D1383141

Acid-Base Balance

CHEMISTRY, PHYSIOLOGY, PATHOPHYSIOLOGY

Acid-Base Balance

CHEMISTRY, PHYSIOLOGY, PATHOPHYSIOLOGY

A. Gorman Hills, M.D.

Professor of Physiology and Medicine
Medical College of Ohio at Toledo

The Williams & Wilkins Company • Baltimore 1973

428 E. Preston Street
Baltimore, Md. 21202, U.S.A.

Made in the United States of America

Library of Congress Catalog Card Number 72-88142
SBN 683-04004-9

Composed and printed at the
WAVERLY PRESS, INC.
Mt. Royal and Guilford Aves.
Baltimore, Md. 21202, U.S.A.

ERRATA

Table I-5.1, page 55: the value listed for pK of water is incorrect. The pK of water is actually ½ pK at 25°, i.e., ½($-\log K_{\text{water}}$). In the footnote to the table cancel though the first sentence of the second paragraph.

Page 230, last line:
for acidosis *read* alkalosis.

Add the following note to sentence (p. 240) beginning "It has been suggested (259) ... "

2. This oversimplified formulation assumes that some of the NH_3 which has diffused into the acidified luminal fluid of the distal tubules is subsequently caused by water reabsorption to diffuse out again. More generally and exactly, net diffusion of NH_3 into the luminal fluid is necessarily reduced by antidiuresis (259), rendering the urine more acidic.

Appendix C, pp. 316–317:

In Equation 3: for a_{H^-}, read a_{A^-}

In Equation 4: for γ_{H^-}, read γ_{A^-}

In Equation 5: for γ_{H}, read γ_{H^+}

In Equation 8: for NH, read NH_3

In line 3 of the bottom paragraph of page 317: for *frictions* read *fictions.*

Appendix E, p. 327:

Bottom line: *for* pK′ − pH *read* pH − pK′.

Appendex F, p. 336:

Line 13 from bottom: for "*ash-TA*" read *dietary less fecal* "*ash-TA*".

Line 12 from bottom: for *milliequivalents* read *milliequivalents, sign reversed*,.

To the memory of:

EDWARDS ALBERT PARK
1877–1969

JAMES LAWDER GAMBLE
1883–1959

WILLIAM CHRISTOPHER STADIE
1886–1959

in cui rampogna
l'antica età la nuova

PREFACE

Luxuriant growth in several branches of any large subject must prompt efforts at an overview. This book is such an attempt, composed in sober recognition of the twin perils of inaccuracy and superficiality inevitably attending such an enterprise. Though written by a physician and directed primarily toward physicians and medical students, it is not a clinical treatise; its central concern is physiology. Pathology is treated summarily, and entirely from the physiological standpoint; there is no intention of providing a compendium of the particular diseases which may disturb acid-base regulation, let alone a digest of therapy. Material of sufficiently specialized character, whether historical, etymological, semasiological, or technical, has been segregated in the appendices in the hope of providing in the body of the text as readable and balanced an account as possible.

Pulmonary control of the arterial CO_2 pressure is briefly considered for the sake of completeness; but the subject of this book, as its title indicates, is the renal contribution to acid-base regulation. The field is examined with the eyes of a physician—focusing, that is to say, upon the intact organism and its adaptive responses to changing normal and abnormal circumstances—whereas renal physiology as an investigative discipline is undoubtedly now "en route to adopting the molecular viewpoint" (619). If intracellular acid-base equilibria, or acid secretion by extrarenal tissues lending themselves readily to *in vitro* study, or related biophysical concerns of present-day renal physiology do not receive much attention here, it is not at all because of any doubt on my part that the nature of the processes occurring inside of and at the boundary of the cell have now displaced the extracellular fluid (including the blood) from the position it had occupied for over a half-century as the cynosure for professional investigators of renal acid-base regulation. But I think it a defensible view that the first fruits of nearly two decades of extensive and intensive probing of these current frontiers of the subject remain as yet fairly few from the standpoint of physicians and other students of the biology of higher animals; accordingly, I have gone no further into the physiology of cells and membranes than I was carried by a sense of its immediate pertinence to the higher levels of biological organization.

On the other hand, by reemphasizing a very early insight—Henderson's perception of the central importance of volatile buffering for acid-base regulation—this book inevitably relates its subject to the most influential theoretical development in renal physiology since the high noon of the clearance era a quarter-century ago: the countercurrent theory of mammalian renal function. The technical revolution of the past two decades in renal physiology, introduced by extensive direct study of the function of mammalian nephron segments by means of micropuncture methods, has given us a topographical description, and the foundation stones for any future explanation, of the acidification of glomerular filtrate—the principal contribution, with ammonia production, of the renal cells to acid-base regulation. But information yielded by micropuncture has also gone beyond the function of the renal cell and the nephron by helping to clarify the mechanisms by which suitable buffers are normally supplied to the urine at rates always appropriate to the acid-base load. Together with water, urea, and oxygen, the buffer gases CO_2 and NH_3, because of their diffusivity, are distributed in the medulla in a manner which cannot be understood in terms of the summated activities of individual nephrons, let alone cells, but only at a higher level of organization, specifically that of a suborgan whose various structures are integrated into an emergent, not an additive, functional ensemble. Like the normal regulation of urinary bicarbonate excretion, a function of the organism as a whole, the medulla's contributions to acid-base regulation defy application of a dogmatic reductionism to the teaching of physiology.

It will hardly be disputed, certainly not by anyone conversant with the clinical literature over the past decade, that messages from chemists to workers in the acid-base fields of medicine and physiology still, on occasion, become shockingly scrambled in transmission; that must be my principal apology for offering a discussion of electrolyte equilibria in solution couched in terms so elementary as to risk the derision of the chemist. The historical approach to acid-base chemistry adopted here is not just an act of intellectual piety or an obeisance to humane scientific cultivation (though I am willing for it to be seen as those, too), but it reflects my adherence to the view of James Clerk Maxwell that the acquisition of a firm grasp of any branch of science is furthered by a vivid awareness of the early stages of its development. The doctrine of electrolytic dissociation, its rapid assimilation into kinetic theory, and the ensuing successful application of thermodynamic treatment to these reactions constitute, after all, one of the grand simplifying achievements in the history of the exact sciences; but nowadays, because of the very power and exactness of the physicochemical deductions, there is a real risk that practical calculation of electrolyte equilibria will become familiar to the student before the character and the magnitude of the problems being so easily solved by the Henderson-Hasselbalch equation have been fully fathomed. Certainly today anyone can gain facility much more rapidly than was possible a generation ago in the manipulation of physiological acid-base equilibria; and this is because "we know so much more" than the chemists and physiologists whose contributions have made our task so much easier. But it is true, too, as T. S. Eliot remarked of all the classics, that "they are what we know."

Toledo, Ohio
Independence Day, 1972

ACKNOWLEDGMENTS

This book could not have been written without assistance of various kinds from many persons. I feel an obligation, as impossible to overestimate as it is difficult to define, to those physicians in company with whom I have had the privilege of investigating aspects of acid-base balance for nearly a decade and a half: to K. A. Woeber, whose alertness in detecting a discrepancy between observation and received opinion originally directed me into research in acid-base physiology; to W. D. Kerr, for his healthy skepticism and energetic commitment; most of all to E. L. Reid, whom Nehemiah might have felt himself fortunate to have recruited. Generous colleagues who have consented to review parts of the typescript, and from whose comments I have benefited, include Drs. James L. Gamble, Jr., Joel B. Mann, and George D. Webster, Jr. More than anyone else, Dr. Giles F. Filley has assisted me, through the most copious allocations of his time over more than 2 years, to clarify my thinking and to articulate my presentation of the subject. Finally, preparation of the typescript for publication would have been unthinkable without the prodigious devotion and high secretarial skills of Mrs. Harold Best and the faultless typing of much of the final draft, under pressure, by Miss Mary M. Teed.

CONTENTS

1. CHEMISTRY

The letter I has not been used to designate an Appendix in order to avoid possible confusion with Roman I.

B. Chemical Notation and Definitions: Chemical shorthand and its shortcomings—examples—instantaneity of ionic reactions—physiological and chemical acid-base terminology.
C. Acidic and Basic Strength: Contemporary theory—acid-base reactions as proton transfers—reciprocal relation between strength of an acid and its conjugate base—quantitative expression of acid-base strength arbitrarily related to H_3O^+-H_2O equilibrium—practical form of the Henderson-Hasselbalch equation.
D. Solutions of Strong Electrolytes: "Leveling" effect of the solvent—calculation of pH of an aqueous solution of strong acid from molar acid concentration—neutralization.
E. Weak Electrolytes (Buffers): α-curves and buffering—range of effective buffering—buffer strength and buffer efficiency—"neutralization" of a weak acid—the isohydric principle—acids and bases ranked according to strength—polybasicity—polybasic buffer curves—amphoteric compounds—isoelectric point of proteins—mixtures of weak electrolytes—their buffer strength—urinary titratable acidity.

6: Volatile Buffer Systems 66

Henderson's recognition of the central importance and special fitness of the CO_2 buffer system to serve blood acid-base regulation—heterogeneous equilibria involving CO_2 and NH_3 in urine—accumulation of H_2CO_3 an efficient way of depressing pH.
A. The CO_2 Buffer System in Extracellular Fluid: A gas-liquid heterogeneous equilibrium with stabilized P_{CO_2}—i.e., an "open system" with respect to CO_2—its unique buffering properties—enhancement by regulatory alteration of arterial P_{CO_2} by the lung.
B. Volatile Urinary Buffering: Stabilization of "effective" urinary pressures of CO_2 and NH_3—provision of economical transport of waste buffer gases into urine as needed (dynamic heterogeneous equilibria)—"excretory buffering"—special fitness of CO_2 and NH_3 for this purpose—buffering efficiency of a volatile system not maximal when $pH = pK'$.
C. Delayed Dehydration of H_2CO_3: Hydration of CO_2 to H_2CO_3—only the first dissociation of carbonic acid relevant to physiology—slow uncatalyzed interconversion of CO_2 and H_2CO_3—high equilibrium ratio CO_2:H_2CO_3—carbonic anhydrase—relatively slow dehydration of H_2CO_3 in absence of carbonic anhydrase—role of nonvolatile buffer—rise of $[H_2CO_3]$ above equilibrium in a kinetic steady state in the distal nephron—large effect of H_2CO_3

accumulation on pH—chemical principles of low disequilibrium pH due to accumulation of H_2CO_3—chemical principles of generation of high urinary P_{CO_2} beyond the papilla—physiological advantages of absence of carbonic anhydrase from distal nephron and muscles—no delayed dehydration of NH_4OH possible—difference between the molecular structures of the hydrates of CO_2 and NH_3.

2. PHYSIOLOGY

1: Homeostasis 85

A. Homeostasis before and after Bernard: Origins—Justus Liebig—physiological function always adaptive to some set of conditions—functional levels in physiological studies—relevance of studies at lower levels to higher levels of organization—Bernard and the constancy of the *milieu intérieur*—its inconstancy—biological rhythms in the higher animal organism—homeostasis an idealized stability of function.
B. Acid-base Homeostasis: Twenty-four hour periodicities—urinary "alkaline tides"—effects of posture and sleep on functions related to acid-base balance—functions related to urinary acid-base composition—glomerular filtration rate—relation of acid-base load to food and fasting—periodicity of arterial P_{CO_2}—postural effects—hemodynamic and endocrine factors—urinary ammonia—experimental design and exclusion of unwanted variables. Intercomparison of the efficiency of homeostatic systems—speed, precision, ruggedness—comparison of renal acid-base regulation with regulation of the volume and osmolality of extracellular fluid—conventional quantification of limits of normal loads of nonvolatile acid and base.

2: Kidney, Lung, and Body Buffers 99

A. Acid-Base Regulation: Collaboration of Kidney and Lung: Extracellular [HCO_3^-] regulated by net balance of nonvolatile acid or base—central importance of plasma [HCO_3^-] lies in its predominance as buffer for nonvolatile acid and base and hence its central role in adjusting acid-base composition of urine—fluctuations of plasma P_{CO_2} do not affect homeostatic regulation of acid-base balance under physiological conditions, but adjust it adaptively when pulmonary regulation deranged.

B. Acid-Base Balance: Rate of glomerular HCO_3^- filtration the principal determinant of the acid-base composition of the urine—continuous massive base generation in peritubular blood, resulting from acid secretion into the nephrons, approximately equals rate of base loss as glomerular HCO_3^- filtration—peritubular HCO_3^- generation plus fluctuating filtered HCO_3^- the renal mechanism for maintaining acid balance as the acid-base load fluctuates—the "alkaline reserve"—urinary HCO_3^- excretion nearly but not exactly synonymous with renal base excretion—nonvolatile buffers contribute above pH 7.4—significance of extra urinary HCO_3^- produced by raised urinary P_{CO_2}—urinary buffers for excess acid (NH_4^+ and titratable acid) reduce pH gradient against which acid must be secreted into the nephrons—acid-base balance and other balances determined by relation between load and excretion rate—special feature of acid-base balance the reciprocal relation between acid and base.

C. Role of the Lung: Stabilization of Arterial P_{CO_2}: CO_2 exhalation and "renal net acid excretion" compared—arterial P_{CO_2} expresses pulmonary contribution to acid-base regulation.

D. Nonvolatile Buffers of Body Fluids: They do not influence extent to which plasma pH must be altered physiologically to achieve a new steady state of acid-base balance—they stabilize pH of venous blood—their importance when arterial P_{CO_2} is abruptly altered—buffers of cell water—cellular metabolic responses in acid-base homeostasis.

3: Arterial P_{CO_2} **109**

A. Stabilization of Arterial P_{CO_2}: Nonhomeostatic changes—homeostatic self-regulation of arterial P_{CO_2} at 40 ± 2 mm Hg—stabilization during exercise—the respiratory center—pulmonary reflexes—the peripheral chemoreceptors—the central chemoreceptors—lesser effect of nonrespiratory changes of extracellular fluid pH as compared with altered P_{CO_2}—stabilization of the spinal fluid pH.

B. Somatic and Renal Responses to Altered Arterial P_{CO_2}: Alterations of blood [HCO_3^-] secondary to altered arterial P_{CO_2}—urinary response to hypocapnia—to hypercapnia—"the displacement phase"—metabolic response to hypocapnia (lactic acid production)—P_{CO_2}—lack of interaction of responses to respiratory and nonrespiratory change—effect of arterial P_{CO_2} on rate of acid secretion into the nephrons.

base generation—constituents of the net acid-base load—the dietary ash—acid-forming minerals (Cl, P, S) and base-forming minerals (Na, K, Ca, Mg)—net dietary ash—correction for urinary neutral S—net metabolic production of organic acids—principles illustrated by lactic acid as parodigm—corrections in calculating contribution of dietary ash to *L*—extrarenal acid and base loss.

C. Quantitative Evaluation of Acid-Base Balance: Focus on L_A—sources of the nonvolatile acid—circularity in evaluating contribution of metabolically produced organic acids to *L*—problematic aspects of reckoning shifts of protons between body water compartments from metabolic balance data—the "sum of the base-sparing mechanisms."

D. Dietary Ash and Therapeutics: Deplorable tendency to minimize counseling concerning dietary ash—usual ash of American diet—uses of dietary ash prescription.

6: The Fundamental System of Renal Acid-Base Regulation.. 142

Renal acid-base regulation requires control of urinary pH and transport of buffer into urine—P_{CO_2} and P_{NH_3} of bladder urine and papillary urine.

A. Urinary pH: Greater range of urinary acidic intensity compared with blood—normal limits of variation of urinary pH—narrower range during water diuresis.

B. Transport of the Buffer Gases and Their Ions into the Urine: Their essentiality—nonionic diffusion of CO_2 and NH_3—origins of the concept—nonionic diffusion as a pH-dependent tubular transport mechanism—antimalarial drugs not excreted to equilibrium by nonionic diffusion—mathematical treatment—not applicable to NH_3. Active transport of CO_2 and NH_3 unlikely—doctrine of their exclusively passive transport falsifiable but not conclusively verifiable—cumulative evidence in the case of ammonia—evidence for diffusion equilibrium of NH_3 throughout renal tissue water—axial papilla-to-cortex NH_3 pressure gradient.

C. The Fundamental System: Most readily discerned during water diuresis—approximation of experimental variations in urinary acid-base composition with varying pH by a physicochemical construct—stabilization of the "effective" pressures of CO_2 and NH_3 in papillary urine—effective urinary buffering by these gases due to their volatility—nonvolatile buffers cannot buffer excreted base during water diuresis—excretory buffering by dynamic gas-liquid heterogeneous equilibria—evolutionary background—urinary buffer strength minimal in midrange (near 6.3) of urine pH—CO_2 limitlessly expendible as urinary buffer—urinary ammonia loss occasionally compromises nitrogen balance but takes precedence. Automaticity of urinary response to changing acid-base load—central role of plasma [HCO_3^-]—counterproductive change in rate of acid secretion into nephrons in response to changing load—a dissenting view.

D. Factors Influencing Urinary pH

7: Acidification (I) and Alkalinization of the Urine.......... 162

A. Introduction: Secretion of "hydrogen ion" chemically vague—"acidification" of the urine used here as purely descriptive term—chemically possible mechanisms.

B. Membrane Transport and Urine Acidification: Diffusion and diffusivity—pressure, concentration, and electrochemical gradients—obstacles to movement of particles—measurements of transtubular potentials—of potentials across other secretory epithelia—relevance of these data to the mammalian nephron uncertain.

C. Alkalinization (Water Reabsorption): Alkalinization of the urine by evolution of CO_2 secondary to water reabsorption—near-equivalence with acidification under all circumstances.

D. Possible Chemical Mechanisms of Acid Secretion: Exchange of H^+ or H_3O^+ for Na^+ versus reabsorption of Na^+ and OH^-—possibility of proton hops—link between Na^+ and Cl^- reabsorption—volume-osmolality regulation and acid-base balance normally independent but become interdependent during Na^+ want—nature of coupling between Na^+ reabsorbed and acid secreted—proximal tubular Cl^- reabsorption believed accounted for by transtubular electrochemical gradient—other possibilities.

E. Topography of Urine Acidification: Fluctuation of plasma [HCO_3^-] the most important influence (at constant flow) on urine pH and hence on the pH of luminal fluid in successive nephron segments—species-specific variation—stabilizing effect of disequilibrium pH in distal portion of nephron—its physiological significance—proportional water reabsorption, and hence effect of water reabsorption on pH, greatest in medullary collecting ducts—is acid secreted into the medullary collecting duct in the mammal?—fraction of filtered HCO_3^- delivered to distal nephron always small, but considerable variation in absolute rate as acid-base load varies.

F. Renal HCO_3^- Clearance: Definition—HCO_3^- a threshold substance—glucose threshold and "titration" curve—*Tm*—

superficial resemblances to HCO_3^- titration curve—differences—glomerulotubular balance (stabilized fractional reabsorption)—the HCO_3^- pseudo-*Tm* due to suppression of acid secretion by the concomitant expansion of expanded extracellular fluid volume—physiological base loads increase renal acid secretion—HCO_3^- clearance increases with increasing base loads—the clearance an *approximate* evaluation of the efficiency of renal elimination of a base load.

8: Acidification of the Urine (II) **180**

A. Transtubular Electrochemical Potentials: Transport across membranes and activities of ions—the Nernst equation—passive versus active transport—mammalian transtubular potentials—technical difficulties—the proximal tubule—the distal tubule.
B. The Role of Intracellular CO_2: Effects of P_{CO_2}, carbonic anhydrase, and carbonic anhydrase inhibitors on urine acidification—effects of enzyme and probably enzyme inhibitors ascribable to effect on luminal fluid—P_{CO_2} affects cellular proton transport—overall process is exchange of Na^+ for proton across cell—"source" of secreted protons conventionally held to be CO_2 (H_2CO_3)—no proton source within renal tubular cell needed for the overall process—nature of primary molecular event initiating transport less an investigative than a semantic problem—profusion of molecular models unfalsifiable by technics being used in renal physiology—pseudo-problem of observed rates of acid secretion in acetazolamide-treated animals exceeding calculated rate of uncatalyzed intracellular CO_2 hydration—mechanism of acceleration of renal acid secretion by increased P_{CO_2} not known.
C. Role of Carbonic Anhydrase: Raised P_{CO_2} of alkaline urines—role of delayed dehydration of luminal H_2CO_3—disequilibrium pH normally present in distal tubular fluid during HCO_3^- diuresis—can be induced in proximal tubular fluid by carbonase anhydrase inhibitors and abolished in distal tubular fluid by intravenous carbonic anhydrase—reabsorption of $NaHCO_3$ excluded as major mechanism of urine acidification—disequilibrium pH in medullary collecting duct—sites of luminal and intracellular carbonic anhydrase catalysis—possible contribution of $NaHCO_3$ reabsorption to urine acidification not excluded.
D. Regulation of the Rate of Acid Secretion: Central role of the acid-base load via the plasma HCO_3^-—inappropriate response of acid secretion—importance of changing transtubular electro-

chemical gradient—prominent effect in higher range of urine pH—observable during HCO_3^- diuresis when carbonic anhydrase infused or urine flow lowered—accessory defenses against alkali excess. Abnormal conditions: stimulation of Na^+ reabsorption during Na^+ want—altered extracellular fluid volume—primary renal disorders—altered arterial P_{CO_2}—K^+ depletion—hormones.

9: Urinary Ammonia Excretion **201**

A. Significance: Relation to N Metabolism and Acidity Regulation: Early studies—relation to acid base regulation—to protein metabolism and acid urine—response to NH_4Cl—to mineral acid—relation to pH of urine and body fluids.
B. Normal Regulation of Urinary Ammonia Excretion: Effect of urine flow less than that of urine pH—ammonia contributed by kidney to blood and urine—steady states described mathematically—nonionic diffusion to equilibrium—mathematical description—empirical approximations—simplified theoretical treatment during water diuresis—discrepancy with observation—agreement satisfactory if allowance made for postpapillary rise in pH and a papilla-to-cortex NH_3 pressure gradient.
C. Renal Sources of Urinary Ammonia: Topography—precursors—L-glutamine the principal precursor in the dog—contribution of arterial ammonia inflow—data in man similar—chemical reactions involved in renal ammonia production—renal glutamine synthetase in herbivorous animals.
D. Ultimate Cellular Sources of the Urinary Ammonia: The circulating ammonia pool—the kidney the principal contributor at rest—the muscles—ultimate source of renal ammonia production in the tissues contributing to the circulating glutamine pool—functions of circulating ammonia questioned—active transport of NH_4^+ questioned. Sources of circulating glutamine—peripheral tissues a major contributor—peripheral glutamine release in fasting animal necessarily derived from protein catabolism of whole body—evidence that local protein catabolism supplies it—evidence that local catabolic N is released as NH_3—thermodynamic—evolutionary—ureotelism an adaptation to land colonization by ancestors of mammals—but did not provide for urinary buffer ammonia—retention of nonionic diffusion of ammonia necessitates intrarenal NH_3 production—circulating glutamine viewed as vehicle for nontoxic carriage of waste NH_3 to urine—incorporation of waste NH_3 into glutamine in peripheral cells presumably catalyzed by glutamine synthetase—this

enzyme recently demonstrated in mammalian skeletal muscle—increased diffusion of NH_3 into peripheral cells results in increased release of N as glutamine—most of extra ammonia N added to cell water appears as extra glutamine released—evolution of ammonia carriage as circulating glutamine deserves investigation by studies in comparative physiology. Kidneys and hepatic-splanchnic circulation take up glutamine—they are metabolically specialized areas of the circulation in that ammonia is efficiently removed from them—release of glutamine in other circulatory regions (head, cardiopulmonary tissues)—glutamine release a general property of tissues not specialized in ammonia metabolism?—net glutamine synthesis and release by brain—significance in relation to cerebral ammonia toxicity—relation of the liver to circulating glutamine pool not well established. NH_3 usually (like CO_2) but not always available in excess for urinary buffer—negative N balance of inadequate protein intake contributed to by urinary ammonia and automatically precipitated when protein intake marginal.

10: P_{CO_2}, P_{NH_3}, and the Medullary Countercurrent Systems... 229

A. Urinary and Intrarenal Pressures of CO_2 and NH_3: Raised P_{CO_2} of alkaline urine—delayed dehydration of H_2CO_3 not the whole explanation—countercurrent systems and axial pressure gradients of diffusible substances—generation of NH_3 in medulla likely—delayed dehydration of H_2CO_3 affects urinary P_{CO_2} by increasing delivery of CO_2 to medulla and to urinary tract—postpapillary H_2CO_3 dehydration and CO_2 reabsorption from urinary tract both raise urinary P_{NH_3}—automaticity of system transporting extra NH_3 buffer into urine as urine pH declines—homeostatic utility of increased renal tissue P_{NH_3} as urine flow declines—necessity of return to systemic circulation of some ammonia produced in the kidney—low P_{CO_2} of very acid urines—probably due to medullary CO_2 fixation—effect of water diuresis.
B. Increased Renal Ammonia Production in Acidosis: Description—renal glutamine extraction increased—determined by some change in the renal cellular milieu—nature unclear—hypotheses—effect of increased renal ammonia production on renal tissue P_{NH_3}—potentiates effectively the enhancement of urinary ammonia excretion by lowered urine pH.
C. Effect of Urinary Flow Rate on Urine pH and Renal Acid and Base Excretion: Antidiuresis lowers low urine pH and raises high urine pH—acid-base effect of decreasing flow rate depends

on relative quantities of urinary HCO_3^- and NH_4^+ as CO_2 and NH_3 are reabsorbed secondarily to water reabsorption—possibility of terminal generation of acid in the more acid urines secondary to NH_3 reabsorption.

3. PATHOPHYSIOLOGY

1: Acidosis and Alkalosis 245

Acidosis and alkalosis (respiratory and nonrespiratory), acidemia and alkalemia defined—compensatory responses by lung and kidney.

A. Whole-Body Buffering: Major role of fluid within nucleated cells in buffering rapidly accumulating acid or base—sequential studies after rapid administration of mineral acid—ionic exchanges—mechanism—disposition of alkali—cell buffering versus cellular metabolic activity contributing to homeostasis—ammonia—lactic and other organic acids—difficulties of analyzing acid-base equilibria within cells and in subcellular water compartments—current orthodoxies about cell water and cellular electrolyte transport and equilibria challenged by some—high mean ionic strength of cell water and nonhomogeneity—demonstration of agreement between theory and experiment desirable.

B. Blood Analysis: Uses—limitations—significance of arterial pH—of arterial P_{CO_2}—exact evaluation of the nonrespiratory component impossible—approximations sought—the plasma CO_2 content—useless for evaluating pulmonary function—"comprehensive acid-base analysis" of arterial blood versus "electrolyte structure of the plasma"—plasma [HCO_3^-] an index of the nonrespiratory component, plasma CO_2 content an approximation to plasma [HCO_3^-]. Analysis of the electrolyte structure of the plasma—value of other components of plasma electrolyte structure for etiological differential diagnosis—reciprocal relation of plasma Cl^- and HCO_3^- reflects renal osmolality and volume regulation—nature of molecular link between Na^+ and Cl^- reabsorption obscure—"chloride acidosis"—due to base loss or excess of HCl or its equivalent—accumulation of other acids—Δ or undetermined ion—mixed abnormalities—other ions. "Comprehensive analysis"—effect of P_{CO_2} on

[HCO_3^-]—relation determined by nonvolatile buffer strength—"CO_2 titration" of plasma or blood—basis of attempts to improve estimates of nonrespiratory component—the CO_2 combining power, whole blood buffer base, standard bicarbonate, base excess (*BE*)—*in vitro* CO_2 titration does not reproduce *in vivo* titration—instrumentation and the *BE*—correction for unsaturation—for deviations of arterial P_{CO_2}—use of directly measured acid-base variables and clinical experience.

C. Effects of Altered pH of Extracellular Fluid: Physiological effects of altering pH inherently complex—effects on heart, lung, and O_2 transport—on cellular metabolism—clinical manifestations of acidemia and alkalemia.

2: Nonrespiratory Alkalosis 269

A. Excessive Base Load: May be due to acid loss (HCl) or alkali excess—effect on kidneys—frequently complicated by K depletion—association with contracted extracellular volume—with aldosteronism—with increased renal acid secretion.

B. Renal: After correction of chronic hypercapnia—cardiac and neurological complications—complexity of clinical and experimental alkalosis—renal tubular alkalosis. Consequences of increased acid secretion into the nephrons (internal environment inappropriate)—"paradoxical aciduria"—complicating factors—mineralocorticoids, K depletion, kaliopenic nephropathy, increased renal ammonia production. Reciprocal relation between acid and K secretion—competition for common pathway?—no competition in proximal tubule—parallel changes in some situations. Ion exchanges in alkalosis and K deficiency in muscle cells—in renal cells—molecular basis of increased renal acid secretion not established—mineralocorticoid hormones—role of Cl^-—relation to volume of extracellular fluid—"saline-resistant alkalosis"—direct effect of hormones on renal acid secretion—reabsorption of filtered inorganic anions other than Cl^-—acid-base balance related to renal volume-osmolality regulation, under abnormal conditions, via Na^+ reabsorption—no simple relation between acid and K secretion discernible.

3: Nonrespiratory Acidosis 279

Two classifications—one based on breakdown of homeostasis is followed here.

A. Excessive Acid Load: Often due to base loss—urinary hyperacidity—complicating dehydration—cholera and other diarrheal diseases. Acid excess—ketoacidosis—lactic acidosis—drugs and toxins—chloride acidosis—importance of distinguishing severity of metabolic disturbance and severity of acidosis—response to treatment—hypocapneic alkalosis—mechanism—classification of lactic acidosis.

B. Renal Acidosis: Due to subnormal renal acid secretion in relation to plasma [HCO_3^-], to insufficient urinary buffer, or both—renal tubular acidosis (RTA) and glomerulotubular insufficiency—tests of renal acidification. Urinary pH—artifactual rise produced by urea-splitting organisms—tolerance testing for RTA unnecessary when nonazotemic acidosis accompanied by bicarbonaturia—subclinical RTA and acid loading—varieties of the NH_4Cl test—shortcomings of pH as criterion of response—proper criterion rate of renal acid excretion—independent appraisal of acid secretion and ammonia excretion—relation of ammonia excretion to renal blood flow and hence to glomerular filtration rate (GFR)—sources of variation—standardization—practical considerations. RTA and glomerular insufficiency—heterogeneity of renal disease hampers general prediction of degree of acidification defect appropriate to deminished GFR. Early descriptions of RTA—chloride acidosis of uncomplicated cases—classification—classical (gradient) RTA and proximal (rate) RTA—primary and secondary forms—complications—genetic and sporadic cases—therapy. Glomerulotubular insufficiency—glomerulotubular balance—renal pathology of azotemic renal failure—functional classification not well developed—glomerular insufficiency reduces delivery of nonvolatile buffer to the urine—mathematical relation of plasma concentration of a freely filtered, nonreabsorbable substance to GFR (sulfate)—approximations (creatinine, urea)—diversion of phosphate to stool—uric acid—displacement of plasma HCO_3^- rather than Cl^- in plasma by accumulating anions of acidic compounds in uremia—[HCO_3^-] and [Na^+] − [Cl^-]—ability of diseased kidneys to lower pH in acidosis well preserved—comparison with normal should take urine flow into account—decreased urinary nonvolatile buffer—decreased renal ammonia production—differences between patients with glomerulotubular failure—stabilization of plasma [HCO_3^-] at different value in patients on same acid-base load—falling HCO_3^- filtration as plasma [HCO_3^-] declines permits acid balance at lower rates of acid secretion into nephrons—is all renal acidosis tubular acidosis?—is ammonia production a tubular function?—glomerular contribution to renal failure—contribution of distorted architecture—diseased kidney still a precision device but capacity and sensitivity reduced.

I

CHEMISTRY

one

CHEMICAL EQUILIBRIA

A. REVERSIBLE REACTIONS (168, 513)

The chemical reactions of central concern to us in this book result from the addition of acid or base to, or loss of acid or base from, blood, extracellular fluid, or urine; all of them belong to the category of *reversible reactions* in *aqueous solution.* Also, with a single exception—the luminal fluid in certain parts of the nephron—we deal in normal physiology exclusively with acid-base reactions which are so very rapid as to be *instantaneous* from the physiological standpoint. The quantitative treatment of such reactions according to physicochemical theory aims essentially at quantifying the acidic intensity produced by the addition of known quantities of particular strong and weak bases and acids to water, and especially at calculating the relation between the kind and quantity of acid or base added to, or withdrawn from, a volume of solution of known acidic intensity and the new point of equilibrium which will be reached in the solution.

Significant quantitative study of reversible equilibrium reactions in solution had to await the second half of the 19th century. This rather late start of an important branch of chemistry seems clearly to have been due in part to the spectacular successes in another department of chemical theory which for a time seemed antithetical. Lavoisier's associate, Claude Louis Berthollet, as a result of his studies of the soda lakes of Egypt during the Napoleonic occupation of that country, had convinced himself that certain reactions may be caused to proceed either forward or back, depending upon the mass of the respective reactants present, a conclusion which implies the possibility of a state of equilibrium resulting from equal velocities of forward and back reactions. Unluckily, however, Berthollet contended that the mass of reactant influenced the composition, rather than just the quantity, of product formed; this idea embroiled him in a controversy with

Joseph Louis Proust which lasted throughout much of the first decade of the last century. Since Berthollet was driven to deny chemical combination in fixed proportion—the crucial concept which was shortly to lead to recognition of the fundamentals of chemical bonding and the structure of molecules—his views fell into disfavor, and interest in reversible reactions was not intensively renewed for a half-century. Meanwhile, new developments in physics, especially the beginnings of thermodynamics and the kinetic theory of gases, were providing concepts which would later be applicable to experimentation in solution chemistry (475, 167, 281).

There are many instances of reversible chemical equilibria other than those occurring in solution, for example, gaseous equilibria such as the reversible dissociation of phosgene:

$$COCl_2\ (g) \rightleftharpoons CO\ (g) + Cl_2\ (g) \tag{1}$$

The physiologically vital physical equilibria between the partial pressure of a gas in solution and the pressure of the same gas in a gaseous phase are dealt with later (I-6). Attention is therefore confined here to the kinetics of reversible chemical reactions in solution, a subject which can be traced from an important contribution made by Ludwig Wilhelmy of Heidelberg in 1850.

B. KINETICS

The "inversion" of sucrose—its hydrolysis in water solution (in the presence of acid as a catalyst) into the two hexoses (dextrose and fructose) of which it is composed—had been known for several decades by 1850. This reaction is slow enough to permit serial observations of its velocity, and it becomes slower as it proceeds. Wilhelmy, a physicist, followed the velocity of hydrolysis with the aid of a polarimeter which recorded the change from dextro- to levorotation as sucrose is replaced by the two hexoses (656). He observed that the velocity v of the reaction was always proportional to the sucrose concentration c:

$$-(dc/dt)\ (=v) = kc \tag{2}$$

The reaction is first-order by definition (Figure I-1.1). The general significance of *velocity* of reactions for the kinetic analysis of equilibrium states did not become fully apparent for more than a quarter-century after Wilhelmy's study, however. As will be noted (I-3B), the dissociation of an acid HA in water was at first to be analyzed kinetically, in terms of Arrhenius's theory, as a monomolecular decomposition; according to contemporary acid-base theory, it is brought about by a bimolecular reaction involving a collision between a molecule each of water and acid. However, a bimolecular reaction in which water is one of the reactants is first-order in dilute aqueous solution because the water concentration is very close to unity and is not appreciably affected during the course of the reaction.

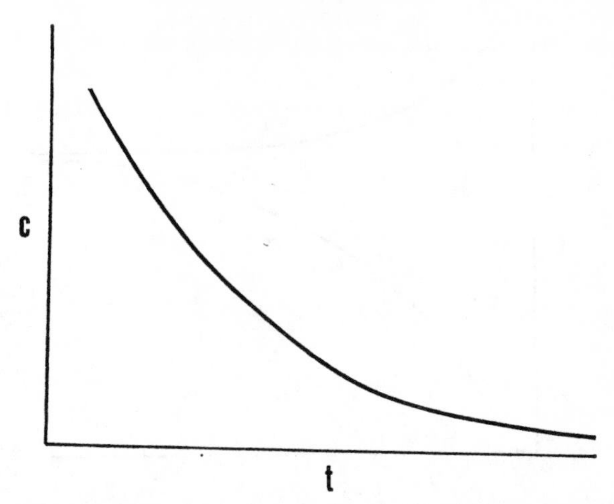

FIG. I-1.1. Graph (schematic) of a first-order decomposition. The instantaneous velocity v, $-dc/dt$ at any moment, can be visualized as a tangent to the curve at that moment. In a first-order reaction v is by definition proportional to the concentration c at every instant; that is, $v = kc$, and so the reaction slows as it goes.

Since $v = kc$, the tangent at any point on the curve divided by the concentration at that point always has the value k:

$$k = -\frac{dc/dt}{c}$$

The value of the velocity constant k is characteristic of the particular reaction under study. The higher its value, the greater the relative tendency of the reaction to proceed to completion.

In 1862 and 1863 Marcellin Berthelot and Péan de St.-Gilles reported very thorough, though still empirical, studies (61) of the reversible slow hydrolysis of esters (e) to yield acid (ac) and alcohol (al):

$$\underset{\text{e}}{\mathrm{R\overset{\overset{O}{\|}}{C}OCH_2R'}} + \underset{\text{w}}{\mathrm{H_2O}} \rightleftharpoons \underset{\text{ac}}{\mathrm{R\overset{\overset{O}{\|}}{C}OH}} + \underset{\text{al}}{\mathrm{R'CH_2OH}} \tag{3}$$

In the synthesis of the ester, which does not go to completion, the proportion of alcohol disappearing increased when more acid was provided; the proportion of acid disappearing similarly increased with the amount of alcohol. The hydrolysis reaction, too, never goes to completion, but can be made to proceed further by adding more water. The French investigators recognized that the failure of the reaction to proceed to completion in either direction represented attainment of a stable state of equilibrium (Figure I-1.2), and clearly asserted the importance of obtaining some general means of treating reactions of this type.

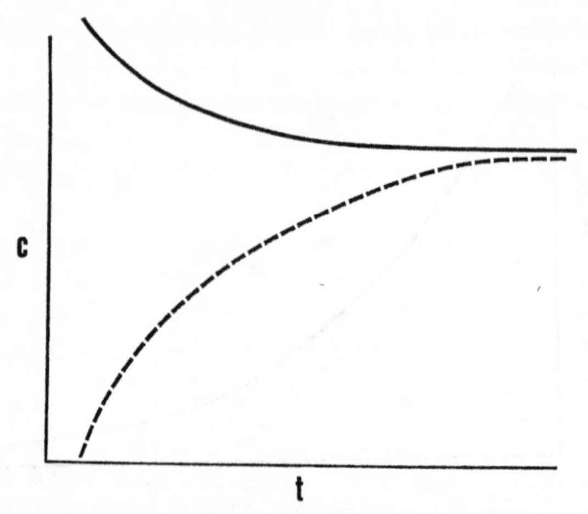

FIG. I-1.2. A state of equilibrium approached from opposite directions of a reversible reaction (schematic). Each *line* represents ester concentration as a function of time; the *solid line* shows progression toward equilibrium starting with a mole of ester, and the *dotted line* shows the reaction of 1 mole each of the constituent acid and alcohol. The equilibrium between ethyl acetate-water and ethanol-acetic acid studied by Berthelot and St.-Gilles was first expressed in 1879 as:

$$K = \frac{[H \cdot C_2H_3O_2]\,[C_2H_5 \cdot OH]}{[C_2H_5 \cdot C_2H_3O_2]\,[H_2O]} = \frac{1}{4}$$

The position of this equilibrium, as represented in Reaction 3, lies to the *left;* esterification has a greater tendency to proceed to completion than does hydrolysis.

C. THE LAW OF MASS ACTION

Such a general treatment was first provided in 1864, and in much more detail in 1867, by Cato Maximilian Guldberg and Peter Waage, mathematician and chemist, respectively, and brother-in-law collaborators, at Christiania (Oslo) (217, 218). They proposed that the "chemical force" with which two (or more) substances react in a reversible reaction can be regarded as equal to the product of their "active masses" and a specific constant characterizing their affinity for each other. Equilibrium represents the dynamic steady state achieved when the "chemical force" in the two directions becomes equal. In 1877 Jacobus Henricus van't Hoff arrived, subsequently but independently, at the same result (631a), basing his treatment, however, on reaction velocities rather than a poorly defined chemical force. At equilibrium, the opposite velocities, determined by the concentrations and a "reaction coefficient" (velocity constant) characteristic of each reaction, become equal. Because of the generality and clarity of his treatment, van't Hoff's paper constituted the principal stimulus to immediate application and further theoretical development of this *law of mass action,* often termed the *mass law;* but Guldberg and Waage are credited with the first statement of this fundamental theorem of the kinetics of reversible reactions in solution.

According to the kinetic theory of matter, a reaction between two molecules of different types can occur only as the result of a collision between one of each type: such a collision is a necessary though not sufficient condition for a synthesis. It is easy to deduce that for a bimolecular reaction the frequency of such a collision, and hence also the velocity of the unidirectional reaction, must be proportional to the concentration of each reactant, that is to say proportional to the product of their concentrations. Therefore, if v_1 and v_2 are velocities of hydrolysis and synthesis, respectively:

$$v_1 = k_1 [e] [w] \tag{4}$$

$$v_2 = k_2 [ac] [al] \tag{5}$$

where k is the velocity constant characteristic of the particular reaction (Figure I-1.2). Clearly the value of each specific velocity constant k expresses the relative tendency of the particular reaction to go to completion. For any equilibrium state, where $v_1 = v_2$, we can write:

$$k_1 [e] [w] = k_2 [ac] [al] \tag{6}$$

and in dilute aqueous solution, where [w] is essentially unity:

$$k_1 [c] = k_2 [ac] [al] \tag{7}$$

Consequently the *equilibrium constant* K at any temperature is given from the concentrations by transposition of Equation 7 as:

$$K = \frac{k_1}{k_2} = \frac{[ac] [al]}{[e]} \tag{8}$$

The value of K tells us what proportion of the ester will have been hydrolyzed after a dilute aqueous solution has been prepared and sufficient time has elapsed for equilibrium to be reached; it also allows predicting how much ester will be derived from any starting quantities whatever of acid and alcohol. It gives, in short, *a general solution*, for any reaction of this type at a given temperature, to the problem of the equilibrium concentrations which will be arrived at after any quantities whatsoever of the respective reactants are added to water. We can now write Equation 8 in a general form which will apply to any equilibrium consisting of a first-order forward reaction and a second-order back reaction:

$$K = \frac{c_1 c_2}{c_3} \tag{9}$$

where c is concentration and the subscripts indicate the three reactants.

Guldberg and Waage, and also van't Hoff, were able to show that application of the mass law to the data of Berthelot and St.-Gilles gave a satisfactory account of the experimental findings; and extensive investigations in many laboratories soon amply demonstrated the applicability of the law to a wide variety of reactions in solution. Nevertheless, it is necessary to stress the crudity with which the theoretical basis of the mass law has just been presented. For example, the order of any reaction cannot be deduced from its apparent molecularity but must in each case be experimentally determined. Again, reaction velocities are in general very much less than would be expected if every suitable collision resulted in chemical change; modern kinetic theory takes into account molecular configurations and especially the activation energies required for reaction, and accounts theoretically in this way for the low ratio of observed velocities to those expected if every suitable molecular collision resulted in a chemical transformation (van't Hoff's "reaction coefficient"). We cannot go further into reaction kinetics here; however, a refinement in the treatment of acid-base equilibria according to the mass law, which takes into account the electrical forces to which ions are subject, will be examined subsequently (I-4), and will afford an illustration of modification of a simple kinetic analysis to render it applicable to real situations without loss of generality.

The chemistry of equilibria in solutions deals with the formation of precipitates, acid-base titrations in nonaqueous solutions, complex ion equilibria and ligands, and oxidation-reduction reactions, as well as simple ionic equilibria in aqueous solution (168, 513). In this book we are concerned only with simple ions in solution whose equilibria all obey the mass law, with the special feature that they are so rapid as to be instantaneous from the physiological standpoint. (The CO_2-HCO_3^- buffer system sometimes constitutes an exception, or rather a pseudo-exception; cf. I-6C.) Originally, simple ionic reactions were thought of as actually instantaneous; but though they are very fast indeed—their half-times are generally of the order of fractions of a millisecond—their velocities have proved to be measurable with the aid of the sophisticated methods of contemporary physical chemistry (140, 225). However, long before such measurements became feasible, the reactions of weak acids and bases were recognized on formal grounds as another class of reversible reactions in solution obeying the mass law (I-3B) and therefore amenable to deductive treatment, according to the same kinetic principles as expressed by Equation 9 which govern equilibria attained by slower processes.

Water reacts with itself, one molecule transferring a proton to another molecule (I-5), in a reversible reaction called *autoprotolysis:*

$$2\,H_2O \rightleftharpoons H_3O^+ + OH^- \tag{10}$$

Ionization occurring in aqueous solutions through which a current is passed was recognized by 1800 (475), and we owe to Michael Faraday the introduction in the 1830's of our terms electrolyte and ion, anion and cation, and cathode and anode (445); but it was not until 1887 that the concept was promulgated that water and its electrolytic solutes are permanently dissociated into ions to a greater or lesser extent (I-3B). The degree of ionization of water is extremely low—$[H_3O^+] = [OH^-] = 0.0000001$ Eq/liter at 25° C—and pure water is in consequence a poor electrical conductor; but its ability to accept or donate protons is of great importance for the acid-base properties of aqueous solutions. Acids donate protons to water, and bases accept them from water (I-5); and when a strong acid or base is added to water, even a dilute solution is a good conductor because the total concentration of ions present becomes considerable. In a N/10 aqueous solution of a strong acid like HCl (0.1 mole/liter or 100 mmoles/liter), for example, the sum of the concentrations of the ions present in solution can be directly calculated since a strong acid is wholly dissociated in dilute solution:

$$HCl + H_2O \rightarrow H_3O^+ + Cl^- \qquad (11)$$

It is seen that there will be 200 mmoles of ions/liter, compared with which the sum of the concentrations of H_3O^+ and OH^- in pure water is negligible. But for reasons to be explained subsequently (I-5), the *product* of $[H_3O^+]$ and $[OH^-]$ in any aqueous solution remains 10^{-14}, as in pure water: in the HCl solution described above, $[H_3O^+]$ will be 10^{-1} or 0.1 Eq/liter, but $[OH^-]$ will have fallen to the almost unimaginably low value of 10^{-13} or 0.0000000000001 Eq/liter.

Before proceeding, in the remainder of Part I, to a systematic consideration of acid-base reactions in aqueous solution, it is desirable first (I-2) to review some more general aspects of water as a solvent, especially for electrolytes.

two

AQUEOUS SOLUTIONS

A. WATER

Life as we know it is the child of water. Living things on our planet which do not (as most do) inhabit fresh or salt water trace their ancestry to aquatic forms. Also—as Claude Bernard taught us (II-1)—all higher animals harbor within themselves an aqueous medium, the internal environment, within which reside the living elements of the body, the cells.

1. Water and Life

Lawrence Henderson (240) was the first to realize that the properties of water, which are in the highest degree anomalous, are of the utmost importance (together with its terrestrial abundance) in providing conditions friendly to the development and support of life (122, 137, 141, 300). The thermal properties of this substance, particularly its high heat capacity and its peculiar expansion as it approaches the freezing point, make it uniquely effective in stabilizing terrestrial temperature by conserving the heat of solar irradiation; oceans and other bodies of water are insulated during winter by the overlying film of ice, and the liquid water beneath is conserved as an immense heat reservoir against the return of the summer sun. Stabilization of the temperature of the internal environment is also promoted by the high heat capacity of water, and the remarkable solvent properties of this substance qualify it to serve with unique effectiveness as the medium of vital exchange of electrolytes as well as of gases and innumerable other solutes involved in metabolism. So strikingly favorable for the support of life is water that many biologists (though not all) share the view of Wald (637–639), that any life which may arise anywhere in the cosmos must do so in an aqueous medium.

2. Polarity

Subsequent to the appearance of Henderson's remarkable book *The Fitness of the Environment* in 1913, it became possible to account for the anomalous properties of water by the polar character of its structure, by its proclivity to form hydrogen bonds, and by its three-dimensional configuration. In terms of the Rutherford-Bohr model of the atom, H_2O is a very stable molecule because the coordinate covalent bond between each H and the oxygen atom satisfies the combining tendency of all three atoms (Figure I-2.1). Each H has completed the first stable shell of two electrons with the aid of an electron provided by O, and O has completed the second stable shell of eight electrons with the aid of an electron provided by each H. But the molecule is a *polar* one, an electrical dipole, because the orbitals of the shared electron pair are displaced away from the outer aspects of the H's, leaving them positively charged, a corresponding negative charge developing on the other side of the molecule in the region of the oxygen atom. The polarity of the molecule entails the development of electrostatic attraction between H_2O and both the cations and the anions which may be present in solution (Section A-4).

3. Hydrogen Bonding

The fact that H can be bivalent through formation of an electrostatic link to electronegative elements—the hydrogen bond—was increasingly

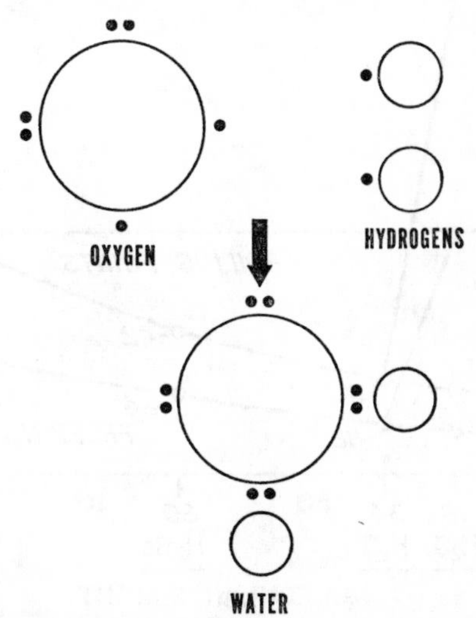

FIG. I-2.1 Formation of the water molecule by the union of one oxygen with two hydrogen atoms. Each hydrogen atom is linked to the oxygen atom by a coordinate covalent bond consisting of two electrons; to each bond both H and O have contributed one electron. The outer aspect of each hydrogen atom is positively charged because the orbitals of the shared electron pairs are displaced toward the O; the H_2O molecule is therefore an electrical dipole.

realized as a result of articles published in 1912 by Werner and Pfeiffer and by Moore and Winmill, and in 1920 by Latimer and Rodebush (583, 449); and the decisive contribution of hydrogen bonding to the anomalous properties of water was vividly emphasized in 1933 in a comprehensive contribution by Bernal and Fowler (57). This form of linkage, though not nearly as strong as the covalent bond, is nevertheless considerably stronger than the weak van der Waals forces of intermolecular attraction. Water is for this reason a highly *associated* liquid: intermolecular attraction is exceptionally great because each water molecule has a strong tendency to form four hydrogen bonds with other water molecules. The high *surface tension* of water results from this strong attraction of one water molecule for another; both this property and the related one of *capillarity*, reflecting the attraction of water to many surfaces ("wetting"), are of great biological importance (608). Some anomalous thermal properties of water (its high melting and boiling points) resulting from its highly associated character are illustrated in Figure I-2.2, which compares H_2O with other compounds constituted of two H's linked to electronegative elements sharing with O the same column of the periodic table. Firm hydrogen bonding between water molecules means that much thermal energy is required to abolish the crystalline structure of ice. Heat energy is also required to reduce the number of the unstable but constantly reforming links between molecules of liquid H_2O as its

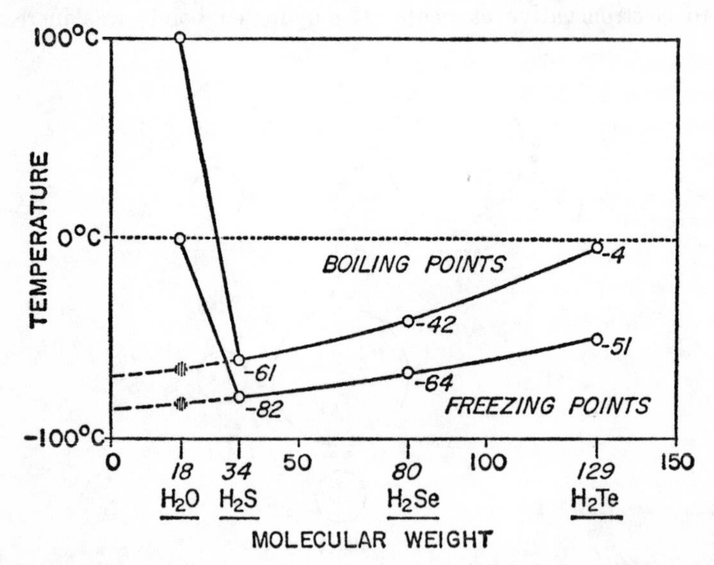

FIG. I-2.2. Anomalous thermal properties of water. Reproduced from Davis and Day (122) with permission of the publisher. The freezing and boiling points are much higher than would be anticipated by extrapolation from the properties of SH_2, SeH_2, and TeH_2, the analogous dihydrides of the three atoms succeeding O in the sixth column of the periodic table.

temperature is raised toward the boiling point, hence the high heat capacity of liquid water.

The ready formation of hydrogen bonds between water molecules can be accounted for stereochemically by the bond angle between the two covalently bonded hydrogen atoms. The "normal" angle of 90° is believed to be widened in H_2O as a result of the repulsion between the two H's, which lie relatively close together owing to the small size of the oxygen atom (Figure I-2.3). The angle in water measures 105.5°, very close to the so-called tetrahedral angle or Maraldi's angle of 109.5° (450, 608). It is called the tetrahedral angle because solid geometry dictates that an atom having four equiangular bonding sites of 109.5° will have its valences directed to the corners of a tetrahedron. The four possible bonds of the water molecule are directed nearly equiangularly, so that the molecules can be fitted together in such a fashion that four hydrogen bonds can form between them (450). Ordinary ice actually has the crystalline structure of packed tetrahedra; nevertheless, tetrahedra cannot be closely packed, and so ordinary ice, the crystalline structure of which is based on tetrahedra, has an open structure and occupies a larger volume than liquid water. That is why water exhibits a paradoxical expansion as its temperature declines from 4° C to the freezing point.

4. Solvation of Electrolytes

Not very many compounds are excellent solvents for electrolytes; all are polar compounds (594) and water is one. The particularly high solubility of electrolytes in water results from its very large *dielectric constant*, which derives not only from the polarity of the H_2O molecule but also from extensive hydrogen bonding between adjacent water molecules. A dipole will rotate about its electrical center when an electrical field is applied, the positive pole being directed toward the negative plate of the field and the negative pole toward the positive plate. By orienting themselves in this manner, molecular dipoles weaken the applied field to a greater or lesser extent, some of the applied energy being expended in maintaining their rotational displacement. The *dipole moment*, or strength of the rotational tendency, depends at any field strength upon the magnitude of the spatial separation of the charges, which is relatively large in H_2O. In addition, clusters of water molecules are at any moment linked together by hydrogen bonds between the molecules; and a given number of dipole molecules independently rotated by an electrical field have much less power to counteract it than do the same number of molecules when two or more are bonded together; the linking of one water molecule to another by hydrogen bonds therefore increases the ability of water to reduce the effect of such an applied field. The dielectric constant of water is approximately 80 to 37° C; this value means that a given voltage applied across an aqueous solution es-

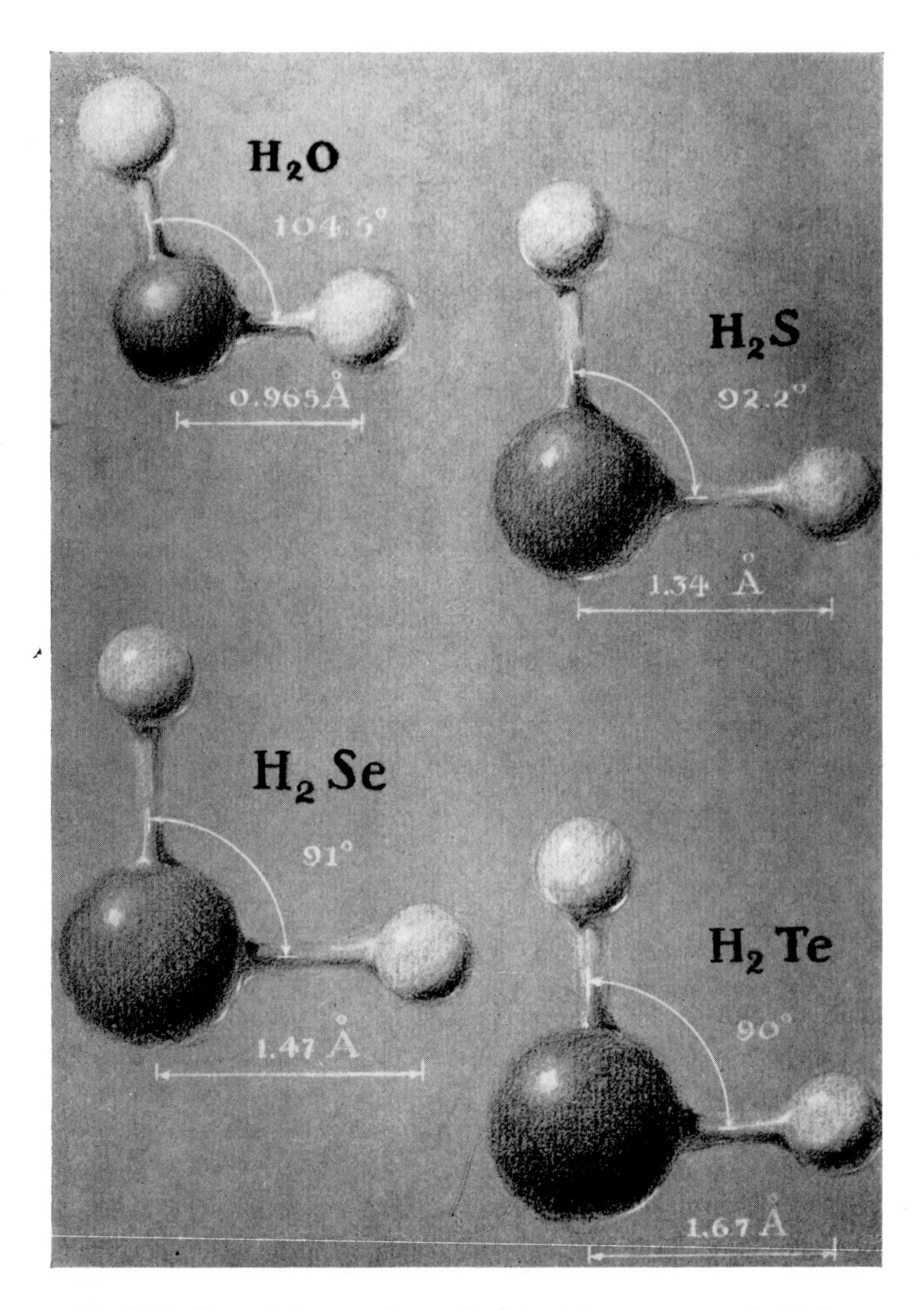

FIG. I-2.3. The angle between the two H—O bonds in water compared with the analogous angles in H_2S, H_2Se and H_2Te. Reproduced from an original in color in Pauling and Hayward (450) with permission of the publisher. Repulsion between the two electropositive H's increases with decreasing diameter of the electronegative element, and effects in H_2O a widening of the bond angle which is sufficient to bring about profound changes in bonding characteristics and in the structure of aggregates of water molecules.

tablishes an electrical field which is only $\frac{1}{80}$ the field generated when the same voltage is applied across a vacuum. The dielectric constant of water is double that which would be found in the absence of hydrogen bonding (122, 137, 141, 300).

The atoms of salts like NaCl exist as ions in the solid state as well as in solution, the crystal lattice of the solid being held together by the electrostatic attraction between the oppositely charged ions. When NaCl is dissolved in water, the electrical fields resulting locally from the regular distribution of positive and negative charges in the crystal are much weakened by the local alignment of dipolar water molecules; the electrostatic force preserving the crystalline structure is reduced to such an extent that it can be overcome by thermal agitation, and the crystal is broken up. Shells of water molecules then surround the ions as a result of electrostatic attraction of anions for the positively charged H's, or of cations for the electronegative O atom; this hydration of the ions hinders the near approach of the oppositely charged ions of the solvent to each other, and so greatly reduces the tendency of the electrolyte crystal lattice to reform (594). The attraction between water molecules and cations, especially those with double or triple charges, is strong enough to result in the incorporation of water molecules into the crystalline solids formed by some of these ions when the solvent water is evaporated.

The dielectric constant of water is sufficiently high that some compounds, not themselves electrolytes, become electrolytes in water; HCl is one, as are a number of other acidic compounds. Liquid HCl is covalently bonded, as indicated by the fact that it is a nonconductor of electricity (see Section C); but in water it clearly exists as dissociated ions, as indicated by the high specific conductance of aqueous HCl and the barely detectable HCl vapor pressure of a 5% solution (594).

B. IDEAL SOLUTIONS AND COLLIGATIVE PROPERTIES

The study of extracellular fluid and the regulation of its composition through respiratory control and alterations of urine composition responsive to the acid-base load have been facilitated immeasurably by the fact that in dealing with these fluids we have to do with aqueous solutions dilute enough to approach *ideality*, and which can in consequence be treated theoretically and deductively with a degree of assurance still exceptional in biology. That is why the chemical physiology of the extracellular fluid, including acid-base equilibria, matured so early among the divisions of life science; "Every science, as it grows towards perfection," as Alfred North Whitehead wrote, "becomes mathematical in its ideas."

Water transport into and out of cells, and the physical chemistry of intracellular water, have been reviewed by Dick (131a). The gross disproportion between our highly satisfactory understanding of chemical events in extracellular fluid and the tentative character of even the most elementary postulates about the behavior of acid-base equilibria in cell water is not due only to the diversity and to the relative inaccessibility of cell water. Even were cell water to be of uniform composition from tissue to tissue, and were it to consist to a large extent within a cell of a single homogeneous phase, quantitative deductive analysis of shifts of acid-base equilibrium would still scarcely be possible in view of the excessive ionic strength of cell water, presumably including the cell sap, as the small, perhaps partially structured phase of cell fluid is termed (I-4B). At the present time, it is the license we enjoy to treat acid-base equilibria in extracellular fluid theoretically, and the minor character of the complication that plasma and interstitial fluid are distinct phases (III-1), which still provide the principal basis for tentative inferences about changes affecting the acid-base composition of cell water in disease.

The desirability of a better understanding, attained through less indirect means, of acid-base equilibria in cell water in various locations is indisputable; but the superior power of the exact sciences has always been related conspicuously to their concerning themselves with problems ready for solution, rather than taking up problems because their solutions are ardently desired.[1] Hopeful projections of a decade or two ago about the outlook for improving our knowledge of the acid-base chemistry of cell water (515) may well be tempered at present by sober assessment of the actual pace of solid advance in this field in recent years, and tempered still more by the paucity of indications that conceptual and methodological developments are imminent which would dramatically improve the accessibility to analysis of acid-base equilibria in cell water.

The concept of an *ideal solution* (475, 167, 166, 382) is an extension of the principle of the "ideal gas," the development of which played such a major role in that great enterprise of the century beginning around 1820 by which chemical combination was progressively understood in terms of the laws of physics, molecule formation, and atomic structure (475, 281). The kinetic theory of matter teaches us that increasing temperature of any substance is the expression of increasing velocity of its molecules; when the temperature of any substance rises to some critical point, intermolecular attraction cannot prevent its indefinite expansion since thermal agitation now overcomes intermolecular attractive forces. This is the gaseous state, and the pressure exerted by a gas against the walls of its container reflects the bombardment of the walls by the molecules; that is why the pressure of a gas increases as the absolute temperature. A gas behaves like an ideal gas when its molecules are sufficiently separated by intermolecular attraction for it to become a negligible constraint on their movement, so that they behave like frictionless point particles. It can be deduced from the assumptions of Newtonian mechanics that such a gas will obey the gas law:

$$PV = nRT \quad (1)$$

where n is moles and R is the gas constant. For every real gas which approaches ideality a mole (6.023×10^{23} molecules) will occupy the same volume V at any given absolute temperature T and pressure P. Most real gases obey the gas law at low to moderate pressure provided the temperature is not very low; as the density of the gas increases with falling temperature and at high pressures, deviations from ideality which are characteristic of the individual gas appear because of increasing attraction between the molecules.

We may define an ideal solution as one the dilution of which does not produce heat effects or modification of its pure constituents, where total volume is equal to the volumes of solvent and solute, and where the solution obeys Raoult's law of vapor pressures. The concept of ideal solutions was very rapidly developed in the 1880's as a result of one of those sudden confluences of seemingly independent lines of investigation into a far-reaching synthesis which from time to time illuminate and accelerate the growth of science. In this case the important general idea which abruptly emerged from experiments in different fields was that which Wilhelm Ostwald, at the suggestion of the philosopher Wilhelm Wundt, was to term the *colligative properties* of a solution, those which depend only upon the number, not the kind, of particles in solution. Colligative properties of a solution comprise the osmotic pressure of a solution, the lowering of its freezing point and vapor pressure, and the raising of its boiling point, compared with those of the pure solvent. The identity of those properties of any solution which depend only upon the number of solute particles present in a stated volume is clearly reminiscent of the identity of the physical behavior of any volume of gas containing the same number of molecules.

Biologically the most important of the colligative properties is osmotic pressure, because (along with capillarity and hydrostatic pressure) it is extensively exploited by organisms for physiological water transport. The phenomenon of osmosis was first observed by the Abbé Jean-Antoine Nollet (423) in the mid-18th century. A bottle containing alcohol, capped with a membrane of pig's bladder and immersed in water, was observed to imbibe water until the membrane bulged and burst; the membrane was more permeable to water than to alcohol. Osmotic pressure affords a most striking macroscopically visible reflection of the magnitude of intermolecular forces; however, quantitative studies had to await the development for research use of artificial truly "semipermeable" membranes—van't Hoff's term for membranes restraining certain solutes with a high degree of efficiency and not just transiently. Such membranes were first described in 1867 by Moritz Traube of Breslau. Wilhelm Pfeffer, a German botanist, then devised the technic of precipitating a copper ferrocyanide membrane on a porous earthenware pot; the pot provided rigid walls which would not yield under the hydrostatic pressure developed as water entered from outside, and with this apparatus Pfeffer made extensive quantitative observations of osmotic pressure of cane sugar solutions (455). He could show that the pressure, measured as the hydrostatic pressure defined by the difference between the level of fluid inside and outside the pot, was proportional to the concentration of solute, and that it also increased with the temperature.

In the early 1880's François-Marie Raoult of Grenoble, studying mostly organic solvents, was establishing that the depression of the freezing point of a solvent by added solute was quantitatively related to the molar ratio of solute to solvent. (The depression of the freezing point of water amounts to 1.86° C per mole per kilogram of water, and measurement of the freezing point is today the standard means of determining osmolality in biological work.) Soon Raoult extended his researches to lowering of the vapor pressure, and by 1887 he could announce his general law (484) that equimolar concentrations of nonvolatile solutes lower the vapor pressure of various solvents by the same fraction. (It will be apparent that the molecular weight of a solute can be experimentally ascertained, insofar as it obeys Raoult's law, from any of its colligative properties.)

At that time the Dutch botanist Hugo de Vries—later, in 1900, to become one of the three rediscoverers of Mendel's findings in plant heredity—had started to use plant cells as osmometers; he prepared aqueous solutions from various pure solutes, using concentrations such that the cells when immersed in them would neither shrink nor swell. Such solutions he termed "isotonic"; higher solute concentrations resulted in transfer of water from the cells, lower concentrations caused water to move into the cells. His measurements showed that the isotonic solutions all had the same freezing point, which was below that of water (130). Being aware of Raoult's work, de Vries, in 1884, happened to tell his countryman, Jacobus Henricus van't Hoff, of his own experiments, which seemed to mean that the relation between the osmotic pressure of a dilute solution and the degree of depression of its freezing point was ascribable to their common dependence upon the concentration of the solute particles.

In the case of the osmotic pressure π we may write this relation at the freezing point as follows:

$$\frac{n}{V} = k\pi \tag{2}$$

where n is the number of moles of any (undissociated) solute, and n/V the concentration of the solute in the solution. Van't Hoff perceived that if π was proportional to T as suggested by Pfeffer's observations, then, by likening the osmotic pressure π to the pressure of a gas P, Equation 2 could be transformed into another which was strikingly analogous to Equation 1, to the equation of state for perfect gases. By 1885 van't Hoff was able to present evidence that the osmotic pressure of a dilute aqueous solution varies as the molar concentration of solute n/V and as the absolute temperature, in formal correspondence with Equation 1:

$$\pi V = nRT \tag{3}$$

Most remarkably, it turned out further that for most solutes R had the same numerical value as in the gas equation. This correspondence between Equations 1 and 3 seemed to imply the closest kinetic analogy between the pressure of a given volume of gas and the osmotic pressure of a given volume of solution; van't Hoff therefore believed that the solvent was incidental. He ascribed osmotic pressure entirely to the concentration of solute particles within the spatial volume occupied by the solution (632). But the solvent can hardly be regarded as irrelevant to osmotic pressure, in view of the subsequently realized importance of solute-solvent interactions; and in fact Equation 3, though valid for dilute solutions, becomes increasingly inaccurate as concentrations rise above M/10.

The "solvent bombardment" concept, which was subsequently developed as a mechanical model to explain osmotic water movement, considered that there is always bidirectional movement of water across a semipermeable membrane, and that bombardment of the pores of the membrane by solute particles diminished egress, but not ingress, of solvent molecules. Osmotic pressure is nowadays related especially to solvent concentration—more precisely, solvent activity (I-4B)—which is of course greater for the pure solvent than for a solution. However, Equation 3 can also be derived by thermodynamic deduction, as applicable in the special case of dilute solutions obeying Raoult's law of partial pressures, from a relation between the osmotic pressure and the lowering of the vapor pressure of a solution by solute. The van't Hoff equation accordingly states a limiting law, approached in dilute solution, governing the osmotic pressure of a solution, in the same sense that the gas law itself is a limiting law (342, pages 202–204).

C. ELECTROLYTIC DISSOCIATION

Van't Hoff's studies of the osmotic properties of dilute aqueous solutions of electrolytes, published in 1886, were of decisive importance for the acceptance of the theory of electrolytic dissociation. He had observed that for certain common substances, notably simple acids, salts, and bases, the value of R in Equation 3 was variably raised above unity. Such substances also lower the freezing point and the vapor pressure of aqueous solutions more than would be anticipated from their molecular weights, as had indeed already been noted by Raoult. Though van't Hoff had no explanation for these findings, he made a table of the values, characteristic of individual electrolytes, of a factor which he called i, by which R in Equation 3 must be multiplied in order to balance his equation.

These observations were of great value to Svante August Arrhenius, who was working in Stockholm on the conductivity of dilute electrolyte solutions, his original results having already been reported in 1884 as his doctoral dissertation to the University of Upsala. This thesis was a systematic study, novel in that measurements were made at considerable dilutions, of many electrolytic solutes and a number of solvents. The key observation was that, though the *specific conductance* of an aqueous solution of any given electrolyte (the conductance of 1 cm^3) increases with increasing *concentration*, *equivalent conductance* (specific conductance times 1000 divided by the equivalence) was found to increase progressively with increasing *dilution* (15, 16). Arrhenius calculated the proportional dissociation (which he called the "activity coefficient" α) for dilute electrolyte solutions. This was defined as Λ/Λ_∞, the equivalent conductance divided by the theoretical value at infinite dilution, at which he assumed that all the constituent ions of the molecules would be free to behave as independent particles. Though he already believed that α gave the proportion of the compound which had dissociated into free ions in solution, he did not at that time venture to claim in print that he had demonstrated it, inferring in his thesis only that α measured the "ionically active" part of the compound, that is, the proportion of ions available to conduct electricity; he feared, quite correctly, that the postulate of dissociation of ions of opposite charge would awaken strong resistance. (Mendeleev, Lord Kelvin, and other eminent persons proved to be tenacious opponents.)

He had written to a number of scientists who he thought might take an interest in his ideas, but got cool responses from all except the young professor of chemistry at Riga, Wilhelm Ostwald. Through Ostwald's enthusiastic support, Arrhenius was able to obtain a fellowship which allowed him to travel and to collaborate with van't Hoff, Ostwald, and others in their own laboratories. Now it is clear that, if electrolytic dissociation occurs in solution, the free ions ought to function osmotically as particles, and that increasing dissociation manifested as increasing conductance should also be reflected in increasing osmotic pressure. For example, if conductance measurements indicate that NaCl is 85% dissociated into Na^+ and Cl^- in a particular solution, then the colligative properties of the solution should reflect the fact that for every mole of NaCl which had been added to make up the solution there should be $0.85 + 0.85 + 0.15$, or 1.85 moles of particles apparently present or—as we would now say, following J. L. Gamble (182)—1.85 *osmoles*.

Since Arrhenius's α and van't Hoff's i could be supposed to offer independent quantitative estimates of the degree of dissociation of an electrolyte in a pure solution of that compound made up to any desired concentration (provided the solution was a dilute one), it remained to compare paired values of α and i for a series of similar dilutions of particular electrolytic compounds. It can be shown (Appendix A) that when this is done:

$$i = 1 + (n - 1)\alpha \quad (4)$$

In 1887 (15) Arrhenius presented comparisons, for similar dilutions of some 88 compounds, of the value of i determined from measurements of osmotic pressure with the theoretical value calculated according to Equation 4 from conductance measurements. With few exceptions, the agreement between the paired values was very satisfactory (Appendix A). Within a few years, therefore, the theory of electrolyte dissociation gained widespread though by no means, at first, universal assent (107, 281, 466, 475).

The principal objection to electrolytic dissociation was not trivial; it was that no explanation was at hand for the rupture in solution of the strong electrostatic link between cations and anions which, so it then seemed, should prevent their behaving like independent particles. Water was at the time regarded as an inert solvent; only much later was it realized that a chemical reaction (hydrogen bonding) occurs between the water and the electrolytic solute (Section A-4). This concept developed gradually, beginning about a generation later, out of the insights of G. N. Lewis and W. Kossel into the nature of valence, and especially out of Niels Bohr's proposals in 1913 concerning atomic structure (475, 281).

NOTES

1. "... the contrast between natural scientists and many social scientists proves instructive. The latter often tend, as the former almost never do, to defend their choice of a research problem ... chiefly in terms of the social importance of achieving a solution. Which group would one then expect to solve problems at a more rapid rate?" (321, page 163).

three

DEVELOPMENT OF ACID-BASE THEORY

A. ORIGINS

Our word *acid* comes from the Latin *acēre,* to be sour, and philology discloses that the archetypal acid of the ancients was vinegar (ὄξος, *acetum:* vinegar; ὄξυς, *acidus:* sour). Souring as a process, as in wine or milk, is a phenomenon so generally familiar that it must from very early times have engaged the attention of persons capable of posing searching questions about the composition of matter. From the time of the alchemists numerous attempts at a chemical definition of acidity are in fact recorded, the most noteworthy names in this connection being those of Paracelsus, Robert Boyle, and Carl Wilhelm Scheele (475, 281, 167, 104, 599, 157, 354). Owing to the want of adequate general chemical theory, however, such efforts were purely speculative; and the concept of acidity remained, down to the chemical revolution of the late 18th century, no more than an empirical generalization, acids being recognized as a class of substances possessing a sour taste and exhibiting the properties of attacking metals, turning blue and violet vegetable colors red, and antagonizing caustics (alkalies).

It is understandable that Lavoisier should have assigned the central role in acid-base chemistry, as in combustion and respiration, to oxygen. He derived our word for this element from Greek roots meaning "creator of acid," and he described O as the "*principe acidifiante,*" considering acids to be oxides of nonmetallic elements like S and P; this concept is fossilized in the German word for O_2, *Sauerstoff*. Supporting experimental evidence was principally the acid produced when S, P, and C are burned (oxidized). Lavoisier's mistake was natural enough at the outset of the chemical revolution before the dawn of organic chemistry; for many acids do contain oxygen, and the oxides of P and S are anhydrides which contain no H, but render water strongly acidic. Alkalies, in his system, were oxides of either metals or alkaline earths—somewhat oddly, in view of the derivation of the word "oxygen." A scientific basis for the difference between metals and nonmetals was not at hand.

It soon became clear, however, that some acids contained no O. Claude Louis Berthollet had presented evidence as early as 1789 that H_2S—a very weak acid, to be sure—is oxygen-free; later it was shown by Davy, by Gay-Lussac, and by Thenard that HCl contains no oxygen, and by Gay-Lussac that HCN contains none (281, 441, 66). By 1830 nearly a dozen oxygen-free acids were known. Electrolytic decomposition of HCl and other evidence had led Sir Humphrey Davy to propose in the first quarter of the 19th century that H is the acidifying principle, a view adopted also by Dulong but not at first generally accepted; an unsatisfactory "dual theory" of acidity was for a time upheld by Berzelius and others (167, 281). Even when H became generally accepted as the element of acidity, after 1840, chemical notation adequate to describe acid-base reactions generally in terms of H^+ transfer could come only very gradually because of the undeveloped understanding, at the time, of chemical radicals and the principles of bonding and molecule formation (281, 167). Above all, the relatively late development of a theoretical basis for understanding equilibrium reactions (I-1) left acid-base chemistry in a very unsatisfactory state until the last quarter of the last century (475, 281, 167).

The dissociation of protons from a polybasic acid, as base is added, offers an avenue to the understanding of acidic intensity; for this reason the studies by Thomas Graham (1833) and especially by Justus Liebig (1838) of such acids represented an important advance in acid-base theory (281, 657a). Since the acidity of hydroacids could not be related to oxygen, Liebig re-interpreted the structure of phosphoric acids in terms of Davy's view that acidity is due to H: "Acids," he wrote, "are hydrogen compounds in which the hydrogen may be replaced by metals" (657a). This concept, essentially correct as far as it goes, was regnant for a generation before the theory of electrolyte dissociation was announced in 1877, and it created a basis for a limited understanding of the behavior of "neutral sodium phosphate" Na_2HPO_4 as an acceptor of H^+ in physiological solutions; indeed, the importance for physiology of exchange reactions between carbonic and phosphoric acids and their salts, for example, was repeatedly invoked after 1840. But of course no quantitative formulation of acid-base equilibria, nor any concept of buffering, was as yet possible (475, 281, 167, 104).

The concept of basicity as the antagonist of acidity has remained remarkably stable since classical times. Very early a variety of compounds including lime, soda, potash, and ammonia were recognized as having in common the ability to negate the acidity of various materials. The alchemists termed such substances *alkali*, borrowed from the Arabic al-qalīy, derived from the word for roasting in a pot; this method of producing "potash" (KOH) from seaweed is of great antiquity. Increasingly, however, from around 1800 the word *alkali* was applied specifically to carbonates and hydroxides of what we still call "alkali metals" and "alkaline earths," and so an increasing need was felt for a more general word indicating the property of negating acidity. Since that time the term *base* has increasingly been invested with this more general sense, which is admirably adapted to contemporary acid-base theory (281, 167, 104).

B. THE IONIZATION THEORY

Arrhenius explained acidity in terms of electrolytic dissociation (I-2C) as a property conferred by the presence of free protons in solution, while basicity was a property conferred by free OH^-. The Arrhenius theory of acidity is obsolete (I-5), but it represented the decisive forward step in our understanding of acid-base chemistry because the reversible dissociation of a proton from an acid is a process which can be treated mathematically as an equilibrium reaction. The kinetic analysis of weak electrolyte equilibria in terms of the mass law introduced by Wilhelm Ostwald (439, 440, 17) is in fact the same as is applied to proton transfers in contemporary acid-base theory (I-5); it allowed the behavior of weak acids and bases in dilute solution to be analyzed deductively, and so led to an exact understanding of *buffering*, that is, of the factors governing the relation between pH change and quantity of acid or base added to or subtracted from solution.

The theory accounted adequately for salt formation and neutralization when equimolar quantities of strong acid and strong base were mixed. Arrhenius held that the strength of an acid HA depended upon the extent to which it dissociates in water:

$$HA \rightarrow H^+ + A^- \quad (1)$$

and the stength of a base B upon the extent of its dissociation in water:

$$BOH \rightarrow B^+ + OH^- \quad (2)$$

This postulate appeared to be confirmed by studies of acidic catalysis by Ostwald which showed good correlation between catalytic activity and the strength of an acid as judged by conductance measurements (439, 440). Except for the fact that the proton in Equation 1 is transferred rather than set free, Arrhenius's formulation of acidic strength was correct, to a near approximation, in the case of weak electrolytes; but the *apparently* incomplete dissociation of strong electrolytes in water cannot be explained in this way. This distinction between the behavior of strong and weak electrolytes in aqueous solution emerged from observations of the effect of progressive dilution upon the proportional dissociation α of an acid in pure water.

It was already well known that the salts of what we call a weak acid yield the corresponding acid when mixed with any strong acid HA; CO_2, for example, is given off by carbonates or bicarbonates which are acidified because they are converted to H_2CO_3:

$$HA + BHCO_3 \rightarrow B^+ + A^- + H_2CO_3(\rightarrow CO_2 + H_2O) \quad (3)$$

This can only mean that the very slight dissociation in water characteristic of a weak acid, as in Reaction 1, is a reversible reaction which can be displaced far to the right by a base which removes H^+, and back to the left when H^+ is then supplied again by adding a strong acid. The ionization represented in Reaction 1 is therefore a reversible one, which according to the Arrhenius theory of acidity could be represented as:

$$HA \rightleftharpoons H^+ + A^- \quad (4)$$

Though the velocities of the forward and back reactions of ionic reactions of this type were far too high to permit experimental evaluation of the velocity constants in his day, Ostwald perceived that if the mass law applied to such reactions they must be characterized by an *acidic dissociation constant* k_A (formerly termed also the ionization constant): this constant would be formally identical to the equilibrium constant of reversible reactions like esterification (I-1C); the value of the acidic dissociation constant would relate $[H^+]$ to the proportion α of acid dissociated. If the forward reaction proceeds at a rate determined at all times by the product of the concentration of undissociated acid and a specific velocity constant, then:

$$v_1 = k_1[HA] \quad (5)$$

In fact, according to contemporary acid-base theory, this is not the case since H_2O is also involved; but the reaction must still be treated in dilute solution as a first-order one, analogous to the hydrolysis of sucrose and of esters in dilute solution, since $[H_2O]$ is essentially unity (I-1C). The back reaction is second-order, like the esterification reaction, and the velocity of recombination is given as:

$$v_2 = k_2[H^+][A^-] \quad (6)$$

At equilibrium, where the velocities are equal:

$$k_1[HA] = k_2[H^+][A^-] \quad (7)$$

and the *acidic dissociation constant* k_A is given as:

$$k_A = \frac{k_1}{k_2} = \frac{[H^+][A^-]}{[HA]} \quad (8)$$

Evidently the acidic dissociation constant gives a quantitative measure of the strength of the acid in aqueous solution. An analogous "basic ionization constant" K_B could be derived for any weak base BOH by substituting OH^- for H^+:

$$k_B = \frac{[B^+][OH^-]}{[BOH]} \tag{9}$$

and would give the strength of the base. This basic dissociation constant is now entirely obsolete for reasons which will become evident (I-5B).

It follows from the mass law that the value of the acidic dissociation constant k_A of an acid will remain unaffected by progressive dilution. Consider 1 mole of a monobasic weak acid dissolved in different volumes of water, and let α be the proportional dissociation of the acid at any volume v in liters. The quantity of each ion present is α equivalent, and the quantity of undissociated acid HA present is $(1 - \alpha)$ equivalent. The corresponding *concentration* of each ion, at any volume of solution v, is α/v Eq/liter and of HAC $(1 - \alpha)/v$ Eq/liter. Substituting in Equation 8:

$$k_A = \frac{(\alpha/v)(\alpha/v)}{(1 - \alpha)/v} = \frac{\alpha^2}{(1 - \alpha)v} \tag{10}$$

or

$$k_A = \frac{\alpha^2}{1 - \alpha} \times \frac{1}{v} \tag{11}$$

This is Ostwald's law of dilution, which states that the value of the equilibrium constant of a weak acid or base obeying the mass law will remain unaffected by increasing the degree of dilution (which varies as $1/v$). Otherwise expressed, the product of the concentration C in equivalents per liter and $\alpha^2/(1 - \alpha)$ should be a constant at all dilutions.

Table I-3.1 presents two sets of experimentally determined values for the acidic dissociation constant k_A of acetic acid.[1] K' is the "apparent dissociation constant" reckoned directly from the concentrations. The relatively small progressive alteration of the value of K' on continued dilution did not prevent early application of kinetic theory by Ostwald and others to weak electrolyte equilibria; later it became possible to correct for the continuous "drift" of K' with dilution by allowance for the effect of ionterionic attraction upon ionic mobilities (I-4B; Appendix C). When this is done and activities are substituted for concentrations in the term $\alpha^2/(1 - \alpha)$, the true thermodynamic dissociation constant K is stable indefinitely in the face of progressive dilution (final column of Table I-3.1).

From conductance measurements and colligative properties (I-2B; I-2C), strong electrolytes appear to be only partially dissociated in concentrated solution; in solutions of the ionic strength of extracellular fluid the apparent degree of dissociation α of Na^+, K^+, and Cl^- is only about 0.85. If the conductance and the colligative properties of strong electrolytes reflect in-

TABLE I-3.1. *The apparent dissociation constant (K') and the thermodynamic dissociation constant (K) for acetic acid at 25° C**

C	$\frac{\alpha^2}{1-\alpha}$	K'	K
0.10000	0.01350	1.85×10^{-5}	1.70×10^{-5}
0.05000	0.01905	1.85	1.72
0.02000	0.02988	1.84	1.74
0.009842	0.04222	1.83	1.75
0.005912	0.05401	1.82	1.75
0.002412	0.08290	1.81	1.75
0.001028	0.1238	1.80	1.75
0.0002184	0.2477	1.78	1.75
0.0001114	0.3277	1.78	1.75
0.00002801	0.5393	1.77	1.75

* Based respectively on data of Kendall (304) and MacInnes and Shedlovsky (370); modified from Maron and Prutton (382, page 444). If C is the molar concentration of the compound, and $1 - \alpha$ is the fraction of the compound remaining undissociated at any value for C, then from Equation 10: the acidic dissociation constant k_A is given as:

$$k_A = C \frac{\alpha^2}{1 - \alpha}$$

This is the mathematical statement of *Ostwald's law of dilution* in its original form based on concentrations; the derivation from activities is similar in principle (104, 370, 382, 482; cf. Appendix C). The law states that, for dilute solutions of electrolytes obeying the mass law, the value of k is unaffected by further dilution.

The values shown for K' are those based on *apparent dissociation*, α being reckoned from conductance data. K' declines about 4% as molarity is reduced from 0.1 to 0.00003. The true thermodynamic dissociation constant K, the derivation of which (I-4B) was not found for nearly a half-century after the postulation of electrolytic dissociation, is stable within less than 2% at all dilutions less than M/20.

Because of its fundamental significance in relation to colligative properties, molality (m) (moles per kilogram of water) is theoretically preferable for many physiological purposes to molarity (C) (moles per liter). In practice we nevertheless generally measure the concentration of particular electrolytes volumetrically as molarity, but determine osmolality by cryoscopy. The difference becomes progressively smaller with increasing dilution, and is small for most biological purposes for extracellular fluid and urine.

complete dissociation, as Arrhenius originally supposed, strong electrolytes should also obey the mass law and its corollaries, including Ostwald's law of dilution; that is, they should exhibit a stable value for k_A as water is added. But they do not do so; on progressive dilution no stable value for k_A is ever found (Table I-3.2). This finding was originally termed "the anomaly of strong electrolytes" (104, pages 250–252). Strong electrolytes are now believed to be completely dissociated even in moderately concentrated solution, their degree of apparent association in solution being ascribed to interionic attraction (I-4B; Appendix C).

TABLE I-3.2. *Effect of progressive dilution upon apparent proportional dissociation of HCl as judged by colligative properties**

m (molality)	i
0.05	1.90
0.01	1.94
0.005	1.95
0.002	1.97
0.001	1.98

* The van't Hoff factor i is the ratio of the colligative effect produced by a particular concentration of electrolyte to the colligative effect of the same molal concentration of non-electrolyte. In the case of an electrolyte of valence Type 1-1 like HCl, the limiting value of i approached at infinite dilution is 2, representing complete apparent dissociation. Data, from Maron and Prutton (382, page 332), are based on observations of freezing point depression but apply to all colligative properties at 0° C; for the body fluids of the mammal, correction for temperature = 37° C must be applied. For a strong electrolyte of valence Type 1-1, apparently 85% dissociated in extracellular fluid, i = 1.85 at 37° C (e.g., Na^+, 0.85; Cl^+, 0.85; NaCl, 0.15).

The ionization theory of Arrhenius is now obsolete, but the contributions of Raoult, Arrhenius, van't Hoff, and Ostwald to solution chemistry remain the firm foundation of our understanding of physiological acid-base chemistry. Before passing in the ensuing two chapters to the later refinements which are part of the contemporary theoretical treatment of the subject, it will be well to place in summary perspective what we owe to the advances already described and the refinements subsequently contributed.

It was the kinetic treatment of the reversible dissociation of weak electrolytes which opened the door to quantitative study of acid-base physiology, because it led to a theoretical formulation of the nature of *buffering* which is still the basis of our mathematical treatment of this biologically vitally important subject. Lacking an adequate theory, Arrhenius's predecessors were not in a position to distinguish clearly, let alone to relate mathematically to one another, the two equally important meanings of the acidity of a solution: its acidic *intensity* and its *total acid content*, i.e., the quantity of acid available, in some volume of the solution, to react with added base and so to retard rise of pH. But it lay on the surface of the ionization theory that the weakness of a weak acid reflected the very low proportion of its total protons dissociating when the acid was added to water, whereas all its potentially dissociable protons are available to antagonize the effect of an added strong base on the acidic intensity of the resulting solution. Nowadays we prefer to say that the extent to which an acid transfers its protons when added to a solution depends upon the quantity and strength of the conjugate bases available in the solution compared with its own conjugate base; nevertheless, the mathematical deduction of the principles of buffering from the mass law in terms of contemporary acid-base theory (I-5D) is identical in principle to that already given.

Titration was used for analytical purposes for many decades before Arrhenius. Titration of the urine to neutrality was applied very early (II-4) in studies of the manner in which men and animals rid themselves of loads of excess nonvolatile acid. Titration involving only strong acids and bases (neutralization in the strict sense) presented no conceptual difficulties; but the remarkable capacity of blood plasma to remain very faintly alkaline in the face of addition of a large quantity of acid was very puzzling to the early investigators of physiological acid-base regulation since they had no concept permitting them to treat *buffering* in a general quantitative manner. The sour taste of acids, the speed with which they attack metals, and the change in color of certain vegetable dyes in solution during the course of (or as the endpoint of) a titration were all recognized as reflections of what we now term acidic intensity; but there was no continuous scale available by which to quantify it, and no concept available to relate it in a quantitative way to the quantity of acid present in solution.

The assimilation of electrolytic dissociation into kinetic theory provided a general rule for relating changing acidic intensity to changing ratios of buffer species—a contribution of enormous practical importance to acid-base chemistry, even though the accuracy and certainty of the calculations would later be improved still further (Table I; I-4B). From the conceptual standpoint it now became necessary to think of electrolyte solutions as containing individual ions, rather than the electrolytic compounds which had been added to water to make up the solution. Figure I-3.1 vividly illustrates the magnitude of the conceptual problem confronting the early analysts of blood plasma. Chemical analysis of the concentrations of substances like glucose or urea, which maintain their identities in water solution, can be expressed in terms of the respective quantities of individual substances which must be added to a volume of water to constitute the solution; such a concept is inadequate for a complex electrolyte solution. When an electrolyte solution is prepared by adding acids and salts to water, independent ions like Na^+, Cl^-, and K^+ are present which are derived from *compounds* like $NaHCO_3$ and HCl, in which the ions are not free to move about independently of each other before the compounds are added to water. Kinetic treatment of acid-base equilibria made it clear for the first time that, after weak electrolytes have been added to water, the particles present as ions cannot be assumed to be those which have been added; a transformation has taken place, as when ester, alcohol, acid, and water are mixed, making it impossible to state on the basis of a quantitative analysis of the solution exactly what reagents were added to water to constitute the solution (Figure I-3.1).

The constitution of an ionic solution prepared from solid reagents and

1000 Gr. Serum[22]).

Wasser	908,84
bei 120° nicht flüchtige Stoffe	91,16
Albumin etc.	82,59
unorganische Bestandtheile	8,57

Chlor	3,565	=	schwefels. Kali	0,283
Schwefelsäure	0,130		Chlorkalium	0,362
Phosphorsäure	0,146		Chlornatrium	5,591
Kalium	0,317		phosphors. Natron	0,273
Natrium	3,438		Natron	1,545
phosphors. Kalk	0,300		phosphors. Kalk	0,300
— Magnesia	0,220		— Magnesia	0,220
Sauerstoff	0,458			

Summe der unorganisehen Bestandtheile 8,574

Dichtigkeit = 1,0292.

FIG. I-3.1. An early analysis of the chemical composition of blood plasma, made by Carl Schmidt in 1850 (542). Reproduced in Gamble (185). A quarter-century before the theory of electrolytic dissociation, salts were thought of as existing in solution entirely as compounds; but it is not evident, in the provenance assigned to the ions in the *right-hand column*, on what principle K was assigned to sulfate and Na to phosphate. (The salts of carbonic acid were overlooked in Schmidt's analysis.)

water is very simply described in terms of the ionization theory, more simply, indeed, than by contemporary theory (see below). Brønsted-Lowry theory (I-5), on the other hand, greatly simplifies the treatment of acid-base reactions occurring in solution by focusing attention exclusively on the proportional acceptance by all conjugate bases present of any protons contributed to the solution by addition of acid, and the proportional donation by the several acids present when a base is added. The theory eliminates the special role accorded to OH^- by Arrhenius, and in consequence its purview does not specifically include the dissociated salts which result, with water, from reactions of acidic and basic compounds (e.g., Example 15). A corollary of this greater simplicity is the elimination of the superfluous "basic dissociation constant" (I-5B). Riggs has written (508, page 258)—very correctly in relation to shifts of physiological acid-base equilibria brought about by stated additions or subtractions of acid or base—that "the custom of writing the [Henderson-Hasselbalch] equation in terms of the concentration of the salt of an acid can lead only to confusion, and (like the equally confusing 'base dissociation constant' K_b) should be abandoned. The transport mechanisms for ions which the physiologist studies commonly deal with single ionic species, and often manifest a quite extraordinary ability to distinguish between such apparently similar ions as sodium and potassium, or bromide and iodide. Aside from having to fulfill the electrostatic requirement that in volumes of any considerable size the sum of the positive charges must equal the sum of the negative charges, these mechanisms know nothing about 'salts.' The physiological scientist should be at least as discriminating as the mechanisms he studies!"

Contemporary acid-base theory describes with admirable succinctness the shifts of acid-base equilibrium resulting from transfer of protons into or out of a solution; but it does not take explicit account of the manner in which electroneutrality is maintained, and the resulting constraints upon ionic independence in extracellular fluid and urine must be separately examined. Equations written in terms of contemporary acid-base theory do not display in full detail the electrical equivalence involved in the shifts of acid-base equilibria resulting from dietary intake and urinary (and extrarenal) loss of electrolytes. For example, we can write (I-5), quite correctly according to Brønsted-Lowry theory:

$$H^+ + NH_3 \rightarrow NH_4^+ \qquad (12)$$

to represent the effect upon the ammonia buffer pair of adding HCl to solution. This notation ignores, as irrelevant, the fact that there must be an anion present to preserve electroneutrality, in this case Cl^-; but it is the necessity for particles of H_3O^+ and other cations always to be balanced by an equal number of anionic negative charges which frequently makes it convenient to speak of the "matching" of a particular cation against a particular anion in the urine and extracellular fluid, or the "covering" of urinary anions by equivalent cations. "Matching" of charges is examined in more detail in connection with the electrolyte structure of the plasma (III-1) and with the significance of the mineral content of the diet in relation to the acid-base load (II-5).

In contrast to contemporary theory, the ionization theory described the acidic intensity of a solution as a ratio of *salt* to acid, that is, as a ratio of *chemical compounds*, and it was in this form that Henderson wrote his equation for the CO_2 equilibrium in extracellular fluid (I-4A):

$$H^+ = k_A \frac{H_2CO_3}{NaHCO_3} \qquad (13)$$

Nowadays we express the last term as a ratio of acid to *base*, corresponding to a rearrangement of Equation 8 (see also I-5):

$$H^+ = k_A \frac{HA}{A^-} \qquad (14)$$

Let use see why the two ratios have the same meaning.

There are two different ways in which we can make up any nonvolatile buffer solution. Assume, for example, that we desire to prepare an acetate buffer consisting of equal parts of acetic acid HAc and acetate Ac^-; we can do so as follows: (i) add to water 0.5 mole of acetic acid HAc and 0.5 mole of sodium acetate NaAc; alternatively, (ii) add to water 1.0 mole of HAc and 0.5 mole of NaOH. The reason why the two procedures produce an identical solution is that in the second case a reaction takes place, the base reacting extensively with the acid to produce the (dissociated) salt, plus water:

$$HAc + NaOH \rightarrow Na^+ + Ac^- + H_2O \qquad (15)$$

Contemporary acid-base theory deals with the acid-base properties of the solution and is not concerned with how it was prepared. Reaction 15, however, directs our attention to the presence in solution of a salt, NaAc, which may either be added as such to the solution or formed quantitatively in solution by the reaction of a weak acid with a strong base.

Inasmuch as the organism can expend energy for selective ion transport, acid or base can be *generated* in extracellular fluid from ingested neutral dietary salts like NaCl; the overall physiological process is in effect a reversal of spontaneous formation of salts by the reaction of acids and bases as in Reaction 15. Consider, for example, what happens when NaCl is added to extracellular fluid, and one of the ions which has gained its independence in solution, the Na^+, is subsequently removed from extracellular fluid in company with a different anion, namely HCO_3^-. Such overall, physiologically linked, processes must very frequently be analyzed by the clinician in terms of their net effect on the composition of extracellular fluid; the example chosen is illustrated every time NaCl solution is administered to a patient during the correction of nonrespiratory alkalosis. Under such circumstances, kidneys whose regulatory powers have been restored by rehydration or any therapeutic means can be expected (II-6) to retain the Cl^- while causing the Na^+ to be excreted in the urine "matched" against some of the HCO_3^- present in excess in extracellular fluid; the result will ordinarily be the desired acidification of extracellular fluid. (Persistence of alkalosis under these circumstances is discussed in III-2B.) The acidifying process will be reflected in the electrolyte structure of extracellular fluid, where some of the HCO_3^- previously "matched" against Na^+ will have been replaced by Cl^-:

$$(Na^+ + Cl^-) - (Na^+ + HCO_3^-) = + Cl - HCO_3^- \qquad (16)$$

In terms of the Arrhenius theory we would represent the overall process as the replacement of the alkaline salt $NaHCO_3$ in solution by the neutral salt NaCl, which is equivalent to addition of acid. Since $[H_2CO_3]$ and $[CO_2]$ are held constant by the lungs, it is as if HCl had been added to extracellular fluid:

$$HCl + HCO_3^- \rightarrow + Cl - HCO_3^- \ (+H_2CO_3) \qquad (17)$$

Whenever some of the HCO_3^- normally present in the plasma has been replaced by Cl^-, we are seeing the reflection, in the electrolyte structure of the plasma, of accumulation of nonvolatile acid. The accumulation may represent simple net addition of the acid to extracellular fluid (Example 17), or it may result from renal regulatory action in which a basic salt (e.g., $NaHCO_3$) is being eliminated in the urine while a neutral salt of the same element (e.g., NaCl) is being ingested (Example 16). Evidently replacement of some plasma Cl^- by HCO_3^- must, conversely, represent accumulation of base in extracellular fluid; this will happen if the kidney excretes Na^+ and Cl^- while $NaHCO_3$ (or NaOH or any other alkaline Na salt which must react with H_2CO_3) is being taken into the body. Renal tubular alkalosis due to potassium deficiency (III-2B) provides an instance where the kidney may permit a pathological accumulation of base in extracellular fluid by inappropriate conservation of HCO_3^- at the expense of Cl^-. It follows, too, that accumulation of Na^+ and Cl^- in equivalent excess above the usual concentrations in the plasma has no acid-base significance, though such plasma is in general hyperosmolal; an equivalent excess of both strong base and strong acid in extracellular fluid is tantamount to accumulation of neutral salt.

Substitution of Cl^- for HCO_3^-, or vice versa, in the electrolyte structure of plasma can also come about by metabolic processes. When an organic salt of Na, like the citrate or the lactate, is ingested, the organic anion will enter the Krebs cycle and be metabolized to CO_2; in the course of this transformation it is necessary that a proton be detached from the buffers of the extracellular fluid from each monovalent organic ion metabolized (II-5). It is as if base has been added to extracellular fluid; the overall effect is that extracellular [HCO_3^-] increases as well as extracellular [Na^+]. For this reason dietary Na is termed a *base-forming* element. Conversely, when the chloride of a dibasic amino acid, like arginine chloride, is ingested and the organic portion is catabolized, the NH_3^+ group of arginine against which Cl^- had been matched must transfer a proton to the extracellular fluid buffers before it can be incorporated into urea (or into NH_3); the net effect is as if HCl had been added to extracellular fluid (II-5). Dietary Cl is accordingly an *acid-forming* element.

It follows that the ultimate effect of the dietary salts upon acid-base balance normally depends principally upon the dietary company kept by the mineral ions present in those salts. No dietary Na salt can ever be acidifying; but whereas dietary organic salts of Na containing a metabolizable anion act in the net like strong bases, NaCl has no such effect, because the base-forming character of Na is negated by the acid-forming character of the Cl. No dietary salt of Cl can ever be alkalinizing; ingested salts of Cl containing a cation metabolizable to a neutral substance (e.g., the NH_3^+ group of a dibasic amino acid) act in the net like a strong acid, whereas NaCl does not contribute to the acid-base load. K, Ca, and Mg may be ingested, like Na, as salts of metabolizable anions; and acid-base analysis of the mineral content of the diet accordingly classifies the alkaline earths and alkali metals as *base-forming elements.* P and S, like Cl, are *acid-forming elements* (II-5B).

C. CATIONS, ANIONS, FIXED BASES, FIXED ACIDS

Considerable confusion has arisen in the literature as a result of overestimation of the implications for our understanding of acid-base balance of redefining acids and bases in accordance with contemporary theory (II-5). The fact that Cl^- is a Brønsted base is not directly relevant to the physiological and clinical acid-base significance of the presence of this ion in the diet and to changes in its concentration in extracellular fluid. Cl^-, once present as an independent particle in extracellular fluid, never under any circumstances participates in physiological shifts of acid-base equilibria as described by contemporary theory, because it is always completely dissociated: Cl^- in extracellular fluid never *functions* as a Brønsted base. But when an increased proportion of the total plasma anion electrically matched against total plasma cation is made up of Cl^-, it is a reflection of accumulation of *acid* in the plasma. The explanation of this seeming paradox has been furnished in Equation 17, which shows that adding HCl, or any net physiological process equivalent to adding HCl, to extracellular fluid will cause HCO_3^- to be replaced by Cl^-, since the HCO_3^- disappears by conversion to H_2CO_3 and CO_2. And addition of dietary Cl^- not matched with Na^+ or another nonmetabolizable alkali cation to extracellular fluid is (from the overall standpoint) equivalent to adding HCl: that is why dietary Cl is an "*acid-forming*" element—the principal one, because of its predominance among the mineral anions of extracellular fluid (III-1B2).

Na is neither a base nor an acid in the Brønsted-Lowry system. But it is physiologically a *base-forming element,* like the three other principle alkali cations of the extracellular fluid, K^+, Ca^{++}, and Mg^{++}. None of these ions is a Brønsted-Lowry base, but their presence in the plasma *reflects* the addition of sufficient base to bring the pH of this CO_2-containing solution to the faintly alkaline range (II-4A; II-5B). In addition, rise of their total plasma equivalence—and this means essentially rise of $\{Na^+\}$, since the concentrations of the other three are small and their capacity to change concentration circumscribed by toxicity—signifies net accumulation of additional base, unless the excess Na^+ is matched by an equivalent excess of Cl^- (III-1B2). This terminology has been criticized on the reasonable enough ground that it will be confusing to those who do not understand that the words "acid" and "base" used in this context do not refer to Brønsted-Lowry acids and bases, or (for that matter) to Arrhenius acids and bases. For any reading of the older physiological literature, however, the meaning of these expressions must be understood, and the physiological relations which they describe are as important as they ever were for understanding acid-base balance (II-5) and for interpreting the electrolyte structure of the plasma (III-1B2).

In this book the terms "fixed base" and "fixed acid" are avoided. "Fixed base" is a term reflecting the venerable concept of plasma as a solution of CO_2 to which (among other solutes) alkaline salts of Na, K, Ca, and Mg had been added. In Gamble's exposition (182) of renal mechanisms conserving base when an acidic load must be eliminated, the "fixed bases," which can only be derived as such from the diet or bodily stores, are contrasted with NH_4^+ which can be derived at the expense of neutral urea in additional quantities as needed to "spare base": NH_4^+, like Na^+, K^+, Ca^{++}, and Mg^{++} can be "matched" electrically in the urine against the anions derived from mineral acids being added to extracellular fluid. The analogous term "fixed acid" is apparently of later coinage, and is radically ambiguous. Many authors—e.g., Smith (577, page 376), Filley (165, page 113), and Lemann and Lennon (333b)—use it as a synonym for nonvolatile acid. Christensen (94b, Chapter 10), however, distinguishes the fixed anions Cl^- and $SO_4^=$, which cannot enter into shifts of acid-base equilibrium, from the buffer anions, including the nonvolatile ones (phosphate and proteinate) as well as HCO_3^-, which can and do participate. Phosphate has special importance among the nonvolatile buffer ions of extracellular fluid in that as crystalloids the ions $H_2PO_4^-$ and $HPO_4^=$ pass freely, except for a portion bound to plasma protein, into the glomerular filtrate and thereby provide the principal nonvolatile buffer of the urine. Plasma pH 7.4 defines partial "neutralization" of phosphoric acid, the ratio $H_2PO_4^-:HPO_4^=$ being 1:4 in extracellular fluid, whereas in well acidified urine practically all the phosphate exists as $H_2PO_4^-$, the filtered conjugate base having been made wholly available as an acceptor of secreted acid.

NOTES

1. The acidic dissociation constant is symbolized by k. This usage is adopted because until the 1920's acid-base equilibria were in fact calculated from apparent concentrations, but the symbol K often used was that now applied to thermodynamic calculations based on activities (I-4; I-5).

four

pH, pK, *K′*, pK′

A. THE HENDERSON-HASSELBALCH EQUATION

Early studies of enzymatic catalysis played an important role in the application, after the turn of our century especially, of the theory of electrolytic dissociation to the elucidation of buffering. Why this should have been so is readily grasped. The velocities of many enzyme-catalyzed reactions are greatly affected by small changes in pH in the vicinity of the enzyme's pH optimum; stabilization of the acidic intensity of the medium by buffers may be a necessary condition for the continuation of the reaction where acid or base is produced, and is a requirement for comprehensive study of the kinetics of such reactions.

Fernbach and Hubert (163) introduced the French word *tampon*, subsequently rendered in English as *buffer* (Appendix B), in explanation of the enhancement by phosphate salts of the velocity of certain reactions catalyzed by enzymes (amylase and diastase) used in the brewing industry. These reactions were found to be inhibited by dibasic sodium phosphate, Na_2HPO_4, alone, but accelerated if acid was also added; yet, as these investigators were able to show by using suitable indicators, addition of any more acid than was required to react with all the dibasic phosphate present retarded the reactions. Since acid converts Na_2HPO_4 to NaH_2PO_4, this appeared to mean that the simultaneous presence of both salts served to maintain reaction velocity. Fernbach and Hubert could show further that the particular phosphate salt used (whether Na or K) was a matter of indifference to these results; and they correctly inferred that the phosphate mixture promoted the velocity of the reaction, not through specific catalysis by the phosphatic ions, but because the presence of the two salts stabilizes the reaction of the solution within narrow limits which define near-optimal conditions for enzymatic catalysis.

The year 1909 was a notable one for acid-base chemistry and physiology. Over the course of almost a century before, beginning with J. J. Berzelius, it had been realized increasingly that interactions between phosphatic salts, carbonic acid, and bases must be important in animal chemistry; and the striking ability of the organism and of mammalian blood to "neutralize" large quantities of acid had been repeatedly remarked by physiologists in the latter half of the 19th century (II-4). In 1877 Walter (645) had reported that massive intravenous administration of acid always killed his experimental animals before the alkaline reaction of the blood had been abolished; he observed the characteristic effects of acidification upon respiration, and arrived at the important insight that voluminous alkaline reserves, including the salts of carbonic acid making up much of the "total CO_2" of the blood, were of crucial importance in the defense against excess acidity. In the first years of the present century Höber introduced potentiometric estimation of the acidic intensity of blood plasma with platinum-coated electrodes (264a); the method was rapidly improved by Höber and others (265, 265a, 177, 176a, 157a) and at once became an important tool for physiological acid-base investigation. The extensive physiological observations of acidity regulation of the last half of the 19th century were reviewed and brought into relation with the then recent clarification of the chemical basis of buffering by the ionization theory, and with precise estimations of the acidic intensity of blood, in a monumental paper written in German and published in 1909 by Lawrence Henderson (237). It is to Henderson himself that we owe not only our grasp of the central importance of the CO_2 buffer system in stabilizing the reaction of extracellular fluid, but also the realization that this special buffering efficiency of CO_2 depends upon its being a gas (I-6). In honor of these contributions, we still refer to the following equation:

$$[H^+] = k \frac{[CO_2]}{[HCO_3^-]} \tag{1}$$

as Henderson's equation. It will be seen to be Equation 8 of Chapter I-3, rearranged and applied to a volatile buffer system. In point of fact Henderson originally wrote it as follows (236):

$$H^+ = k \frac{H_2CO_3}{NaHCO_3} \tag{2}$$

bases being thought of at that time as salts; the practice of symbolizing $CO_2 + H_2CO_3$ as H_2CO_3—the distinction had not then been recognized—lasted well into the 1920's and constitutes one source of confusion for contemporary readers of classic articles by Van Slyke, Gamble, and others. "Total CO_2" generally is understood to include HCO_3^-, and we still have no entirely satisfactory designation or symbol for $CO_2 + H_2CO_3$, which indeed is needed only for specialized purposes: since H_2CO_3 is only about 1/400 the total at 37° C (I-6), $[CO_2 + H_2CO_3] \cong [CO_2]$.

In the course of his studies of enzyme kinetics, S. P. L. Sørensen of Copenhagen explicitly formulated the distinction between the quantity of acid present in a solution and its acidic intensity. In a paper published also in 1909, in both French and German (581, 582)—one which will always be a landmark in the development of acid-base theory—Sørensen examined the relation between the "*total acidity*" of a given volume of solution, determined by titration, and its acidic intensity, which he expressed as its pH (in actual fact the symbol he used was $P_{H}\cdot$). The pH was defined as the reciprocal of the logarithm of $[H^+]$ (or practically, $-\log [H^+]$), and colorimetric and the potentiometric methods then available for its determination were discussed in detail. Clark, whose textbook of 1920 (102) was influential in disseminating and systematizing the new concepts of acid-base equilibria, has paid tribute (104, pages 265 and 269) to the originality and power of the new acid-base parameter.

The past decade has witnessed a surprisingly extensive reversion in clinical circles and in the clinical literature to the use of "$[H^+]$" for expression of the acid-base intensity of extracellular fluid and other aqueous solutions of weak electrolytes. This is a retrogressive tendency which can only be deplored (260, 123, 165). It seems to have arisen originally on the basis of the bizarre notion that the original introduction of the pH scale had no justification other than a lazy man's disinclination to transform potentiometric data into more "natural" units (278), and to have been reinforced, in part by treacherous analogies between protons and the fully dissociated solute ions, especially Na^+, of physiological solutions (338, 278), and in part by a disposition, not unfamiliar in current educational doctrine, to reduce the difficulties of a subject for students by eliminating the conceptual preparation required for its proper understanding (338, 170).

The analogy between $[H_3O^+]$ and other particles in extracellular fluid has not ceased to be urged. In a recent rear-guard action on behalf of "H^+," Rector (487a) pointed out reasonably enough that important physiological functions like rates of particle movement are directly related (as in first-order kinetics) to concentration, and inquired why the fact that the chemical potential of a substance is a logarithmic function of its concentration should cause us to express only acidic activity in logarithmic unitage. The reason is surely that solute concentrations in extracellular fluid are of prime physiological importance because of their chemical significance. They give a clinically acceptable approximation to activities, which in turn exactly express each solute's contribution to the colligative properties of the solution (I-2B), which are of major significance in physiology and medicine. In contrast, the H_3O^+ and OH^- present in water are not solutes; nor do $\{H_3O^+\}$ and $\{OH^-\}$ ever contribute appreciably to the osmolality of extracellular fluid or other buffer solutions of weak electrolytes. An impression gained from the recent clinical literature that $\{H_3O^+\}$ is a concentration in exactly the same sense as the concentration of the solute cations of extracellular fluid will scarcely promote understanding or even realization of the unique behavior of H_3O^+ and OH^- in physiological solutions. Students and physicians so schooled may (for example) find themselves at a loss to understand, if indeed they are aware of, the fact that the "H^+ concentration" of the plasma or urine does not decline like that of Na^+ and of solutes generally when water is added; nor, of course, does $[OH^-]$ (p. 9).

Rector appears further to suppose (487a) that a choice should be made between $\{H^+\}$ and pH on the basis of experimental data which will indicate whether the distribution of random samplings of the blood pH, or that of blood $\{H^+\}$, more closely approaches a normal distribution. It is not easy to grasp how the statistical distribution of blood pH, a biological variable, can touch the physicochemical question under discussion, nor why the distribution of randomly sampled values of blood pH should be of any greater interest from the purely chemical standpoint than those of the urinary pH. One has only to think of the effect of the diet—and, via the diet, the influence of the species—upon the distribution of values of urinary pH in mammals (II-4A) to perceive the irrelevance to chemical theory of histograms of the urinary acidic intensity observed in a normal mammalian population of any type; for the type of frequency distribution obtained will depend principally upon the distribution of acid-base loads in the individuals studied. So also, for that matter, must the distribution of values of blood pH (II-6); but the very narrow range of normal values seems likely to inhibit if not frustrate statistical studies of the type recommended by Rector for whatever purpose they might serve.

Sørensen termed pH "the hydrogen ion exponent"—pH 7.0 representing, for example, the exponent, with sign changed, of the molar concentration of H_3O^+, 10^{-7} Eq/liter—and defined it as the logarithm of $[1/[H^+]$, or (practically) $-\log [H^+]$. According to contemporary acid-base theory, the pH scale is the thermodynamically preferred means of expressing acid-base intensity, of which it is the prime operational estimate (Section B2); $[H^+]$—or better $[H_3O^+]$—is a derived expression which cannot in practice be obtained with perfect accuracy from pH even in dilute aqueous solution. The operational primacy of pH measurements in practical acid-base work has given rise to an inclination opposite to the vogue, just considered, for expressing acidic intensity as "$[H^+]$," a tendency which has been pushed to the point of questioning the propriety of deducing the Henderson-Hasselbalch equation from the mass law (338). As H. N. Christensen (96) has very justly said: "It is possible to emphasize the minor uncertainties to the point where we make pH very difficult for the medical student. If for the sake of rigor, we prevent the student from being able to conceive of the pH as approximately the [negative] log of the hydrogen ion concentration, I think we have gone too far."

It was soon seen to be advantageous to derive a new variable, pK, from the dissociation constant K of an acid or base. This was defined as the logarithm of $1/K$, in analogy with the derivation of pH; this variable is not only convenient for calculations of electrolyte equilibria based on pH values, but is the thermodynamically preferred expression of the strength of an acid (Section B2). Henderson's equation (Equation 1) could then be transformed by taking negative logarithms of all terms, as follows:

$$\text{pH} = \text{pK} + \log \frac{[HCO_3^-]}{[CO_2]} \qquad (3)$$

This equation was introduced in 1917 by K. A. Hasselbalch of Copenhagen (230). The general equation (I-3, Equation 6) similarly becomes:

$$\text{pH} = \text{pK} + \log \frac{[\text{base}]}{[\text{acid}]} \qquad (4)$$

This is the fundamental relation of physiological acid-base chemistry, the renowned Henderson-Hasselbalch equation. It is not exact until activities are substituted for concentrations; the derivation of activities from concentrations is explained in the ensuing section.

B. THERMODYNAMICS

Thus far we have considered chemical equilibria only from the standpoint of kinetic theory. However, like the gas law, the law of mass action is applicable only under well defined, limited conditions. Real gases approach ideal behavior—that is, they obey the gas law (I-2, Equation 1)—at low to moderate pressures and ordinary temperatures; but their behavior deviates markedly at higher pressures or very low temperatures, i.e., at high densities, when the molecules are close enough to be attracted to one another by intermolecular forces. In this sense the gas law is a "limiting law," applying insofar as gas molecules can be treated as point particles exerting negligible attraction for one another. Similarly, dilute solutions of non-electrolytes, which we may define roughly as M/5 or less, approach ideal thermodynamic behavior in that their colligative properties observe simple relations which can be deduced with a high degree of accuracy from their molecular weights, and vice versa.

Dilute electrolyte solutions, however, including plasma and extracellular fluid, are affected by electrostatic attraction between dissociated ions bearing opposite charges which cause these to behave as if they were partly associated, when in reality it is their freedom to move independently which is partially constrained. Such forces affect the behavior of solutions of strong and (to a lesser extent) of weak electrolytes, causing deviation from behavior predictable from the mass law. In the case of weak electrolytes, interionic attraction can of course affect only the very small proportion of total compound which is dissociated, and so causes relatively small deviations from the equilibrium states which would be predicted according to the mass law; nevertheless, the deviations are large enough to prevent us, in precise physiological work with extracellular fluid and urine, from dealing directly in concentrations when using the thermodynamic dissociation constant K. Peter Debye and Erich Hückel, however, found a theoretical way (see below and also Appendix C) of accounting quantitatively for the interionic attraction responsible for the "anomaly of strong electrolytes"—their failure to obey the mass law—which is applicable also to interionic attraction in dilute solutions of weak electrolytes such as extracellular fluid and urine. Before considering their theory in more detail, we shall briefly review its background in the thermodynamics of solutions.

1. Thermodynamic Treatment of Nonideal Solutions

The most general theoretical analysis of chemical equilibria is that developed by deduction, particularly by J. Willard Gibbs (193), from the first and second laws of thermodynamics. The thermodynamic approach, as it does not depend upon any postulations about molecular events for its validity, furnishes the most authoritative, as well as the most abstract and general, account of the equilibrium state. Gibbs deduced a new thermodynamic quantity, the "free energy" now generally symbolized G in his honor, which allows us to determine whether any particular spontaneous change, including reversible chemical reactions in solution, can occur within a system. Spontaneous change can by definition occur only if it is accompanied by a decrease in the free energy of the system, ΔG; accordingly, equilibrium is the state in which the free energy of the system is at a minimum. We have already noticed (I-2B) the very striking analogy between a dilute solution and an ideal gas when the osmotic pressure π is treated as the analog of the gas pressure P, which led van't Hoff to deny to the solvent any essential contribution to osmosis. It is now considered that van't Hoff's equation, though an approximation useful in dilute solutions, cannot be applied generally because osmotic pressure depends rather on the chemical potential of the solvent than on the concentration of solute.

It was Gilbert Newton Lewis (343, 344) who succeeded, a generation after van't Hoff proposed his equation, in treating nonideal systems thermodynamically by introducing two additional thermodynamic quantities, *fugacity* and *activity*. The fugacity f is a measure of the *escaping tendency*, a perfectly general property of constituents of any phase; and equilibrium obtains when f is equal for all constituents throughout the system. For example, water vapor over a solution is in equilibrium with the liquid phase when f for water is equal in vapor and liquid phases. Evidently any evaporation or condensation of water involves a change in free energy. The total free energy of G of any system is always unknown, whereas the *change* in free energy ΔG of a system in passing from one state to another, as for example in ionization of an electrolyte, can be experimentally determined. Similarly, f is an undefined quantity, but it can be evaluated in relation to an arbitrary reference state. Lewis showed that this relation is given by:

$$G = G^{o}\ RT \ln \frac{f}{f^{o}} \quad (5)$$

where G^{o} and f^{o} are values for the reference state. The *activity* $\boldsymbol{a}$ is then defined as f/f^{o}, and for any transformation, such as a shift of chemical equilibrium, the change in free energy is given as:

$$\Delta G = nR\mathrm{T} \ln \frac{\boldsymbol{a}_2}{\boldsymbol{a}_1} \quad (6)$$

Equation 6 permits thermodynamically exact deductions, because the evaluation of activities takes into account mutual attraction between molecules, interactions between solute and solvent in a solution, and the effects of electrical fields on charged particles.

Table 1 of the preceding chapter showed that Ostwald's law of dilution is obeyed only approximately during progressive dilution of solutions of weak electrolytes when the dissociation constant is calculated from apparent concentration data. The drift of K' with continuing dilution was not enough to discredit the early kinetic treatment of electrolyte equilibria, and it is eliminated by substitution of activities for concentrations in the Henderson-Hasselbalch equation. The activity, a thermodynamic property, is derived in the case of any ion in solution by the equation of Debye and Hückel, which takes into account electrostatic attraction between ions of different charge. Their theory is discussed below; it is also applicable to the behavior of strong electrolytes which do not obey the mass law (Table I-3.2).

Evidence that it is the attraction between the oppositely charged ions which accounts for the apparently incomplete dissociation of strong electrolytes, as judged by their conductance and colligative properties, was presented in 1907 by Niels Bjerrum, who made use for this purpose of what he called the *osmotic coefficient*, g, defined as the ratio of an observed colligative property to what would be found were the electrolyte fully dissociated. Evidently the osmotic coefficient of a completely dissociated electrolyte exhibiting no interionic attraction would have a value of 1. Fractional values for g would express the degree of proportional dissociation according to the Arrhenius concept; $1 - g$ would be zero for a completely

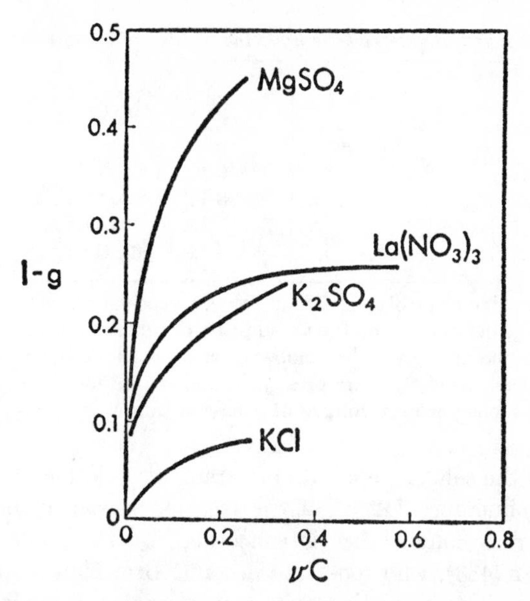

FIG. I-4.1. Plot of $1-g$ against the ion concentration, νC, for strong electrolytes Modified from Prutton and Maron (482, page 219). ν is the number of particles yielded by the dissociation of one molecule of electrolyte; $1-g$ is apparently undissociated electrolyte, judged by colligative properties, expressed as a proportion of the total compound present. The curves are representative of compounds of similar valence type at the dilutions shown; at higher concentrations individual differences between compounds of common valence types appear.

dissociated compound, and unity for a wholly undissociated one. When for a variety of strong electrolytes $1 - g$ is plotted against νC, the concentration of ions in solution assuming complete dissociation, families of curves are found, segregating within the range 0 to 0.2 molar according to the valence type of the compound (Figure I-4.1); at greater concentrations individual differences between compounds belonging to the same valence group appear. All curves originate at 0, for at infinite dilution complete dissociation is postulated. As the concentration of ions is increased to around 0.2 molar, the influence of the valence type becomes evident (Table I-4.1); from the samples shown it may be seen that $1 - g$ increases as a function of the number of charges in a unit volume of solution containing the same number of ions (Table I-4.1). This influence upon colligative properties exerted by the number of charges present in volumes of solution containing the same number of ions cannot be explained by Arrhenius' concept of partial dissociation of strong electrolytes.

In 1923 Debye and Hückel, working in Zürich, deduced from Coulomb's law and from evaluation of the probable distribution of ions in solution an expression for the magnitude of the effect exerted by the "ionic

TABLE I-4.1. *Effect of valence type upon the number of charges per ion in solution**

	Particles per unit volume of solution†	Number of charges per unit volume of solution†
KCl	6 K^+, 6 Cl^-	12
K_2SO_4	8 K^+, 4 $SO_4^=$	16
$La(NO_3)_3$	3 La^{+++}, 9 NO_3^-	18
$MgSO_4$	6 Mg^{++}, 6 $SO_4^=$	24

* Representative electrolytes of four valence types are listed in order of increasing mean charge per particle on the assumption of complete dissociation in solution. The right-hand column gives the number of charges which will be present if 12 ions derived from each electrolyte are present in a given volume of solution.

† Assume 12 ions per unit volume of each solution.

strength" of the solution upon its apparent dissociation constant and also upon its conductance (123a). Their treatment was further improved in 1926, insofar as conductance is concerned, by the Norwegian physicist, Lars Onsager (433), who took into account Brownian motion of the ions, as well as the electrophoretic effects and relaxation of the ionic atmosphere produced by an applied current which had not been considered by Debye and Hückel. The attraction between oppositely charged ions may be visualized as restraining the freedom of independent movement which the particles would have if they carried no charge; such constraint on their mobilities produces an effect upon the colligative properties of a solution of strong electrolyte similar to that of incomplete dissociation, and causes the values of g to fall below unity. These findings, which Arrhenius sought to account for in terms of incomplete dissociation, are given a satisfactory, quantitative explanation by the Debye-Hückel theory. At least in dilute and moderately concentrated solution, strong electrolytes are now believed to be wholly dissociated.

The theory of interionic attraction permits calculation of activities from concentrations and therefore allows us to correct in a satisfactory way for the apparent failure of dilute solutions of weak electrolytes to conform with a high degree of accuracy to behavior predicted by the mass law (Table I-3.1). Estimated concentrations can be converted into activities by multiplying by γ, the "activity coefficient" appropriate to the ion of interest to us:

$$a = \gamma C \tag{7}$$

Equilibrium calculations can then be carried out with the thermodynamic dissociation constant (Appendix C, Equation 3). In practice it is more convenient in physiology to calculate the apparent dissociation constant K' from K and the ionic strength μ of the solution; using K' we can work directly with the concentrations of compounds and pH to calculate shifts of ionic equilibria. For physiological purposes K' may be calculated according to an approximation to the Debye-Hückel limiting equation:

$$K' = K \pm 0.52\sqrt{\mu} \tag{8}$$

(see Equations 14 and 15, Appendix C).

The apparent acidic dissociation constant of an acid or base may be greater or less than the thermodynamic constant, depending on whether or not the acidic congener is ionized.[1] We cannot follow the derivation of the theory of iterionic attraction from Coulomb's law, but an understanding of the manner in which the equation of Debye and Hückel is applied to solutions of weak electrolytes is important for us because it is the basis of all precise work with physiological electrolyte equilibria. In Appendix C is presented the application of the theory of interionic attraction to the two volatile buffer systems of cardinal importance in physiology, and practical calculation of the activities of the ions of principal importance in urine and extracellular fluid are reviewed in Appendices C and D.

2. pH versus $[H_3O^+]$

Any substance contributes to the free energy of a heterogeneous equilibrium (I-6) in proportion to its *chemical potential*, which, as may be seen from Equation 6, is therefore a logarithmic function of its activity. The principal chemical relations of interest to us in acid-base physiology are thermodynamic functions defined by differences in the chemical potential of H^+ in two related systems (such as luminal fluid in the nephron and peritubular blood) or in two successive states of a system (such as the effect on blood of infusing a given quantity of base). Such differences are proportional to the logarithm of the ratio of activities, or (what is mathematically equivalent) to the difference between the logarithms of the activities, of H^+ in two different fluids or in the same fluid before and after the position of equilibrium of the buffer systems present has been displaced. For this reason, the pK′ of an acid offers the preferred comparative estimate of the strength of acids in actual solution; and the difference between the pK′s of two acids is proportional to the change of free energy per mole of proton transfer between them. Similarly, pH units are those which express on a linear scale such thermodynamic properties of acid-base equilibria in solution as the voltage required if an H^+-secreting cell is to establish and maintain a difference of H^+ activity across the cell or its wall, or the potential difference between two solutions of different H^+ activity which will be recorded by a hydrogen electrode. For the same reason, titration of weak acids or bases yields symmetrical curves when pH is plotted against equivalents of acid or alkali added, and these curves are formally identical for all single acidic dissociations over almost the entire range of pH (I-5). Most of our chemical questions as physicians concerned with the analysis and treatment of acid-base disorders, and a large proportion of the chemical concerns of the acid-base physiologist, have to do with relating the quantity of acid or base added to or removed from solution to the resulting change in acidic intensity and to change in the ratios of particular conjugate pairs.

The form in which we pose such inquiries in order to get a precise answer is generally: what is the difference between the position of equilibrium of a buffer pair in solution (e.g., urine or blood plasma) in State 1, the initial state, and in State 2, after a given quantity of acid or base has been added or removed? The physician asks this kind of question when (for example) he calculates the quantity of intravenous base which will be needed to correct a "negative base excess" disclosed by acid-base analysis of blood (III-1); and investigations of the renal or pulmonary regulation of acid-base balance repeatedly take this form. Again, it is the ratios of activities of H^+ of the two states, before versus after, which govern the change in acidic intensity; and the most appropriate unit to express any such change in acidic intensity is the difference in pH between State 1 and State 2.

As has been noted, the errors and uncertainties involved in the interconversion of pH and $[H_3O^+]$ are of very small magnitude compared with physiological variability and, sometimes, compared with the imprecision of the values of physical constants such as absorption coefficients of buffer gases. From what has been said in the preceding paragraphs, however, as well as in Section A, it will be clear that the concentration or activity of H^+ is not in any case the most suitable function for the expression of the acidic intensity of physiological fluids. Contemporary acid-base theory emphasizes the level of availability of protons from all potential proton donors, rather than only the very small amount of H_3O^+ present, as the basis of the acidic intensity of a solution.

The theoretical and practical uncertainties of determining the H^+ activity, a_{H^+}, through potentiometric estimation of pH, and the further errors of converting activity to concentration—that is, of determining the value of the activity coefficient, γ_{H^+} (Equation 7)—have been discussed succinctly by Davis (123) and in detail by Bates (39, 40). In brief, because electroneutrality must be preserved in considerable volumes of solution, net ionization always results in the appearance of equal numbers of positive and negative charges, and it is therefore impossible to determine the activity coefficient γ of a single ion. For this reason a conventional activity coefficient of unity is generally assumed for H^+ in reference solutions used for standardizing the measurement of pH differences: the pH scale is to this extent arbitrary. In addition, junction potentials and related technical aspects of potentiometry introduce errors which become very appreciable except under suitably controlled conditions. For such errors to remain negligible, the solvent composition of the reference solution and the unknown must be similar, and the unknown must contain only simple ions and molecules in total concentration not exceeding 0.2 molal. Furthermore, measurements are not reliable if the pH of the unknown is beyond the limits 2.5 to 11.5. Extracellular fluid, being less than 0.2 molal in respect to its principal solute, NaCl, normally lies within the limits within which we are entitled to relate concentrations and pH theoretically in evaluating acid-base equilibria. The same is true of most urines, with some reservations in regard to the most concentrated ones.

NOTES

1. Where the conjugate base carries more charge than the acid, as in:

$$[H^+] = K' \frac{[H_2CO_3]}{[HCO_3^-]}$$

or

$$[H^+] = K' \frac{[H_2PO_4^-]}{[HPO_4^{=}]}$$

its mobility is more constrained than is that of its congener acid by the electrical field produced by the ionic environment. The effect on reaction kinetics is as if the base concentration were reduced, and so the apparent acid-to-base ratio is increased: K' therefore exceeds K, and pK' $<$ pK. But where the acid carries more charge than its conjugate base, as in:

$$[H^+] = K' \frac{[NH_4^+]}{[NH_3]}$$

the apparent acid-to-base ratio is decreased; $K' < K$, and pK' $>$ pK.

five

CONTEMPORARY ACID-BASE THEORY

A. PROTON TRANSFER

Arrhenius' formulation of acidity represented the decisive advance in the theoretical treatment of acid-base equilibria in dilute aqueous solution (I-3) and provided the conceptual basis for the principal advances in acid-base physiology over almost a century past. But it was based entirely on water solutions and ascribed acidity to the ionization of H, whereas ionization is not in fact a prerequisite to the reaction of acids with bases. Hantzsch (226, 227), in particular, has pointed out how experimental evidence runs counter to the ionization theory of acidity; these arguments have been summarized concisely by Bates (38, page 173). In brief, acids which appear to possess equal strength in water are found not to be equally strong in organic solvents; furthermore, the stronger of such acids appear to be more strongly acidic in nonaqueous solvents, in which they are little ionized, than in water, where they are highly ionized. (Their intense acidity in solvents in which they are ionized little if at all is demonstrable by salt formation in the presence of basic indicators and by their catalytic activity in reactions such as the inversion of sucrose.) For these reasons, contemporary acid-base theory conceives both the *strength of acids and bases* and the *acidic intensity of solutions* in terms of *proton availability*—the ease with which protons can be detached from existing combinations—not in terms of the degree of dissociation of protons from any particular conjugate base in solution. Specifically, it is realized that for practical purposes a solution of a strong acid in water contains no free protons; those dissociating from the acid are transferred to the weak base H_2O to form H_3O^+, termed the hydronium (or oxonium) ion. Bell (44, pages 19–24) has discussed the evidence, which is derived from magnetic resonance data and infrared and Raman spectrography, for the existence of a stable hydronium ion. Additional water molecules are bonded, probably with lesser and varying degrees of firmness, to the hydronium ion, as they are to other cations which may be present in aqueous solution (I-2A).

Two theories have been developed which are in better accord with the observed manifestations of acidity in a variety of solvents (38, 44, 625). The more comprehensive is that of G. N. Lewis (345, 346), in which a base is viewed as a substance capable of furnishing an electron pair, and an acid is a substance able to accept such a pair to form a coordinate covalent bond; in Lewis' view the proton itself is an acid, but not the only one. This theory, because it is not very convenient for the treatment of aqueous acid-base equilibria, is seldom applied in physiology. For this reason we shall not consider it further here; it is presented in systematic detail by Luder and Zuffanti (362). J. N. Brønsted (75, 76) originated the definition of acidity which governs the usage followed in this book, which we shall refer to as *the contemporary theory* of physiological acid-base equilibria; his views are very similar to those independently evolved in somewhat less detail by T. M. Lowry (360, 361).

An acid is defined, according to Brønsted, as a proton donor, and a base as a proton acceptor. These definitions evidently imply that in every acid-base reaction at least four chemical entities, consiting of two pairs of reactants, must participate. In the first place, an acid, in the act of donating a proton, must itself be converted to a compound identical in all respects except for the absence of the proton; this compound is called the *conjugate base* of the acid. Examples are:

$$NH_4^+ \rightarrow NH_3 + H^+ \quad (1)$$

$$H_2PO_4^- \rightarrow HPO_4^= + H^+ \quad (2)$$

and the separation of a proton from an acid is called *protolysis*. However, H^+ does not, for practical purposes, exist free in aqueous solution, so that in the second place the protons which are shown above as dissociating from the acids at the left have to be accepted by the conjugate base of some other acid; otherwise no reaction could take place. Two "conjugate pairs," at least—each pair consisting of an acid with its conjugate base—are therefore involved in every acid-base reaction, which might perhaps better be thought of as a transaction. As an example:

$$HCl + NH_3 \rightarrow NH_4^+ + Cl^- \quad (3)$$

In water solution, however, more than two conjugate pairs must participate. Water is termed an *amphiprotic* solvent, meaning that it can function both as an acid and a base.[1] Indeed, pure water reacts to a very small extent with itself by *autoprotolysis*, a transaction in which one water molecule, acting as an acid, donates a proton to another, acting as a base:

$$2\ H_2O \leftrightharpoons H_3O^+ + OH^- \quad (4)$$

Autoprotolysis is quantitatively insignificant, as shown by the value 10^{-14} of the equilibrium or autoprotolysis constant of water (K_w) at 25° C:

$$K_w = \frac{[H_3O^+]\,[OH^-]}{[H_2O]} = \frac{10^{-7} \times 10^{-7}}{1}\ \text{Eq/liter} \tag{5}$$

but it is the capacity to react with itself and with other acids and bases which confers upon water its amphiprotic character as a solvent and assures its participation in every acid-base reaction in the body. The value of K_w at 25° C is conventionally used as the basis for exposition of the principles of physiological acid-base chemistry because of the convenience of the round numbers; the custom is continued in this book for the same reason. It should be kept in mind, however, that K_w at 37° C is in fact approximately $10^{13.6}$ Eq/liter, and that variation of the value with temperature alters the relation of the acidic intensity, expressed in absolute terms, to the acidic intensity relative to water. Neutrality is by definition equality of $[H_3O^+]$ and $[OH^-]$, and at 37° C, pH at neutrality is given as:

$$\text{pH} = \sqrt{\text{p}K_w} = 13.6 \div 2 = 6.8 \tag{6}$$

The effect of temperature upon K_w is of theoretical and practical importance in comparative acid-base physiology, and has become of importance in clinical medicine as a result of the introduction of hypothermia in surgery.

The uncharged molecule and both ionic forms of water are always present in aqueous solution; consequently in theory H_3O^+ and H_2O always participate as proton donors when any base is added to aqueous solution, and OH^- and H_2O as proton recipients when an acid is added. However, donation and reception of protons is a competitive affair involving (respectively) all acids and all bases present, so that the quantity of protons donated by a particular donor or accepted by a particular acceptor may be trivial. Physiological solutions always contain a substantial number of conjugate pairs over and above those furnished by water. The participation of many such pairs in any overall acid-base reaction in aqueous solution can be represented adequately, as Filley (165) has emphasized, only by rather complex symbolism. For example, if the base NH_3 is added to extracellular fluid, the consequences might be represented as follows:

$$\begin{array}{ccc}
NH_3 & \xrightarrow{(+H^+)} & NH_4^+ \\
 & \uparrow & \\
H_2PO_4^- & \longrightarrow & HPO_4^= \\
\text{uric acid} & \longrightarrow & \text{mono-urate} \\
H_3O^+ & \longrightarrow & H_2O \\
H_2O & \longrightarrow & OH^- \\
 & \vdots & \\
 & \text{etc.} &
\end{array} \tag{7}$$

Example 7 shows that the addition of the single conjugate base NH_3 to the solution has resulted in a shift of the position of equilibrium of every one of the conjugate pairs present as a result of donation of protons to the added base by all the acids present.

The relative quantity of protons donated or received, respectively, by the various acids and bases when base or acid is added depends in part on the concentration of the particular buffer pair but in part on the acidic strength of the conjugate pair in relation to solution pH. It will be seen (Sections C and E) that, in spite of the abundance of water, the solvent does not contribute, for practical purposes, to exchanges of protons when acid or base is added to extracellular fluid because of the abundance of *buffers*—conjugate bases stronger than H_2O and conjugate acids stronger than H_2O. And, though H_3O^+ is the strongest acid and OH^- is the strongest base which can exist in aqueous solution, and both are always present in extracellular fluid, their concentrations are always too low, at any physiological pH, for them to play a quantitatively significant role in proton exchanges.

B. CHEMICAL NOTATION AND DEFINITIONS

For convenience this book often uses conventional chemical shorthand to describe physiological events which initiate shifts in the position of electrolyte equilibrium in body fluids. This shorthand is never intended to suggest that the position of equilibrium of one conjugate pair of body fluids can be altered without affecting the equilibrium of all pairs present. For example, if we wish to describe the effect on the ammonia buffer system of reabsorption of NH_3 from luminal fluid in the distal nephron down an NH_3 concentration gradient created by water reabsorption, we shall write:

$$NH_4^+ \rightarrow H^+ + NH_3 \rightarrow \text{(diffuses off)} \qquad (8)$$

This symbolization does not imply that a proton is set free in the luminal fluid by the reaction; it simply focuses attention upon the partial replacement of the lost NH_3 by protolysis of NH_4^+. Example 8 makes no attempt to depict associated shifts of the position of equilibrium of the other urinary buffer systems; in fact, of course, a decline in pH will result which will be attended by donation of protons by all acids present and acceptance of protons by all bases present. But to indicate all the associated reactions would require reaction notation much more elaborate than Example 7. Multiple unseen acceptors for protons donated by any acid which is added to the solution (or whose conjugate base has been removed from the solution) and multiple unseen donors of protons to any base added to the solution (or whose conjugate acid has been removed from the solution) are always presupposed by chemical shorthand of the type used in Example 8.

Observe also that the notation exemplified in Example 8 does not in itself make clear that removal of NH_3 from solution, rather than addition of NH_4^+, is the primary process being described; accordingly, the shorthand notation is incomplete unless the primary event is also specified. Similarly, the effect of passive reabsorption of CO_2 from luminal fluid, as water is reabsorbed from the terminal nephron segment, will be represented by movement to the right of the reaction chain:

$$H^+ + HCO_3^- \xrightarrow{a} H_2CO_3 \xrightarrow{b} CO_2 \xrightarrow{c} \text{(diffuses off)} \quad (9)$$

The "H^+" at the left represents protons available for donation by all acids present other than H_2CO_3. Note again that this notation does not of itself specify that the primary event responsible for the displacement of the equilibrium is removal of CO_2; addition of HCO_3^- or transfer of protons to the system will also cause its position of equilibrium to shift to the right. If $NaHCO_3$ is added to the solution as the primary event, or CO_2 is removed, all the nonvolatile conjugate acids present would donate protons, and the same notation would be used. The decrease in concentration of all the conjugate acids which donate protons, and the corresponding increase in concentration of their conjugate bases, is presupposed but not represented. Evidently in either case the pH of the system must rise; formation of additional H_2CO_3 when protons are transferred to HCO_3^- would tend to cause CO_2 concentration to rise, so that CO_2 will diffuse off if it is free to do so. The same symbolization would again be used as a shorthand description of the effect of adding an acid (other than CO_2) to the system. In this case the added acid would be the only proton donor, whereas all bases present would accept H^+; the pH would decline since acid is being added, rather than rising as in the other two cases where acid was lost or base was added.

The acid-base reaction chain represented in Example 9 is a unique one physiologically in that it does not necessarily proceed instantaneously. The delay is not related to the electrolyte reaction proper (Reaction a), which is physiologically instantaneous; but the uncatalyzed hydration of CO_2 or dehydration of H_2CO_3 (Reaction b) takes appreciable time, after the equilibrium is displaced in either direction, to attain a new point of equilibrium, unless the reaction is catalyzed by the enzyme carbonic anhydrase. The one location known to us in the body in which enzymatic catalysis of this reaction is normally lacking is the terminal nephron segment, and this is a matter of considerable physiological significance (I-6C). However, with this single exception, all acid-base reactions with which we deal in normal physiology are (practically speaking) instantaneous. This means that the notation used in Example 8 (and in Example 9 when enzymatic catalysis is assumed) expresses in abstraction from its context an event which, though causally prior to the associated shifts of equilibrium of other conjugate pairs, in practical fact occurs simultaneously with the other reactions with which it is linked.

Filley (165) has emphasized the pressing desirability of unambiguous terminology in acid-base chemistry and physiology. The following distinctions are observed in this book.

1. The term "*H^+ transfer*" is restricted to the shifts of position of the acid-base equilibrium of a defined fluid phase resulting from addition to the solution, or subtraction from it, of an acid or base. Such a reaction occurring in urine, blood plasma, or interstitial fluid can be treated deductively with confidence.

2. The term "active transport" of H^+ or H_3O^+ (or other ions) designates exclusively local (microscopic) energy-requiring processes which specifically propel the particle across a membrane. The term does not include carriage of an ion through a membrane in consequence of osmotic or hydraulic water movement. "Active transport" is required for selective transport of an ion across a membrane (e.g., capillary wall or nephron), against an electrochemical gradient, but could also assist passive movements of ions down such gradients.

3. "Diffusion" is used for movement which is effected by a difference in chemical or (in the case of ions) electrochemical potential. It includes, but is not restricted to, movement of this type across membranes.

4. "Movement" of an ion or other particle is used where no stand is being taken as to whether active transport or diffusion is involved.

5. "Transport" includes all relatively nonspecific types of carriage, whether macroscopic (circulatory) conveyance (convection) or microscopic movement of ions through membranes by fluid filtration, osmotic water movement, or pinocytosis.

6. "Secretion" may be active or passive. In the case of solutes it always implies movement of the transported particle in the direction of secretion, but "acid secretion" is used (II-7; II-8) to designate an acidifying process not necessarily involving movement of H_3O^+ or H^+ across the secretory membrane in the direction of acid secretion.

C. ACIDIC AND BASIC STRENGTH

In a previous chapter (III-3) it was shown how the postulation of electrolytic dissociation by Arrhenius led to the quantitative analysis of acid-base equilibria as reversible reactions of the form:

$$HA \rightleftharpoons H^+ + A^- \qquad (10)$$

which obey the mass law. According to contemporary theory, however, the dissociation of an acid A_1 to yield its conjugate base B_1, as it transfers a proton to a base B_2, takes instead the form:

$$A_1 + B_2 \rightleftharpoons B_1 + A_2 \qquad (11)$$

where A_1-B_1 and A_2-B_2 are each a conjugate pair, the numerical subscript indicating the conjugate pair and the letters indicating whether the pair is present in its acidic or basic form. Example 11 is *the general expression for acid-base reactions* in terms of contemporary theory (cf. Example 3).

The completely reciprocal character of acid-base reactions means that we have no theoretical basis for distinction between the strength of an acid and the weakness of its conjugate base. On the other hand, it is clearly possible to determine from mass-law relations whether a mixture of equal parts of conjugate Pair 1 in Example 11 has greater acidic strength than a mixture of equal parts of Pair 2. For example, if equal equivalent quantities of A_1, B_1, A_2, and B_2 are mixed, and Pair 1 is more strongly acidic than Pair 2, then the reaction must move to the right to establish a new position of equilibrium at which Pair 2 has accepted protons from Pair 1. There will then be more of A_2 than A_1, and less of B_2 than B_1 at the new position of equilibrium.

Since weak electrolyte equilibria obey the mass law (I-3B), we have available a means of quantifying *relatively* the acidic strengths of any two conjugate pairs in terms of the equilibrium constant (I-1C, Equation 9) or more precisely the acidic dissociation constant (I-3B, Equation 8) for their equilibrium as represented by Example 11; we can express the velocities of proton transfer in each direction as:

$$v_1 = k_1 \frac{a_{B_1} a_{H^+}}{a_{A_1}} \tag{12}$$

$$v_2 = k_2 \frac{a_{B_2} a_{H^+}}{a_{A_2}} \tag{13}$$

and at equilibrium, where $v_1 = v_2$:

$$k_1 \frac{a_{B1} a_{H^+}}{a_{A_1}} = k_2 \frac{a_{B_2} a_{H^+}}{a_{A_2}} \tag{14}$$

$$\frac{k_1}{k_2} = \frac{a_{B1} a_{H^+}}{a_{A_1}} \div \frac{a_{B_2} a_{H^+}}{a_{A_2}} \tag{15}$$

$$K = \frac{k_1}{k_2} = \frac{a_{B1}}{a_{B_2}} \times \frac{a_{A_2}}{a_{A_1}} \tag{16}$$

The acidic strength of conjugate Pair 1 in Example 16 varies directly as the equilibrium ratio of the acids A_2 and A_1, and inversely as the equilibrium ratio of the bases B_1 and B_2. These two ratios are necessarily equal, and quite obviously there is no way by which they can be evaluated independently. All we can evaluate is the relative affinity for protons of any conjugate pair compared with that of some reference pair.

The solvent furnishes a convenient reference. In the case of water we choose quite arbitrarily to determine the *acidic* strength of solute relative to the solvent and so the pair H_3O^+-H_2O is conventionally chosen for reference; we could equally well express acid-base strength as basic strength, in terms of a "basic dissociation constant," but this would only represent a duplication of variable for a single experimentally determinable entity. The strength of an acid or base in water solution is then given (using Equation 16) in terms of its *acidic dissociation constant* as:

$$K = a_{B_1}a_{H_3O^+}/a_{A_1}a_{H_2O} \tag{17}$$

However, since the activity of water in dilute solution approaches unity and is not detectably affected by shifts of the above equilibrium, the term H_2O can be dropped, and the acidic dissociation constant becomes:

$$K = a_{B_1}a_{H_3O^+}/a_{A_1} \tag{18}$$

Equation 18 allows us to express the strengths of any acid and base in aqueous solution in a quantitative manner on a single continuous scale. Evidently any substance which will transfer protons to pure water is an acid, and any substance accepting protons from pure water is a base; accordingly H_2O is the weakest acid and the weakest base which can exist in aqueous solution. If addition of a compound causes $[H_3O^+]$ to rise above 10^{-7} Eq/liter at 25° C the compound is an acid; evidently if K of a compound exceeds 10^{-7} Eq/liter, i.e., pK is <7.0, it is an acid, and the lower the pK, the stronger the acid. Any compound whose pK exceeds 7.0 at 25° C is a base, and the higher the pK the stronger the base. Table I-5.1

TABLE I-5.1. *pK at 25° C of some compounds of physiological importance**

Compound	pK
Lactic acid	3.86
Acetic acid	4.76
H_2CO_3(pK_1)	6.35
Water	7.00
$H_2PO_4^-$-$HPO_4^=$	7.20
NH_4^+	9.25

* Values are taken from Edsall and Wyman (137, pages 452–453). It is really a buffer pair rather than a compound for which a dissociation constant can be written. The first entries are acidic compounds for which an unambiguous pK can be shown because they have only one dissociable proton and hence can be a member of only one buffer pair. NH_4^+ is not a compound, but can be listed since it, too, can be a member of only one buffer pair (NH_4^+-NH_3). This is not true of $H_2PO_4^-$, so the buffer pair must be specified.

Water is a special case, pK being identical whether H_2O acts as conjugate base or acid. It is only in relation to water that we can restrict acidic compounds to substances like lactic and acetic acid whose pK is less than 7.0 at 25° C. Glucose, pK 12 to 13, is an acid in more general terms: it can donate a proton reversibly when the medium is sufficiently alkaline (cf. Appendix G).

ranks some acids and bases of physiological interest according to pK on a continuous scale of acid-base activity.

Since it is convenient to work directly with concentrations rather than activities (I-4B1; Appendix C), we will rewrite equation 18 in terms of the apparent dissociation constant:

$$K' = [B_1][H_3O^+]/[A_1] \tag{19}$$

This expression is mathematically identical to the relation deduced originally in terms of the ionization theory from the mass law (I-3B, Equation 8) except that the effect of interionic attraction is also taken into account. The logarithmic form of Equation 18 (cf. I-4A, Equation 4) is:

$$pH = pK' + \log \frac{[\text{base}]}{[\text{acid}]} \tag{20}$$

This is the form of the Henderson-Hasselbalch equation used for practical calculations.

D. SOLUTIONS OF STRONG ELECTROLYTES

Very strong acids and bases cannot display their full strength in water because of the amphiprotic character of this solvent; this phenomenon has been termed by Hantzsch (226, 227) the "levelling effect" of the solvent. The basic character conferred on water by its capacity to accept protons causes the dissociation of many strongly acidic compounds (such as HCl) to proceed so nearly to completion in water that for all practical purposes the conjugate base (such as Cl^-) can then never accept protons; i.e., it can under no circumstances function as a base or enter into acid-base equilibria in aqueous solution. Since such a compound, when added to pure water, immediately transfers essentially all its protons to H_2O (which as solvent is always present in great excess), H_3O^+ is then the only acid in solution. (OH^- also accepts protons in theory to become its conjugate acid H_2O, but since in pure water, and even more in acidic solution, the concentration of OH^- is vanishingly small, its influence on the new position of equilibrium is negligible.) Accordingly, no acid can be stronger than H_3O^+ in water. Similarly, strongly basic compounds added to pure water immediately accept enough protons from H_2O to be converted entirely to the conjugate base, leaving OH^- as the only base present in appreciable concentration; and so no base can be stronger than OH^- in water.

Whether an acid is 99, or 99.9, or 99.99% dissociated in water is difficult to ascertain directly, since the concentration of the undissociated portion is becoming too small to allow accurate measurement. But in strongly *protogenic* solvents, where the acidic properties of the solvent are sufficiently strong to suppress proton transfer by the acids to some extent, dissociation constants of such acids become measurable, and large differences in the values of K become apparent between acids appearing equally strong in water. Similarly, the basic character of strongly *protophilic* solvents allows differences in the strength of strong bases, inapparent in water, to become manifest. And in *aprotic* solvents, which can neither accept nor donate protons, acidic or basic strength can be fully manifested by reaction or catalysis, which do not require ionization.[1]

The acidic intensity of a very dilute aqueous solution of a strong acid can be deduced directly from a knowledge of its concentration with very little error. In solutions as concentrated as extracellular fluid, a correction for interionic attraction would have to be made when deducing pH from concentration; in the ensuing exposition interionic attraction is ignored, since it does not affect the principles which are to be made clear. The central purpose of what immediately follows is to show that, where concentrations of strong acid or base are present which are at all appreciable in physiological terms, solution pH is far away from neutrality; consequently the pH of an aqueous solution consisting of a mixture of strong electrolytes only is extremely unstable in the middle pH range, around 5 to 9, within which the pH of extracellular fluid constitutes a very narrow subrange.

The pH of an aqueous solution of a strongly acidic compound can be calculated if the molar concentration of the acid is known; the reason is that such an acid, when it is added to pure water, instantly transfers all its protons to H_2O. These transferable protons are now all present as H_3O^+; $[H_3O^+]$ can be calculated from the molar quantity of the acidic compound which has been added, the number of detachable protons it possesses, and the volume of solution to which it has been added, and pH can be calculated (ignoring interionic attraction) from $[H_3O^+]$. In the case of a monobasic acid like HCl, $[H_3O^+]$ is simply the molality. When an acid transfers more than one proton to water, $[H_3O^+]$ will be the molarity multiplied by the valence; M/10 H_2SO_4 will contain 100 mmoles of the compound/liter, but 200 mEq of H_3O^+/liter.

It is also possible to calculate the pH of a mixture of strong acids and bases from the quantities added, since 1 equivalent quantity of strong base always "neutralizes" exactly 1 equivalent quantity of strong acid, i.e., removes an equivalent quantity of H_3O^+ from solution. Hence in theory, if we have acidified water by adding to 1 liter 10 mEq of strong acid, pH will be returned to 7.0 if we then add precisely 10 mEq of strong base.

The reason for the extreme instability of pH in the middle pH range of a solution containing only strong electrolytes is simply that pH close to 7.0 at 25° C implies extremely low equivalent concentrations of either added acid or added base ($[H_3O^+] = [OH^-] = 0.0000001$ Eq/liter at pH 7.0). A tiny excess concentration of either a strong acid or a strong base carries pH far toward the acid or alkaline extreme, as illustrated in Table I-5.2. The table gives the course of events during the titration of a liter

TABLE I-5.2. *Titration of aqueous HCl**

$H_3O^+ + Cl^- + Na^+ + OH^- \rightarrow Na^+ + Cl^- + H_2O$

HCl solution: 1 liter of N/1000 (1.0 mEq/liter)

Titrate with: N/10 NaOH (0.1 mEq/ml)

NaOH added		$[H_3O^+]$	Acid contributed by HCl remaining	pH
ml	*mEq*	*Eq/liter*	*mEq*	
0	0	0.001	1.0	3.0
5.0	0.5	0.0005	0.5	3.3
7.0	0.7	0.0003	0.3	3.5
9.0	0.9	0.0001	0.1	4.0
9.9	0.99	0.00001	0.01	5.0
9.99	0.999	(0.000001)	0.001	(6.0)
10.0	1.00	0.0000001	0	7.0
10.1	1.01			9.0

* The pH shown in the right-hand column is the value calculated (ignoring interionic attraction) from the concentration of the strong acid originally added (top row) or from the H_3O^+ remaining as the originally added acid is progressively neutralized by NaOH.

The value pH 6.0 is an approximation, like the corresponding value for $[H_3O^+]$ (in parentheses) because between pH 6.0 and pH 7.0 the H_3O^+ contributed by dissociation of H_2O ceases to be negligible in relation to the H_3O^+ formed by transfer of protons from HCl to H_2O.

Though the product of $[H_3O^+]$ and $[OH^-]$ is constant in any dilute aqueous solution whatsoever, the *sum* of $[H_3O^+]$ and $[OH^-]$ is extremely small only near neutrality. At pH <3.0 (cf. top row) $[H_3O^+]$ is measurable in milliequivalents per liter, as is $[OH^-]$ at pH >11.0: such solutions are accordingly good conductors of electricity, whereas pure water is not.

of 0.001 molar HCl ($[H_3O^+] = 10^{-3}$ Eq/liter) by N/10 alkali. Starting pH is evidently 3.0, and exactly 1 mEq of alkali, or 10 ml of N/10 NaOH, will be required to return pH to 7.0. As alkali is added progressively (left-hand columns), $[H_3O^+]$ and pH are calculated from the acid not yet neutralized, i.e., the acid originally present less that which has reacted with the added NaOH.

It will be seen that 90% of the alkali required to neutralize all the acid must be added before pH rises from 3.0 to 4.0. In order to reach pH 5.0, 90% of the remainder of the neutralizing quantity of alkali must be added, or 99% of the total. We have now reached a point where the pH is extremely sensitive to the addition of a drop of alkali. From here on we cannot calculate $[H_3O^+]$ or pH exactly from the strong acid remaining; when less than 0.01 mEq of strong acid is present, the 0.0001 mEq/liter of $[H_3O^+]$ present as a result of the ionization of water can no longer be neglected. But it is evident that it will be extremely difficult to add a small enough quantity of alkali from the titration buret to arrest the pH change at exactly pH 7.0. Instead, addition of a single drop of alkali is likely to

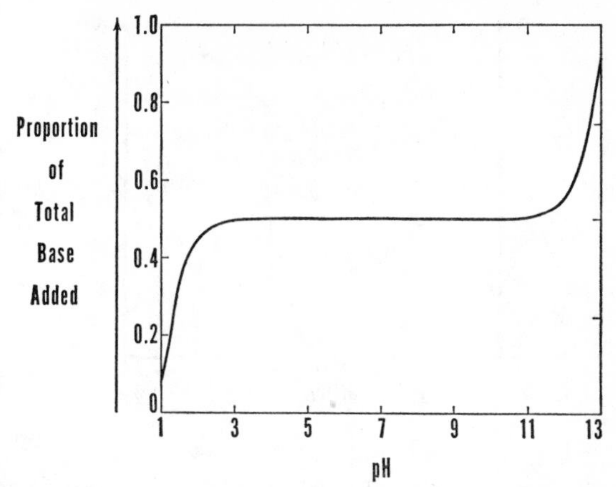

FIG. I-5.1. Titration curve of a strong acid by a strong base.

push the pH fairly well over into the alkaline range. Addition of only 0.1 ml of excess alkali over the neutralizing quantity will already give an excess of 0.01 Eq/liter of OH^- (bottom, left-hand columns); $[H_3O^+]$ will be 0.000000001 Eq/liter and pH 9.0.

Figure I-5.1 gives a graphic representation of the addition of strong base to strong acid; it differs from the titration represented in Table I-5.2 in only two respects: (i) alkali is added until there is an excess of base corresponding to the original excess of acid (2 equivalent quantities for 1 equivalent of acid originally present), so as to yield a symmetrical curve relating pH to alkali added; and (ii) the original acid solution is more concentrated (pH 1.0, corresponding to 100 mEq/liter). Here 0.1 Eq of alkali (say 100 ml of normal alkali) is required to neutralize the acid, and 99.99% of this quantity would have to be added to bring the pH up to 5.0. The slightest further addition of base is likely to raise the pH to 9.0 or beyond.

E. WEAK ELECTROLYTES (BUFFERS)

Figure I-5.2 shows by contrast the relation between pH and acid remaining when any nonvolatile weak acid is titrated by strong alkali. This type of plot is termed an "α-curve" because the ordinate expresses the proportion α of compound present as the conjugate base, i.e., the proportional acidic dissociation of the compound. Note well, in contrast to the comparable curve (Figure I-5.1) yielded by titration of a strong acid by a strong base, the following points.

1. The midpoint of the buffer curve is designated pK (or, taking interionic attraction into account, pK′). Its value is the pH at which the concentrations of conjugate base and acid are equal ($\alpha = 0.5$), a fact which can be deduced from the Henderson-Hasselbalch equation, since then the log of the ratio becomes zero and the term drops out.

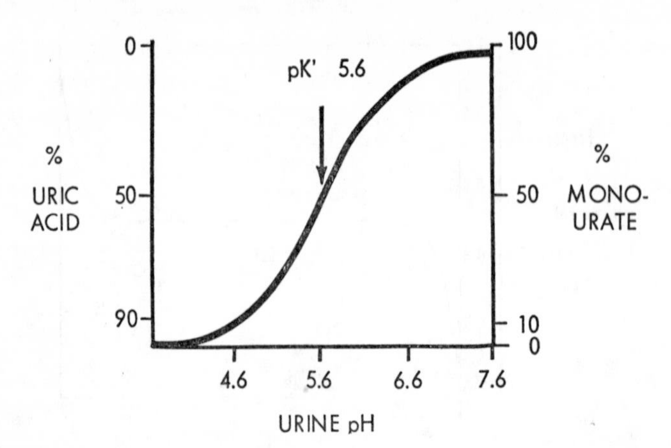

FIG. I-5.2. Titration of 1 mole of a nonvolatile weak acid (uric acid) by 1 mole of strong base. In theory, there is never total conversion of one species of a buffer pair to the other within the pH range in which the mass law applies, but nonvolatile buffering ceases to be effective for practical physiological purposes beyond the limits of pH = pK′ ± 1.5. This is not true of volatile buffers (I-6).

2. It is not possible for pH to vary more than about 3 units (pH = pK′ ± 1.5) as long as any appreciable quantity of both conjugate pairs of any weak acid is present. When one conjugate pair is $\frac{1}{10}$ the total, and buffer capacity is therefore 90% used up, pH must (according to the equation) be pK′ ± 1; pH = (pK′ ± 1.5) comprises the range of ratios from 1:30 to 30:1, beyond each of which limits only 3% of compound is still available for buffering.

3. Buffer strength—the quantity of strong acid or base which must be added to a volume of solution to effect a given change in pH, usually expressed as μEq H^+ pH^{-1} ml^{-1}—naturally increases with the concentration of buffer compound. But buffering *efficiency* (buffer strength per concentration of compound) depends upon the relation of solution pH to compound pK: the shape of the α-curve shows that buffering efficiency for a nonvolatile buffer is maximal when pH is in the neighborhood of pK. It follows that the buffer strength of a solution of a weak electrolyte depends upon two factors: the concentration of compound and the range of pH over which buffering occurs in relation to pK of the compound.

4. When an acidic compound (e.g., uric acid in Figure I-5.2) is titrated progressively with strong alkali, some of the conjugate acid (uric acid) is still present at the neutral point, by definition where $[H_3O^+] = [OH^-]$. A weak acid is therefore not "neutralized" by an equivalent quantity of strong base in the sense in which the word applies to strong electrolytes. The weaker the acid, the closer its pK to neutrality, and the more conjugate acid remains at the neutral point: the acidic intensity of the solution must be reduced further if a weak acid is to give up all its protons. Con-

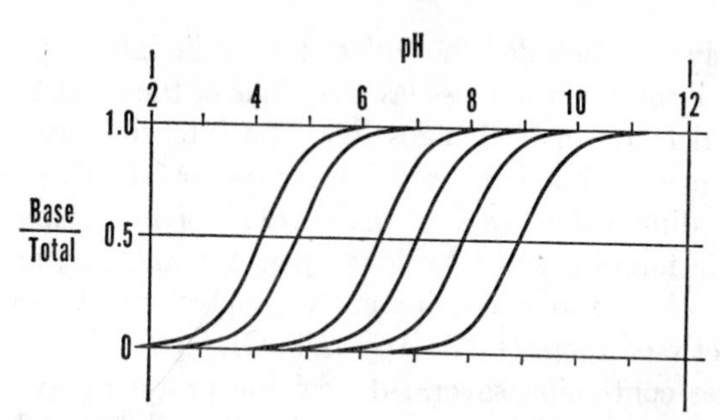

FIG. I-5.3. Ratios of base to total compound for selected weak electrolytes as a function of pH. Modified from Clark (104, page 273); from *left* to *right*, the titration curves are those of bromphenol blue, bromcresol green, bromcresol purple, bromthymol blue, phenol red, and ammonia. The function of indicators depends on the differing pigmentary properties of the two species of a weak acid or base. Given sufficiently intense color or colors of one or both members of the conjugate pair, the dye can be present in solution in trace amounts and so itself not contribute appreciably to the position of the equilibrium it is used to study.

versely, of course, the more weakly basic a compound is, the more the acidic intensity of the solution must be raised if the base is to accept all the protons which it can accept.

Figure I-5.3 shows a series of buffer curves of individual weakly electrolytic compounds. The *form* of the α-curve is *perfectly general* for all weak electrolytes; the difference between the acid-base behavior of the different compounds resides entirely in the location of the curve on the pH scale, or (otherwise expressed) on the value of pK. Evidently Figure I-5.3 ranks acids and bases graphically in respect to acid-base strength on a continuous scale according to the same principle used in Table I-5.1; it shows the order in which the various compounds will effectively yield protons as pH rises, or accept protons as pH declines. The dissociation constant of a weak acid or base in water, like its molecular weight or melting point, is a property characteristic of the compound; pK is the best expression of the inherent acid-base properties of a compound.

When acid is added to a solution containing several buffer pairs, the distribution of the new protons between the various conjugate bases present is determined by the several dissociation constants and by the relative concentrations of the respective conjugate pairs. Each pair contributes in this way toward buffering the acid by accepting protons. And when base is added each pair contributes by donating protons, the position of equilibrium of each pair being always fixed by solution pH according to a rearrangement of Equation 20:

$$\mathrm{pH} - \mathrm{pK}' = \log \frac{[\mathrm{base}]_1}{[\mathrm{acid}]_1} = \log \frac{[\mathrm{base}]_2}{[\mathrm{acid}]_2} \text{ etc.} \qquad (21)$$

This is called the isohydric principle.

Progressive dilution does not affect the acidic intensity of a solution of a buffer pair insofar as the mass-law relations of the reactants are concerned. In practical fact, small changes in acidic intensity are usually observed when buffers are diluted: pH may either rise or fall. These changes are due entirely to diminution, as a physical result of increasing dilution, of the electrostatic forces exerted by ions upon one another; according to Coulomb's law these forces are inversely proportional to the square of the distance between them (I-4B; Appendix C).

Molecules containing several dissociable protons may resemble H_2SO_4, which is a strong acid in respect to the dissociation of both its protons; polybasic acids of this type behave like other strong acids in the physiological range of pH save that equivalence is twice (or more) the molality. By contrast, in the case of some compounds containing two or more transferrable protons, including substances of physiological importance like carbonic, uric, and phosphoric acids, there is a difference in the ease with which the several protons can be detached. When such a compound is titrated with strong base, a double or multiple buffer curve results (cf. Figure I-5.4). *In vivo* buffering is not ordinarily affected by this type of

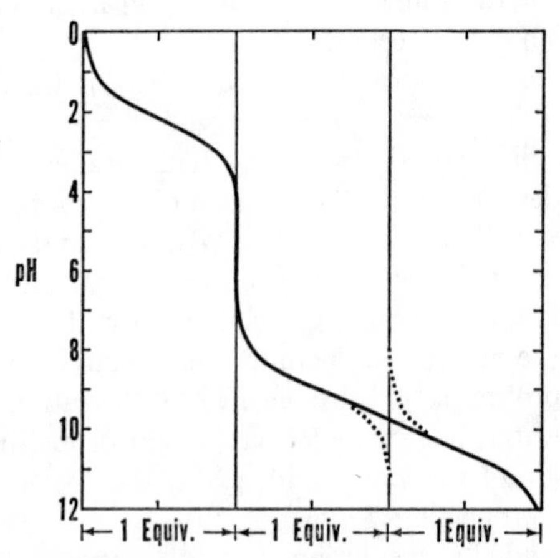

FIG. I-5.4. Titration curve of lysine. Modified from Clark (104, page 278). Lysine in very acid solution has three associated protons, one as COOH and two as NH_3^+ groups. As strong base is added these protons are progressively transferred, the order of detachment, as pH rises, being governed by the dissociation constant of each group. There is no significant overlap between the first dissociation of lysine (K_1, $COH \rightleftharpoons COO^- + H^+$) and the second ($K_2$), but pK_2 and pK_3 (for ionization of the two amino groups) are close enough for the titration curve to become nearly linear between pH 9.0 and 11.0. Solutions containing in appreciable quantity a number of nonvolatile buffers whose dissociation constants do not greatly differ will exhibit titration curves which, like that of lysine between pH 9.0 and 11.0, are more or less linear. This is often nearly true of urine in the pH range of interest (4.5 to 8.0) (see Figure I-5.5).

discrete polybasicity of physiological acids—except in the important special case, considered below, of amphoteric compounds like amino acids and proteins—because of the limited range of variation of the pH of extracellular fluid, and even of urine. We never deal physiologically (for example) with a solution pH at which the *first proton dissociation* of phosphoric acid (which has a dissociation constant termed K_1) or the second of carbonic and uric acids (K_2) affects the equilibrium materially. However, titration of a fluid containing multiple acids (or bases) with neighboring values for pK′ will of course reflect the superimposition of the curves for the individual acids present in proportion to their concentrations. The sigmoid shape of the simple titration curve is accordingly lost when pH is sufficiently adjacent to the pK of several weak electrolytes present in solution (Figure I-5.4).

Multiple buffer pairs in solution do not necessarily presuppose multiple compounds. As originally proposed by Niels Bjerrum (166, page 349), proteins and amino acids carry both positive and negative charges as a result of the reversible ionization of their carboxyl and amino groups:

$$R\langle\begin{matrix} COO^- \\ NH_4^+ \end{matrix} \tag{22}$$

Such compounds are called ampholytes or Zwitterions and are described as *amphoteric*. The reversible dissociation of a proton from each substituent obeys the mass law and is characterized by its own particular dissociation constant. Figure I-5.4 exhibits the titration curve of lysine; the pK's of the two amino groups are close enough so that the proton dissociations overlap materially as pH is varied, producing a nearly linear titration curve in the far alkaline range.

There is no pH at which proteins do not bear multiple charges; the "isoelectric point" of the protein is the pH at which the number of positively charged groups is the same as the number of negatively charged groups. At pH 7.4, there is a net excess of negative charges in the plasma proteins and hemoglobin, which accordingly figure among the compounds contributing to the total net anions of blood. Whole blood, because of its high protein content, is more strongly buffered than any other physiological fluid; hemoglobin is a particularly efficient buffer in the pH range 7 to 8, owing especially to its numerous histidine residues with values for pK′ in this neighborhood.

Figure I-5.5 shows a titration curve of urine; the function approaches linearity because of the overlapping pK's of the nonvolatile buffers present in the urine in substantial concentrations. It is not necessary to know the nature of the particular buffer pairs present in order to determine by titration the buffer strength of the nonvolatile urinary buffers, i.e., the quantity

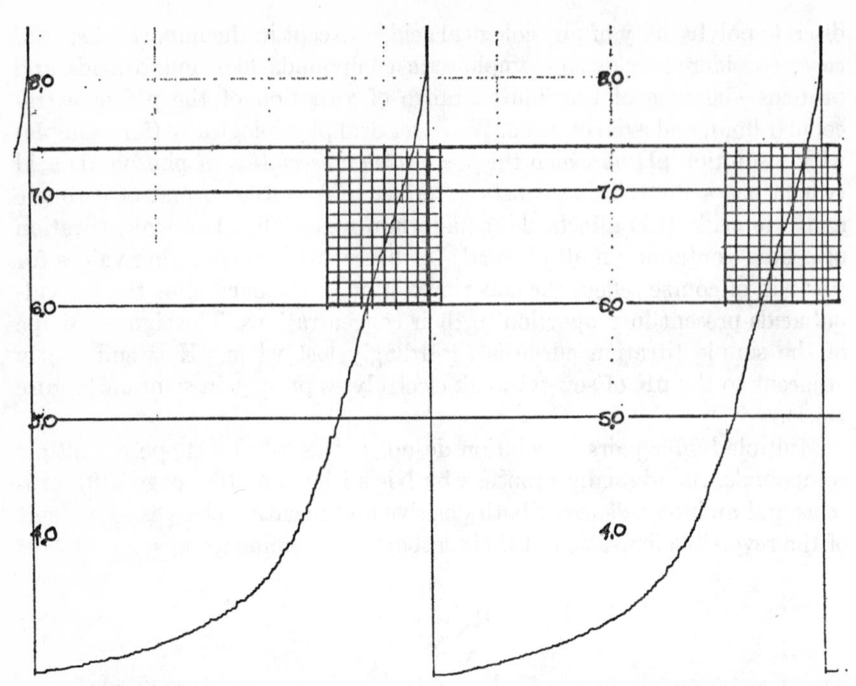

FIG. I-5.5. Titration curve of urine (two successive 5-ml aliquots acidified with 1 ml of N/10 acid diluted to 25 ml and aerated). Original pH's 6.78, 6.77; flow rates 12.2, 12.5 ml/minute. The tracing is one made by an automatic recording titrator. The blue cross-hatching on the read-out paper does not reproduce, and has here been reinforced in ink between pH 6.0 and 7.4: divisions on the *vertical scale* represent 0.1 pH and those on the *horizontal scale* represent 0.02 ml of added N/10 NaOH, or 2 μEq. The nonvolatile buffer strength of the urine over the pH range between that of the collected urine (or any other pH of interest) and plasma pH is generally estimated as the slope of the curve, which usually approaches linearity, relating alkali added to observed pH change; this function is conventionally expressed and symbolized $dH^+/dpH\ ml^{-1}$ of urine.

It is sometimes desirable, notably for calculating transfers of protons among the urinary buffers as H_2CO_3 is dehydrated beyond the papilla (495), to calculate nonvolatile buffer strength as an excretion rate: dH^+/dpH per (volume of urine secreted in 1) minute.

of acid or base which will be accepted by these buffers per change of pH unit per milliliter over any range of pH of interest. The "titratable acidity" of urine, which estimates the protons accepted by the urinary nonvolatile buffers during acidification of the glomerular filtrate, is therefore directly measurable on a urine sample provided its pH is low enough so that it contains essentially no HCO_3^-. The rate at which protons were being transferred from extracellular fluid to the urinary nonvolatile buffers is given by multiplying the titration figure (the alkali required to return the pH of 1 ml of urine to plasma pH) by the rate of urine flow in milliliters per minute. When appreciable HCO_3^- is present, it can be caused to evolve as CO_2 gas by adding measured acid to the urine sample; after aerating, the "residual titratable acidity" or "titratable-acidity-minus-CO_2" can then be evaluated by titration, the titration figure being the alkali required to return the pH of 1 ml of urine to plasma pH less the acid previously added to 1 ml of urine. The physiological significance of this variable, and related considerations, are discussed later (II-4C; Appendix F).

NOTES

1. *Aprotic* solvents can neither accept nor donate protons. Solvents which can donate protons but have little or no capacity to accept them are called *protogenic.* Those which can accept protons but donate them little, if at all, are called *protophilic.*

six

VOLATILE BUFFER SYSTEMS

When one member of a buffer pair is a gas, the buffer system may under certain circumstances function much more efficiently to stabilize pH than is possible in the case of the nonvolatile systems heretofore considered. The singular properties of the CO_2-HCO_3^- equilibrium as it exists in extracellular fluid were first perceived by Lawrence Henderson (240) to represent vital exploitation of acid-base behavior possible only under the following two conditions: (i) one member of the buffer pair must be a gas dissolved in solution at stabilized gas pressure as base or nonvolatile acid is added, and (ii) the ion-to-gas ratio must be kept high. It follows that evolutionary selection of pH 7.4 at 37° C for mammalian extracellular fluid, a value which establishes the 20:1 $HCO_3^-:CO_2$ ratio, has permitted especially effective defense of plasma pH (Section A-1).

The characteristics of heterogeneous gas-liquid equilibria with high, or potentially high, ion-to-gas ratios and stable gas pressures are exploited by the organism for buffering purposes not only in extracellular fluid, but in urine as well. Transfer of protons between extracellular fluid and urine at a rapid rate, i.e., effective renal excretion of nonvolatile acid and base (II-2; II-4; II-6), is thereby made feasible, notwithstanding the rather limited range of possible variation in urinary pH (about 1.2 units during water diuresis); still, this variation is large in comparison with the range of fluctuation in plasma pH. The manner in which the CO_2-HCO_3^- system in the glomerular filtrate supports renal acid-base regulation is closely related to the use of the same system in the plasma, since fluctuation of plasma [HCO_3^-], in response to the normal variation of acid-base load, determines the variation of the rate at which HCO_3^- is filtered at the glomerulus. The high $HCO_3^-:CO_2$ ratio in higher-pH urine allows renal base excretion as HCO_3^- to rise markedly for a small rise in urinary pH; the high ion-to-gas ratio of another volatile buffer system, NH_4^+-NH_3, in lower-pH urine allows renal acid excretion as NH_4^+ to rise markedly for a small decrease in urinary pH (Section B).

Finally, if one member (whether gas or not) of a buffer pair is present in low concentration relative to its conjugate acid or base, a small absolute change in that low concentration, without change in the concentration of the congener, exerts a relatively large effect on the ratio of buffer species and hence on the pH (cf. Figure I-6.1). This principle is exploited in extracellular fluid when the lung moderates the deviation of pH produced by non-respiratory acidosis or alkalosis by causing arterial P_{CO_2} to deviate from its normal value (Section A-2). Accumulation of H_2CO_3 above its equilibrium value with dissolved urinary CO_2 in the distal nephrons when higher-pH urine is being secreted is a physiological process unique to the kidney by which the relatively stable CO_2 pressure of the luminal fluid ceases to stabilize pH. In effect, the heterogeneous equilibrium CO_2-HCO_3^- ceases to exist as soon as H_2CO_3 is no longer equilibrated with CO_2. Because of the extremely small normal ratio H_2CO_3:HCO_3^-, accumulation of a tiny absolute quantity of H_2CO_3 has a relatively enormous effect upon the ratio H_2CO_3:HCO_3^-, and hence upon the pH of the luminal fluid (Section C).

A. THE CO_2 BUFFER SYSTEM IN EXTRACELLULAR FLUID

1. Stabilization of Arterial P_{CO_2}

Henderson likened the extracellular fluid to a solution *in vitro* in which dissolved CO_2 is held at a constant molar concentration which is low relative to HCO_3^- by equilibration with a voluminous gaseous phase containing CO_2 gas (240); such a two-phase physicochemical system is one type of *heterogeneous equilibrium*. The general thermodynamic treatment of such equilibria was first developed, rigorously and very abstractly, by J. Willard Gibbs (193); another case important for the physiology of the body fluids is the *Gibbs-Donnan equilibrium* (133, 193) which governs the distribution of ions between two liquid phases separated by a semipermeable membrane, or subject to some other constraint upon the distribution by diffusion of solvent or solutes. Here we are concerned with a different type of two-phase system, a gas-liquid one in which the essential features are that the gas concentration is *stable* and *low* relative to that of its congener ion.

The partial pressure, or tension, of carbon dioxide, P_{CO_2}, in blood passing through the alveolar capillaries is very rapidly equilibrated with that of the alveolar air (II-3); and according to Henry's law the [CO_2] of a fluid is proportional to the P_{CO_2} of the gaseous phase with which it is equilibrated (Appendix D). Thanks to pulmonary regulation, alveolar P_{CO_2}, and hence arterial plasma [CO_2], is not ordinarily affected at all by the rate of metabolic CO_2 production, i.e., by the rate at which CO_2 is passing into, through, and out of the extracellular fluid and lungs, but is stabilized by reflex regula-

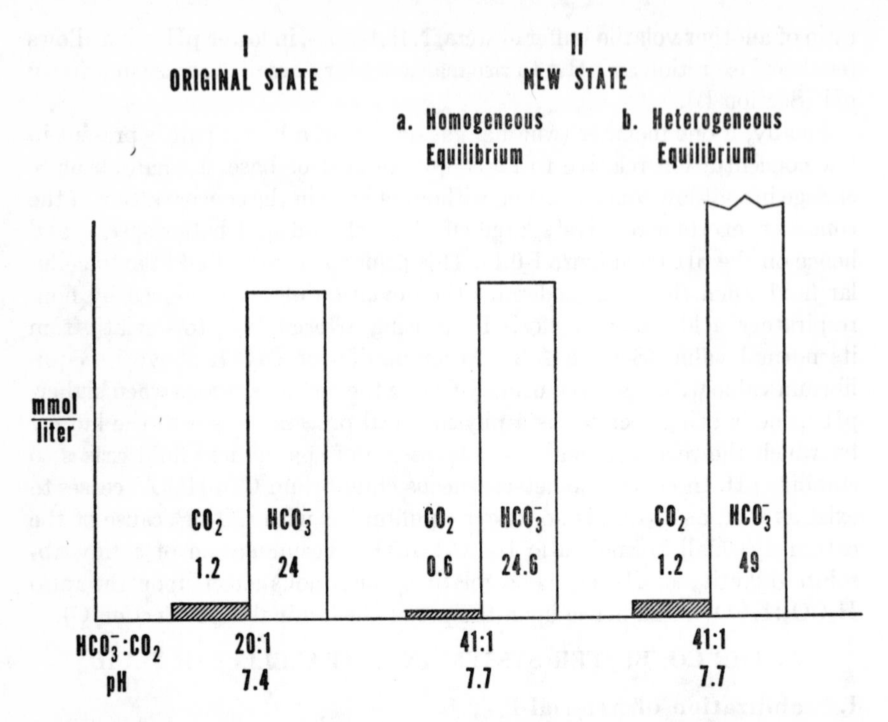

FIG. I-6.1. Effect of stabilization of $[CO_2]$ upon the amount of alkali required to raise the pH of a bicarbonate buffer solution from 7.4 to 7.7. The diagram applies to a bicarbonate buffer system similar to that of extracellular fluid but containing no other buffers for the added alkali; see text. Modified from Hills and Reid (259).

tion of the respiratory rate (II-3A). In turn, arterial plasma P_{CO_2} is equilibrated in all the tissue capillaries with tissue P_{CO_2}, so that P_{CO_2} throughout the body depends primarily on pulmonary action. Most tissues, since they are in general producing CO_2 in bulk, have a CO_2 pressure a few millimeters greater than alveolar P_{CO_2}; kidney tissue P_{CO_2} is scarcely higher than arterial P_{CO_2} because renal blood flow is so large relative to net local tissue CO_2 production.

Where the CO_2-HCO_3^- buffer system in extracellular fluid differs essentially from any nonvolatile buffer system (I-5) or from a homogeneous CO_2 equilibrium (one in which there is no possibility of equilibration with P_{CO_2} external to the liquid phase) is simply that CO_2 will freely diffuse into or out of solution, when base or acid is added, in such a fashion as to hold $[CO_2]$ constant:

$$H^+ + HCO_3^- \rightleftharpoons H_2CO_3 \rightleftharpoons CO_2 + H_2O \quad (1)$$

Any acid added to the CO_2 buffer system in extracellular fluid necessarily generates additional CO_2, but P_{CO_2} does not rise; instead CO_2 diffuses off.

When base is added, dissolved CO_2 will be converted to HCO_3^-, but [CO_2] does not fall; instead, CO_2 diffuses in. These movements of the gas into and out of the fluid phase account for the remarkable stabilization of the pH of the solution observed as long as the ratio of ion to gas is high.

The unique buffering properties of such a system are illustrated in Figure I-6.1, whose three bar pairs have been constructed according to the Henderson-Hasselbalch equation, each bar representing concentration in millimoles per liter. The actual state of the CO_2 buffer system (pK′ = 6.1) in normal extracellular fluid is represented by the *left-hand pair;* at pH 7.4 the ratio HCO_3^-:CO_2 is 20:1, since for normal arterial blood at P_{CO_2} = 40 mm, dissolved [CO_2] is 1.2 mmoles/liter, while [HCO_3^-] = 24 mmoles/liter (II-2). The two companion bar pairs represent the change from this "original state" of the system when pH has been raised to 7.7 by adding alkali to 1 liter of solution; the change in concentration of both dissolved gas and ion depends on whether we are dealing with a homogeneous equilibrium (*middle diagram*) or a heterogeneous equilibrium with a gaseous phase sufficiently voluminous for the CO_2 pressure in both phases to be held essentially constant (*right-hand diagram*). In each case it is assumed that the only solute present is the CO_2-HCO_3^- buffer system. The equilibrium is homogeneous if there is no gaseous phase; such an equilibrium responds to addition of alkali exactly as would a nonvolatile buffer system (I-5). In the illustrative instance chosen for Figure I-6.1, buffering by the homogeneous equilibrium would naturally be inefficient, since pH is already 1.3 above pK′. After 0.6 mEq of alkali is added to the liter of fluid present, the pH has moved up to 7.7, defined by the HCO_3^-:CO_2 ratio 41:1; the ratio has approximately doubled because [CO_2] has been halved with hardly any proportional change in [HCO_3^-]. The large rise in pH occurs because it takes only 0.6 mEq of alkali to halve [CO_2].

Where there is a gaseous phase containing CO_2 equilibrated with dissolved CO_2, and sufficiently voluminous so that absorption of CO_2 into the liquid does not materially affect the CO_2 tension, much more alkali must be added in order to effect the same change in pH; this is illustrated by the *right-hand diagram.* Evidently the reason for the difference is that the small absolute change in [CO_2], which so greatly affected the ratio in the homogeneous equilibrium, is here entirely prevented. The only way the HCO_3^-:CO_2 ratio can rise now is for [HCO_3^-] to increase. In order to attain the HCO_3^-:CO_2 ratio of 41 which was reached after addition of only 0.6 mEq of alkali in the homogeneous equilibrium, enough alkali must be added to double the already relatively large concentration of HCO_3^-. As 25 mEq of alkali is added, CO_2 continues to diffuse into the system and to react with the alkali until [HCO_3^-] is 49 mEq/liter; only then has the HCO_3^-:CO_2 ratio doubled, and the pH been raised to 7.7. The unique and spectacular buffer power of this type of heterogeneous equilibrium is apparent; (25 ÷

0.6) or 41 times more alkali is required, compared with the homogeneous equilibrium, to effect the same rise in pH. In whole blood, in spite of the high concentration of the exceptionally strong nonvolatile buffer hemoglobin, the major buffer *in vivo* is still the CO_2-HCO_3^- system; and in plasma and interstitial fluid almost all the excess base or nonvolatile acid accumulating in extracellular fluid must be buffered by this system.

Not every gas-liquid heterogeneous equilibrium would stabilize dissolved $[CO_2]$ as in Figure I-6.1; were the gaseous phase of small volume relative to the liquid phase, P_{CO_2} of the whole system would rise markedly when acid was added and fall when base was added, and its buffering characteristics would differ very little from those of a homogeneous equilibrium. Pulmonary stabilization of arterial P_{CO_2} is as effective as equilibration of the P_{CO_2} of extracellular fluid with a large gaseous phase would be, but control is exercised by active physiological means, and is brought about by sensitive reflex control of respiration. Extracellular fluid buffering accordingly takes place in an *open system* (II-2; III-1), the essential feature of which is that $[CO_2]$ in the liquid phase is stabilized by free flow of the gas into or out of the liquid phase as base or acid are added, respectively. From the chemical standpoint the system resembles the heterogeneous equilibrium with a voluminous gas phase as represented in Figure I-6.1.

2. Regulatory Modification of Arterial P_{CO_2}

In addition to stabilizing arterial P_{CO_2} in health, the lungs make a vitally important contribution to acid-base homeostasis, principally under pathological circumstances, by modifying the normal value of P_{CO_2} (40 mm) in a regulatory manner; and because of the low $CO_2:HCO_3^-$ ratio in extracellular fluid, relatively small changes in the absolute arterial P_{CO_2} are very effective in modifying plasma pH. When acidosis or alkalosis develops because the kidney is incompetent or overwhelmed by large acidic or basic loads, the resulting deviation of plasma pH is normally moderated by (respectively) decrease or increase in arterial P_{CO_2}; these changes are brought about through reflex control of the respiration, and they provide pulmonary "compensation" for the effect of the acid-base imbalance on the pH of extracellular fluid. Though it is hardly possible in practice for the lungs to reduce arterial P_{CO_2} to much less than a third of the normal value, a reduction of this magnitude markedly affects the pH. The logarithm of ½ being 0.3, it follows that extracellular pH would be 0.3 higher, at arterial P_{CO_2} = 20 mm Hg, than at 40 mm, if $[HCO_3^-]$ remained unchanged. This is not the case in a solution like plasma, which contains nonvolatile buffer, because solution $[HCO_3^-]$ varies with $[CO_2]$ (III-1); but depression of arterial P_{CO_2} is nevertheless an important defense against nonrespiratory acidosis. Similarly, when the normal arterial P_{CO_2} is doubled by the normal pulmonary response to severe nonrespiratory alkalosis, plasma pH would be 0.3 lower than it would otherwise be, if [HCO_3^-] did not change, and is in fact substantially lower. The regulatory function of the lung accordingly provides a powerful accessory defense against nonrespiratory acid-base imbalance (Table III-1.1).

B. VOLATILE URINARY BUFFERING

The principle of the "open system" based upon gas-liquid buffer equilibria is exploited by the organism for acid-base homeostasis not only in the extracellular fluid, but also in the urine. The "effective" pressures of CO_2 and NH_3, i.e., the tissue pressure with which the urine leaving the kidney last equilibrates, exhibit relatively little variation as urinary pH varies, and with it the urinary concentration of the congener ions HCO_3^- and NH_4^+ (II-6C1; II-10A). There is good evidence that the "effective" gas pressures are those of papillary tissue and urine (II-10A). The chemical principles governing the changing composition of urine as its pH changes are accordingly similar to those governing the acid-base composition of extracellular fluid discussed in the previous section: the change in concentration of either urinary HCO_3^- or NH_4^+, as urinary pH varies, is large relative to the urinary pH change. However, the manner in which gas-liquid heterogeneous equilibria are exploited for vital purposes in the urine is somewhat different from that in extracellular fluid, because the physiological objective is not completely analogous: urinary heterogeneous equilibria serve primarily to provide adequate concentrations, and hence *excretion rates*, of the right kind of urinary buffer at any acid-base load. From the standpoint of teleonomy, urinary pH varies sufficiently for adequate concentrations of the urinary buffer ions to be achieved at the actual effective urinary gas pressures (254).

There is a corresponding difference between plasma and urine in respect to the manner in which the CO_2 equilibrium supports the physiological functions of these fluids. The renal system can be seen as providing a system for continuous and inexpensive transport of the buffer gases into the urine. CO_2 transport is of course also involved in the stabilization of plasma pH; Figure I-6.1 illustrates (for example) how rise in pH is moderated, when base is added to extracellular fluid, by ingress of CO_2 into the fluid with conversion of the gas to HCO_3^-. However, once a new steady state of acid-base balance has been established at the increased base load, net transport of CO_2 gas into or out of extracellular fluid is not involved in maintenance of that state. This is not true in the case of the urinary response to changing acid-base load. When the load is alkaline, waste CO_2 is being converted continuously to urinary HCO_3^-; the fact that the conversion actually occurs in extracellular fluid does not alter the fact that the organism can draw indefinitely on its limitless supply of waste CO_2 gas to provide buffer for the base being eliminated in the urine. Similarly, when large acid loads are being eliminated, waste NH_3-N provided by protein catabolism of the organism as a whole is detached within the kidneys from blood glutamine, the chemical vehicle transporting urinary buffer NH_3 in the circulation (II-9D), and is made available to buffer excess protons; this process, too, can continue indefinitely.

While, therefore, Figure I-6.1 can also serve to exhibit the chemical basis of the relation between concentration of HCO_3^- in the urine and urinary pH, it does not bring out the special feature which distinguishes the urinary heterogeneous equilibria: the fact that they are *dynamic* ones. In other words, the buffer gases, in any steady state of acid-base balance, flow continuously out of the body, partly as gas but also (and at very variable rates) as their ions. Hills and Reid (259) described the overall role of the volatile buffer systems in the urine as "excretory buffering" in order to direct special attention to this unique feature of urinary buffering; the same usage is followed in this book. The term *excretory buffering* applies equally to the nonvolatile urinary buffer compounds which must also be transported continuously into the urine, and is convenient from the general biological viewpoint because it emphasizes that ample urinary buffer for excesses of acid and base alike must be continuously available, though these two categories of buffer substance are of course not in general simultaneously required.

This requirement that urinary buffers be transported continuously into the urine, sometimes at high rates, contributes to the special fitness of CO_2 and NH_3 to serve regulation of acid-base in the mammal, in vertebrates generally, and doubtless in invertebrates as well (254). The nonvolatile urinary buffers, even though some of them (like uric acid and creatinine) are end products of metabolism, are available as wastes only to a limited extent. The principal nonvolatile urinary buffer, phosphate, is available in the steady state only to the extent to which it is absorbed from the diet, while the nonvolatile end products of metabolism make a very small contribution at best to the total urinary nonvolatile buffer. CO_2 and (fundamentally) also NH_3 are by contrast bulk metabolic wastes (see below). Moreover, in order for bulk transport of buffer gases into the urine to be brought about by diffusion with little change of gas tension, they must be highly diffusible; i.e., they must move rapidly down very small pressure gradients. CO_2, and NH_3 even more so, are small molecules which are extremely diffusible, in accordance with Fick's law of diffusion, which states that diffusivity is (as a first approximation) inversely proportional to the square root of the molecular weight. Their exceptional diffusivity is another shared property which renders CO_2 and NH_3 especially suitable to serve as urinary buffer gases.

Henderson (240) demonstrated the unique fitness of CO_2, by virtue of its physicochemical properties, to serve as the principal buffer of extracellular fluid. The group of properties of CO_2 to which Henderson drew special attention—volatility, aqueous solubility, and dissociability—also characterize NH_3. It is the difference in the value of the dissociation constants of the two compounds—one lying above plasma $[H_3O^+]$ and the other below—which enables them to collaborate as urinary buffers, one serving to buffer excess base and the other serving to buffer excess acid. In addition, in these small molecules, the carbon and nitrogen atoms are at or close to their lowest energy level; their valences are fully satisfied in stable configurations, all the available bond energy of the more complex dietary compounds containing C and N atoms having been tapped by the time they have been metabolized to CO_2 and NH_3. That is why these gases are the natural bulk end products of animal metabolism, though for special reasons (II-9D) large terrestrial animals must transport and excrete waste NH_3-N by means of chemical vehicles. No other dissociable compounds could so freely be made available to serve in bulk as urinary buffers because no other compounds in the cosmos possess the needed combination of properties (240, 254).

Figure I-6.2, taken from Gamble (182), is the classic exposition of physiological exploitation of dynamic heterogenous equilibria for purposes of excretory buffering—in this case through donation of protons by H_2CO_3 to excess base. The proton donation actually takes place, not in the kidney, but in the extracellular fluid; in excreting the HCO_3^- salt of Na^+ (or any other alkali cation) the kidney merely regulates the moiety of the HCO_3^- filtered at the glomerulus which is allowed to escape in the urine (II-4B). However, the diagram is not concerned with mechanisms, but simply with the influence which alteration of urinary pH must have, according to the laws of chemistry, upon urinary HCO_3^- concentration, assuming that urinary P_{CO_2} is stable at 40 mm Hg. If the urine pH is close to plasma pH, urine will have about the same concentration of HCO_3^- as the plasma, since the urinary P_{CO_2}, more particularly the "effective urinary P_{CO_2}" (II-6C1) does not greatly differ from that of blood. This means that when the urine is faintly alkaline considerable base is being eliminated, namely around 25 mEq in every liter of urine; for urinary HCO_3^- represents elimination of base which has been added to extracellular fluid and which necessarily reacts with H_2CO_3 to form HCO_3^- (II-4B). A small rise in urinary pH above 7.4 greatly increases urinary base excretion because of the exponential relation between $[HCO_3^-]$ and pH at stabilized P_{CO_2} (Figure I-6.2); only when the urine has been acidified markedly to around pH 6.0 does urinary base excretion as HCO_3^- become insignificant (cf. Figure II-6.1).

Figure I-6.3 shows similarly how decline in the pH of papillary urine will necessarily increase the urinary total ammonia concentration if the urine and tissue are equilibrated in respect to P_{NH_3} at a constant value. The pK′ for NH_3 in body fluids has a value (around 9.0) which is high relative to

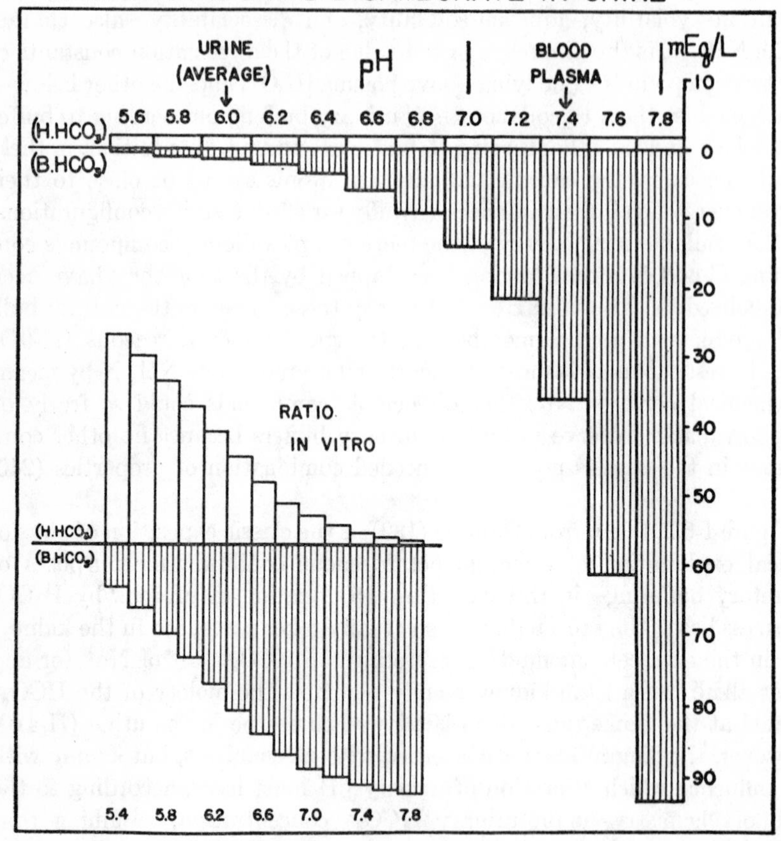

FIG. I-6.2. Excretory buffering of excess base as urinary HCO_3^- represented as a function of urinary pH. From Gamble (182, page 75). According to the original but now obsolete usage, "carbonic acid" and "H_2CO_3" were used here to mean dissolved $CO_2 + H_2CO_3$, or practically speaking $[CO_2]$. The *large upper diagram* shows the relatively large absolute variation in $[HCO_3^-]$ produced in the vicinity of pH 7.4 with relatively little change in pH by addition of base or acid, provided $[CO_2]$ is held constant at about 1.2 mmoles/liter. Note that urinary $[HCO_3^-]$ is negligible from the physiological standpoint when urinary pH is 6.0 or less. The *small lower diagram* applies to a homogeneous equilibrium which is identical at pH 7.4 with the heterogeneous equilibrium depicted. The *small diagram* illustrates that in a homogeneous equilibrium, over the range of pH represented, $[HCO_3^-]$ is much less influenced when the pH is changed by adding base or acid.

urinary or blood pH; and in acid urine the value of (pK′ − pH), and hence the ratio $NH_4^+ : NH_3$, is high enough so that the total ammonia concentration of markedly acid urine is comparable to the concentration of HCO_3^- (or total CO_2) in higher-pH urine, in spite of the fact that the urinary con-

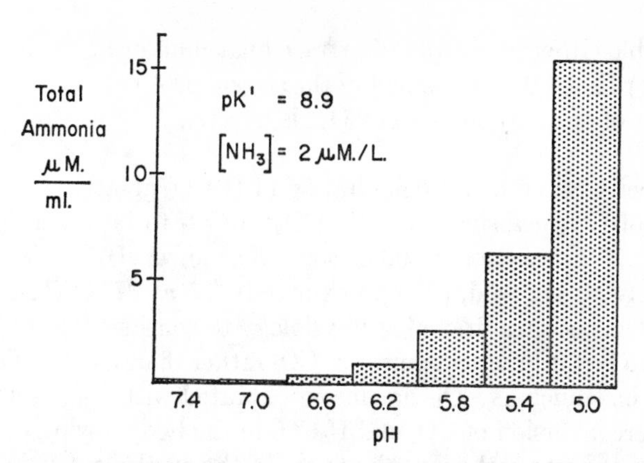

FIG. I-6.3. Effect on concentration of total ammonia of varying the pH of a solution in which $[NH_3]$ is held constant at a value near that of urine. From Hills and Reid (258).

Total ammonia is practically synonymous with NH_4^+ in this pH range because of the high pK′ (about 9.0) of ammonia. At urinary pH 7.0, for example, pK′ − pH is 2.0 and hence approximately 99% of total ammonia is present as NH_4^+; at pH 6.0 about 99.9% is present as NH_4^+. In acid urine virtually all the ammonia present has served to buffer excess acid by accepting protons; the properties of a heterogeneous equilibrium with a low gas-to-ion ratio are being exploited by the kidney for excretory buffering of excess acid.

centration of NH_3 gas is very low compared with that of CO_2 gas (cf. Figure II-6.1). Note that, when the basic congener of a buffer pair is a gas whose concentration is stabilized, the buffering effectiveness of the system increases indefinitely as (pK′ − pH) increases; when the gas is the conjugate acid, as in the case of CO_2, buffering effectiveness increases indefinitely as (pH − pK′) increases. This behavior of heterogeneous gas-liquid equilibria differs fundamentally from that of homogeneous buffer equilibria, where buffering is maximally effective when pH = pK′ (I-5); this physiologically quintessential physicochemical principle is still much underemphasized in relation to renal regulation of acid-base balance and is indeed not invariably recognized by writers on the subject (255).

C. DELAYED DEHYDRATION OF H_2CO_3

Carbon dioxide gas is an acid anhydride, one of a class of compounds which at first constituted a barrier to acceptance of the idea that acidity was conferred upon a compound through its containing one or more hydrogen atoms (I-3A). Dissolved CO_2 renders water faintly acidic because its reaction with the solvent forms carbonic acid:

$$CO_2 + H_2O \rightarrow H_2CO_3 \qquad (2)$$

Though H_2CO_3 possesses two detachable protons, only the first dissociation is appreciable either at the pH of extracellular fluid or at the highest possible urinary pH of 8.0; detachment of the second proton to yield an appreciable amount of the carbonate ion $CO_3^{=}$ is effected only in strongly alkaline solutions.

The velocity constant for dehydration of H_2CO_3 greatly exceeds that for hydration of CO_2, causing the ratio $CO_2:H_2CO_3$ to be very high at equilibrium; at 37° C it is about 500 or somewhat lower. H_2CO_3 is accordingly a moderately strong acid, pK approximately 3.5 at 37° C (II-6; Appendix D), but it has been traditional in physiology to consider the conjugate acid of HCO_3^- to be for practical purposes CO_2 rather than H_2CO_3. This practice is justified inasmuch as CO_2 remains equilibrated with H_2CO_3 at almost all sites of interconversion of CO_2 and H_2CO_3 in the body, owing to the availability of the enzyme carbonic anhydrase. In the erythrocytes, in renal cells, and in organs elaborating an acid or alkaline secretion (stomach and pancreas, respectively), reversible hydration of CO_2 is rendered physiologically instantaneous by the presence of the enzyme, a situation permitting the participation of H_2CO_3 to be ignored:

$$CO_2 + H_2O\ (\rightleftharpoons H_2CO_3) \rightleftharpoons HCO_3^+ + H^+ \qquad (3)$$

Because of the high $CO_2:H_2CO_3$ ratio, CO_2 is in effect a much weaker acid than H_2CO_3, indeed one of the weakest of all acids.

The discovery of carbonic anhydrase resulted from investigations in Copenhagen of the rate of evolution of CO_2 from shaken and acidified bicarbonate solutions. O. M. Henriques (249) found that in the presence of hemoglobin CO_2 was given off more rapidly than would have been expected in the light of the rather slow rate of dehydration of H_2CO_3 found earlier, especially in the very thorough studies by Faurholt (158). Henriques went on to demonstrate carriage by hemoglobin of some CO_2 as carbaminate (II-3); however, the rapidity of immediate CO_2 liberation in similar experiments in several laboratories appeared to be greater than could be accounted for entirely by blood carbaminate. Separation of carbonic anhydrase from erythrocytes was accomplished independently by Meldrum and Roughton (392), who christened it, and by Stadie and O'Brien (587), whose initial report included kinetic study of the properties of the partially purified enzyme. The properties of the enzyme and its distribution in nature have subsequently been studied extensively, and are reviewed by Maren (379, 380). It is a cellular enzyme widely distributed in nature which is present in erythrocytes and in those mammalian tissues where secretion of nonvolatile acid or base, or CO_2 transport, necessitates rapid interconversion of H_2CO_3 and CO_2.

Reaction 3 provides a pseudoexception to the rule that ionic reactions can always be considered instantaneous from the physiological standpoint. Although the ionic reaction proper (at the right) is extremely rapid, the left-hand reaction is relatively slow. Equilibrium is still attained rapidly, by physiological standards, after a shift of position of a pure HCO_3^--CO_2 buffer system; but this situation does not occur in the urine or in the body fluids, where nonvolatile buffer is always present also. As pointed out originally by Kennedy, Eden, and Berliner (304a), the half-time for attainment of equilibration is increased progressively as the concentration of nonvolatile buffer increases, as well as by increasing concentrations of HCO_3^-, and it may well be upwards of a minute under conditions corresponding to some attained in the terminal portion of the nephron and the upper urinary tract. As early as 1948 Ryberg pointed out that the kinetic studies of Faurholt (158) made it likely that delayed dehydration of H_2CO_3 formed by acidification of bicarbonate-rich glomerular filtrate was responsible for the raised P_{CO_2} of higher-pH urines; and it has recently been proven directly that in the luminal fluid of the distal nephron (distal tubules and collecting ducts), H_2CO_3 accumulates above its equilibrium relation with CO_2 during bicarbonate diuresis (II-8C).

Though pharmacological agents which inhibit the catalytic effect of carbonic anhydrase have been extensively studied and applied for therapeutic use (378, 55, 379, 380), there is, it appears, only one site in the body where under normal circumstances Reaction 3 is not at all catalyzed by carbonic anhydrase; this is in the renal tubules and collecting ducts (II-8C). The resulting accumulation of H_2CO_3 in the lumen contributes to the defense of the organism against a base load, because it lowers the pH of the luminal fluid, which in turn suppresses acid secretion into the nephrons by increasing the transtubular pH gradient against which acidification must proceed (II-8D).

A major contribution to the striking elevations of urinary P_{CO_2} above tissue P_{CO_2} is made by continuing decomposition in the urinary tract of H_2CO_3 present in papillary urine above its equilibrium relation into CO_2 (II-10A). The elevated $[H_2CO_3]$ provides the driving force for continued dehydration, which would yield very little CO_2 were it not for the presence of the nonvolatile buffers. These retard the rise in urinary pH, which in a pure solution of HCO_3^- and CO_2 would occur so rapidly as to arrest almost at once the generation of CO_2 from HCO_3^-, since the only significant proton donor then would be H_2O. The nonvolatile buffers, by decreasing the rise in pH produced by a given quantity of acid, permit conversion of HCO_3^- to CO_2 to proceed further; rise in P_{CO_2} increases in relation to both the nonvolatile buffer strength and the $[HCO_3^-]$ of the urine.

The chemical principles which result in disequilibrium lowering of the pH of bicarbonate-rich urine throughout the distal nephron are exhibited in Figure I-6.4. The nephron is likened to a tube freely permeable to CO_2 gas, through which is propelled a solution of concentrated $NaHCO_3$ from

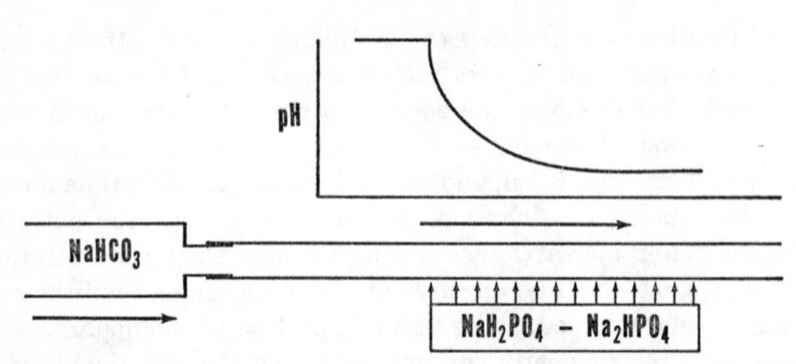

FIG. I-6.4. Chemical processes which result in lowered disequilibrium pH of bicarbonate-rich urine throughout the distal nephron. Schematic.

a syringe. A dilute solution containing equal parts of Na_2HPO_4 and NaH_2PO_4 (pH 6.8) is injected through multiple nozzles into the flowing $NaHCO_3$ solution as indicated; the result is continuous generation of H_2CO_3 at a rate faster than it can be dehydrated by the relatively slow spontaneous decomposition reaction proceeding in the absence of enzymic catalysis. $[H_2CO_3]$ in luminal fluid accordingly rises as the fluid progresses along the tube (*small upper graph*), and pH necessarily declines progressively. However, the progressive rise in $[H_2CO_3]$ results in a steady increase in the rate of the first-order decomposition of H_2CO_3, so that the net rate of synthesis of H_2CO_3 slows; when the rate of dehydration of H_2CO_3 is equaled by the increasing rate of H_2CO_3 formation, the concentration of H_2CO_3 stabilizes as the result of *dynamic equilibrium* between its formation and its decomposition. The *small upper graph* shows the curve obtained when the pH of the fluid within the tube is plotted as a function of the distance along the tube. The low stable disequilibrium pH indicated at the *right* of the *small upper graph* then continues essentially constant (*right-hand portion* of *upper diagram*) as long as the rates of synthesis and decomposition of H_2CO_3 are in approximate balance. Provided that the fluid within the tube continues to contain a great deal of HCO_3^- relative to the protons being transferred to this base by the phosphate solution, the rate of H_2CO_3 synthesis will not fall off; this relation is approximated in the distal nephrons when alkaline urine is being secreted (II-8C).

Figure I-6.5 represents events which may occur after a low disequilibrium pH has already been effected, as in the distal nephron during bicarbonate diuresis; represented in the figure is the synthesis of H_2CO_3 in excess of its equilibrium relation to CO_2 as effected by instantaneous mixing of a bicarbonate solution with phosphate buffer at pH 6.8. The situation at the point of mixing is the counterpart of the low disequilibrium pH existing in the fluid within the medullary collecting duct as it passes through the renal papilla (II-10A); it is the result of a concentration of H_2CO_3 in excess of its equilibrium relation to CO_2. This excess $[H_2CO_3]$ provides the driving force

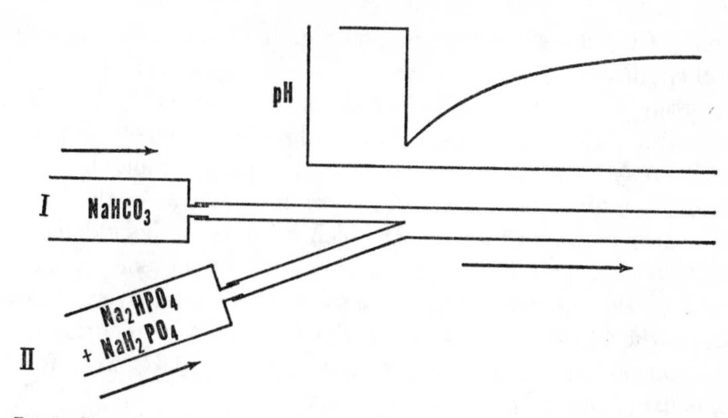

FIG. I-6.5. Spontaneous dissipation of lowered disequilibrium pH. Schematic.

to move the bicarbonate buffer system to a position of stable equilibrium as the fluid proceeds through the tube:

$$H^+ + HCO_3^- \rightarrow H_2CO_3 \rightarrow CO_2 + H_2O \quad (4)$$

The H^+ at the left represents protons necessarily transferred from the nonvolatile buffers to HCO_3^- according to the isohydric principle as H_2CO_3 is removed from solution by dehydration and pH rises. If the wall of this tube is poorly permeable to CO_2 (like that of the urinary tract beyond the papilla) the CO_2 tension of the fluid (P_{CO_2}) must rise as CO_2 gas is formed in solution; P_{CO_2} as well as pH (*right-hand side* of *upper small diagram*) will progressively rise along the length of the tube until the bicarbonate buffer system of the fluid has reached stable equilibrium. A moving column of fluid in which H_2CO_3 had been generated at a concentration above equilibrium, as shown in Figure I-6.5, was adapted by Roughton (Appendix D) to the evaluation of the velocity constant of the dehydration reaction. Serial sampling along a fluid column proceeding through the tube at a uniform rate allows the rate of decrease in $[H_2CO_3]$ to be measured and kinetic analysis to be undertaken.

Accumulation of H_2CO_3 in any physiological solution above its equilibrium relation with CO_2 immediately converts a gas-liquid heterogenous equilibrium to a homogeneous equilibrium; a very well buffered solution is instantly converted into a relatively very poorly buffered one. Because of this, and because H_2CO_3 is a moderately strong acid, a tiny accumulation of this substance is highly effective in depressing the pH of luminal fluid in the nephrons, in which now only the nonvolatile buffers and ammonia are available to resist pH change. The essential role of carbonic anhydrase in acid-secreting and base-secreting cells must surely be its ensuring that steep pH gradients, tending to develop rapidly in small volumes of cell water at the sites of acid secretion and transport, are minimized by rapid diffusion of CO_2 into those fluid volumes and by enzymatic preservation of the local equilibrium relation between CO_2 and H_2CO_3 (II-8B).

Conversely, as pointed out by Edsall and Wyman (137), the absence of carbonic anhydrase from the muscles, where enormously increased CO_2 production may result from sudden exercise, facilitates rapid diffusion of the CO_2 into the interstitial fluid, and its prompt removal in the blood minimizes the drop in muscle cellular pH which would result if sudden accumulations of CO_2 were instantly hydrated. Similarly, the absence of the enzyme from the plasma allows transient passage of CO_2 from erythrocytes to alveoli, as the blood passes through the pulmonary capillaries, with less acidification of the plasma than would result if H_2CO_3 were formed instantaneously from CO_2 dissolved in plasma.

H_2CO_3 is synthesized from CO_2 by a true chemical reaction involving alteration of existing chemical bonds. There is no possibility of delayed dehydration of NH_4OH, in analogy to delayed dehydration of H_2CO_3. NH_4OH does exist in solution, but only as the result of the incessant formation of and rupture of hydrogen bonds between NH_3 and H_2O (583, 137). The atomic structure of NH_3 may be represented as:

$$\begin{array}{c} \cdot\cdot \\ H : N : H \\ \cdot\cdot \\ H \end{array} \qquad (5)$$

each hydrogen atom forming a coordinate covalent bond with N. The resulting displacement of the orbitals of the electrons of N toward the H atoms imparts a marked polarity to the molecule, the positive pole being located on the side of the N atom occupied by the lone electron pair. H_2O is a polar molecule for similar reasons (cf. Figure I-2.1); pure liquid ammonia and pure water are highly associated liquids because of the hydrogen bonds between their molecules (I-2), and hydrogen bonds are present in aqueous ammonia solutions between molecules of NH_3 and H_2O.

The lone electron pair of NH_3 is able to accept a proton through the formation of a dative bond; the molecule thereby becomes positively charged. Once the ion is formed, all four H atoms are attached by a double bond to N and become indistinguishable; the electronic formula for ammonium is therefore:

$$\left[\begin{array}{c} H \\ \cdot\cdot \\ H : N : H \\ \cdot\cdot \\ H \end{array}\right]^{+} \qquad (6)$$

The attraction of the N for a proton makes NH_3 potentially a strong base; but it is evident that, if the fourth bonding site of N is occupied by a hydrogen bond linking the molecule to water, a proton can no longer be accepted. Hydrogen bonding of NH_3 to H_2O to form NH_4OH explains why ammonia is a much weaker base in water than would be predicted from the structure of NH_3; the effective concentration of the base is reduced by hydration. The same phenomenon occurs when NH_3 is dissolved in any polar solvent; and whereas quaternary ammonium compounds like tetramethyl ammonium, in which all four bonding sites of the N are occupied, are strong bases comparable to OH^-, the mono-, di-, and tri-substituted compounds of ammonia, which, like NH_3, can form hydrogen bonds, are also for that reason only moderately strong bases like ammonia itself (583).

II

PHYSIOLOGY

one

HOMEOSTASIS

A. HOMEOSTASIS BEFORE AND AFTER BERNARD

"It is a well-known observation, that healthy individuals very promptly pass urine after they drink fresh spring-water. If one drinks at rapid intervals 10 glasses, each 6 to 8 ounces, of water containing less than .34% salt, urine passed 10 minutes later will be normally colored; but then over the next 1½ hour one will see 8 to 9 voidings, of which the later samples are as clear and colorless as spring-water and scarcely differ from it in salt content. There are persons who can in this way drink 6 to 8 glasses of water one after another without any discomfort.

"Things go altogether differently with water whose salt content is similar to blood. Add only a gram of salt per 100 cc. spring-water, and you will find after 2 hours, even when 3 to 4 glasses have been drunk, that no urine has been passed. It is almost impossible to drink more, because the salt water oppresses the stomach, as if the blood vessels absorbed it poorly, evidently because the fluid inside the vessels (blood) and outside (salt water) exert no physical effect on one another through endomosis or exosmosis.

"Water containing somewhat more salt than blood exerts a third kind of effect . . . for in this case not only is urine secretion not provoked, but water enters the gastrointestinal tract from the blood . . . ; purging ensues, which will be accompanied by thirst if the salt solution is quite concentrated.

"If one assumes that a certain salt content is absolutely required of the blood, one can conclude from these experiments, which anyone can easily confirm on himself, that the constitution of the tissues or the blood vessels opposes an obstacle to any enhancement or diminution of the salt content of the blood, so that blood is not allowed to pass beyond certain limits in becoming richer or poorer in salt. I have ascertained by further investigation that the salt content of the urine passed shortly after a copious drink of water always exceeds that of the water, whereas in the later samples its content of salts, including phosphates, in vanishingly small. . . .

"It follows, that all salts present in the urine, without exception, must be regarded as transient blood constituents, which are excreted precisely because they are no longer appropriate to the normal blood constitution" (349, pages 179 and 180, author's translation).

Like evolution by natural selection, the central theorem of the physiology of the intact higher animal organism, homeostasis, seems to verge on tautology. It is apparent to prescientific, commonsense observation that higher animals, during the period between adolescence and senility, closely approximate constancy in their form and composition; and from this it follows that the normal functions of the body and of its various organs, especially assimilation and excretion, can be fully understood only by recognizing that they are unceasingly adaptive. Everyday events like the taking of food or water and fasting, sleep and wakefulness, rest and physical activity, changes of position, responses to changing ambient temperature and humidity—all these normal circumstances, and many others, challenge the stability of the functional state of the organism and call for incessant modifications of the character of and rate of physiological processes.

Justus Liebig's account, reproduced above, of his observations of the renal responses to ingested water or to saline illustrates how inevitably homeostatic theory entered into the very origins of a systematic chemical physiology. It is simply not meaningful to ask for a description of the normal chemical composition of urine, or of the internal exocrine secretions like gastric juice or bile, without subjoining a description of the conditions the inquirer has in mind. The functioning of these organs (and all others) is in reality always a response to *some* condition, and cannot be quantitatively defined, in any ideal sense, when abstracted from a particular situational context; accordingly, our definition of normal conditions—for example, normal acid-base loads—is inevitably to some extent arbitrary or conventional. The remarkable stability of the chemical composition of the blood and of the extracellular fluid generally is the reflection of the innumerable continuous physiologically adaptive responses of the tissues, suborgans, organs, and the organism itself to continually changing everyday circumstances.

The higher animal organism, as the most complex of living creatures, exhibits a hierarchy of physical and chemical systems, and its physiological functions can and must be studied at various functional levels. Modern biology is intensively concerned with investigating membranes, cells, organelles, and macromolecules. The triumphs of molecular biology in our time are universally admired, and studies at this level possess a generality of biological reference to which the traditional physiology of higher animals can hardly aspire. Furthermore, as the investigator of the operation of the mammalian organism proceeds down the scale from organism to perfused organ or tissue, to surviving tissues *in vitro* (whether as tissue slices, cells, or membranes), and on to homogenates, organelles, extracts, and isolated enzyme systems, he gains progressive freedom in regard to the conditions under which it is possible to study the chemical bases of physiological functions. An isolated organ can be studied under conditions incompatible with its functioning in the intact organism, and cells and organelles can be studied under conditions never encountered in the intact organ. This freedom in respect to imposed conditions is a major asset in the search for a better understanding of the molecular basis of the fundamental processes which give rise to life and upon which the biological systems progressively developed by evolution in higher organisms are necessarily superimposed.

On the other hand, everyone primarily interested in the organism in its full concreteness, including physicians, must be concerned above all with understanding what actually happens in the intact body, including the functions of organs and cells and organelles *in situ;* with defining the range of variation in "normal" conditions; and with characterizing responses to altered physiological conditions, in quantitative terms and as a function of time. A major problem confronting the student of organism at the present time is deciding in what manner and to what extent the great profusion of facts being contributed by studies at lower levels of biological organization, with their great freedom in respect to conditions, is relevant to the highest level in which he is primarily interested, with its extreme restriction of possible conditions of study and its profusion of unwanted, but given, situational influences. It is safe to anticipate that the effort to bridge the interpretations gap between biological levels of organization will continue to require the attention of students of organism for decades to come.

A generation after the publication of Liebig's paper quoted at the beginning of the chapter, Claude Bernard divined that the physicochemical stability of the higher animal organism is *required* by the vital units, the cells, of the body, and is *provided* primarily by the extracellular fluid, the true internal environment (*milieu intérieur*) of the cells, and by those functions which determine its composition. "Divined" is a more accurate verb here than "perceived," because, as Lawrence Henderson noted, the doctrine of the stability of the internal environment was a comprehensive generalization which was based on relatively few facts. The most striking instance of the *chemical* constancy of the internal environment was provided a generation after Bernard by the full realization of the extraordinary stability of the acidic intensity of extracellular fluid. Bernard's evidence, aside from considerations such as those earlier adduced by Liebig, had come above all from his own studies of heat regulation, as we read in one of his lecture-demonstrations of the hypothermia produced in the rabbit by spinal cord transection: "The first fact which strikes us in the study of bodily temperature is that it is a necessary condition of the internal environment, an important attribute of the blood plasma which bathes all the body's anatomical structure. . . . Living creatures have the power to produce heat. They are not abandoned to external conditions, like minerals whose temperature follows the variation of the environmental temperature. Through being alive they possess in themselves a source of heat which permits them to react to their environment and resist it" (58, page 10, author's translation).

Part and parcel of the doctrine of the constancy of the internal environment is, of course, definition of its degree of inconstancy. Change is continually being imposed on the physicochemical characteristics of the extracellular fluid by variations in the external world, or by the changing relation of the organism to it (e.g., recumbency and assumption of the erect posture). Furthermore, we know now what Bernard did not, that there are important *internal* causes for fluctuations in the function of the cells, and hence in the composition of extracellular fluid. Circadian rhythms, though they can be brought into synchrony with natural or artificial environmental cycles lasting 24 hours, or somewhat more or less ("entrainment"), express activity of an innate "Zeitgeber" or biological clock (81, 78). Though these and other biological rhythms no doubt represent, at least in an evolutionary sense, adaptation to external periodicities, they are non-homeostatic phenomena in the sense that it is difficult to view them as expressing a clear-cut response to an easily definable external challenge to the stability of the organism. These rhythms, as well as the immediate responses to ordinary recurrent conditions like sleep, recumbency, and the taking of food and water, complicate the task of the student of the physiology of the intact organism. They are conditions which the experimenter is to a greater or lesser extent powerless to control, yet whose influence he must nevertheless define and allow for in the interpretation of data obtained in experiments which have to take place over some definite interval of time.

We must therefore always detach homeostasis—the ideal underlying stability of the internal environment maintained by "continual and delicate compensation established by the most sensitive of balances" (cf. legend to Figure II-5.1)—from the superimposed fluctuations and rhythmicities which will to some extent affect studies of any physiological function in the intact animal. In addition, every regulatory response to imposed conditions has its own characteristics: its own speed of response, absolute and relative precision (or homeostatic efficiency), and vulnerability to damage. A purely descriptive but quantitative account of these characteristics is needed in the case of every homeostatic system, quite aside from deeper inquiries into the mechanisms of the responses. "Physiology," wrote Claude

Bernard, "must direct its investigations to the internal environment, determine its composition, and grasp its nature, both considered independently and in its interrelations with the intrinsic vital units of the body" (59, page 7, author's translation).

B. ACID-BASE HOMEOSTASIS

1. Nonhomeostatic Phenomena

Characteristic rhythmicity of the acidic intensity of the urine (Figure II-1.1) and of the rate of renal excretion of acid or base has attracted attention for the better part of a century. Kayser (302), Kleitman (312, pages 63–65), and Brunton (79) have reviewed the earlier literature dealing with this subject; the two most prominent features of the usual 24-hour periodicity of urinary acidity are the low pH and increased acid excretion during nocturnal sleep and the "alkaline tides" observed after meals, especially after breakfast. According to Brunton (79, page 380), Dr. Bence Jones communicated to the Royal Society in 1845 a phenomenon to which his attention had been directed by "Dr. Andrews of Belfast," the "alcalescence" of the urine collected about 2 hours after breakfast. Bence Jones suggested that the increased pH of postprandial urine, subsequently known as the alkaline tide (79), might be due to gastric secretion of HCl, an opinion which has in the main been sustained. Acid withdrawn from extracellular fluid and temporarily sequestered in a component of "transcellular water" (the lumen of the gastrointestinal tract) must naturally temporarily deplete extracellular base reserves. Persons with achlorhydria do not exhibit postprandial rise in urinary pH; however, the correlation between gastric HCl secretion and rise in urinary pH is not especially impressive quantitatively, in part because other digestive secretions similarly sequestered postprandially are alkaline; in addition, arterial P_{CO_2} may rise enough, as a result of taking food, to exert a compensatory, and indeed a transiently acidifying effect on the urine (535).

Because of the intimate relation of urinary acid-base composition to the renal handling of Na^+ and K^+ (II-7; II-8) and because of the influence of the urinary flow rate upon the excretion of urinary titratable acidity and of ammonia (II-10), it is impossible to consider the periodicity of urinary excretion of acid or base without reference to the urinary excretion rates of other electrolytes and water as these are influenced by situational factors and circadian rhythms. The literature on this subject was reviewed many years ago by Piéron (457) and recently by Wesson (652, pages 584–587) as well as by Kayser (302) and Kleitman (312). Analysis of the causal factors responsible for observed circadian excretory patterns is complicated by the multiplicity of possible influences (intrinsic rhythms; rhythms related to light and darkness, activity and rest, and sleep and waking; effects of

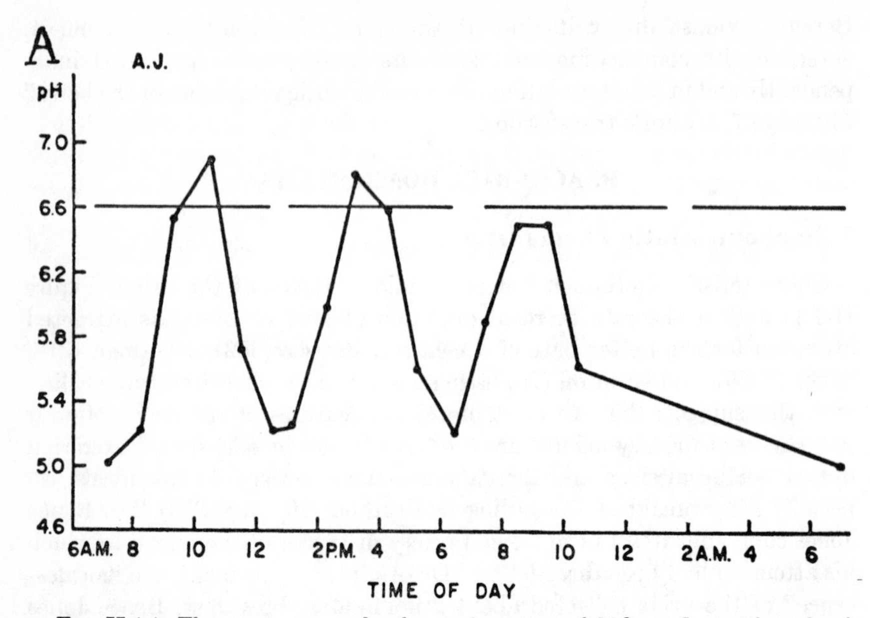

FIG. II-1.1. The most commonly observed pattern of 24-hour fluctuation of urinary pH. Reproduced, with permission, from Elliott, Sharp, and Lewis (150). The rate of acid or base excretion will vary in general concordance with the urinary pH, but not exactly; the most important of the nonvolatile urinary buffer compounds, phosphate, generally has a different circadian pattern, with maximal excretion toward late afternoon and minimal values during sleep (312, 457). Accordingly, circadian change in the rate of acid excretion tends to be less prominent than circadian change in urinary pH, since lowest pH during sleep is associated with minimal nonvolatile buffer excretion. (Relatively little dissociation of urinary pH from volatile urinary buffer excretion is possible since normally HCO_3^- excretion largely determines pH and urinary pH largely determines urinary NH_4^+.)

The marked morning rise in urinary pH is much the commonest feature of the circadian rhythm; it can be related to at least three factors in many persons: (i) increased glomerular filtration rate after the nocturnal low; (ii) increased sensitivity of the respiratory center on awakening; and (iii) gastric HCl secretion upon anticipation or consumption of breakfast. The subsequent postprandial rises in urinary pH shown in the figure are both less constant and less prominent in most persons than the matutinal one. Considerable variation in the general pattern is encountered in those who sleep at night and take food at conventional times; deviations from ordinary habits of sleep and mealtimes further increase the variability of the excretory pattern, as do some clinical disorders (652). It is hazardous to relate the rate of urinary acid or base excretion in a particular subject either to imposed conditions or to such normal determinants as plasma pH and [HCO_3^-], sampled once or several times daily, without taking circadian variations into account (cf. II-6C_2).

posture, endocrine activity, and emotion; etc.). The unavoidable presence of at least some of these factors in any experiment carried out in the intact subject, and modifications of the responses to them as related to the age of the subject, necessitate particular care both in experimental design and in regard to interpretation when the effect of any imposed condition upon the excretion rates of acid or base, HCO_3^-, or ammonia and titratable acidity by intact persons (or experimental animals) is to be studied.

From Piéron's review (457) of early investigations by 16 authors, it was already evident that the urine volume decreases at night in the majority of human subjects, whereas recumbency increases the urine volume (604, 605). Probably the inhibitory effect of sleep on urine formation suffices to overcome the simultaneous and opposite influence of recumbency (42); also the positive water balance at night under these circumstances appears to exhibit an impressive correlation with the normal nocturnal decrease in body temperature (571). Simpson (571–573) and Norn (425) confirmed the usual nocturnal decrease in urine volume and also documented a concurrent decrease in the rate of urinary excretion of Na^+, K^+, and Cl^-. The low nocturnal urinary flow rate persists when equal aliquots of water are drunk hourly over 24-hour periods (571, 652). These prominent features of the circadian cycle of urinary water and electrolyte excretion have been confirmed many times subsequently (377, 522, 591, 400, 651); the electrolyte cycle makes its appearance within the first half-year of life, is less prominent in the elderly, and is reversed in persons who sleep regularly during the day instead of at night (652, 591). Despite some difference in emphasis between British and American students of the subject, it is in general clear (652) that circadian fluctuations of the glomerular filtration rate (GFR) (651), with lower values at night, parallel the observed circadian changes in urinary water and electrolyte excretion and are to a large extent responsible for them, though specific renal tubular influences also play a role (591, 400). The cause of the rhythmical fluctuation of GFR itself has not been identified; adrenal hormones appear to play little or no part, while recumbency causes GFR and the excretion of water, Na, K, and Cl to increase.

Since urinary pH declines (572, 153, 320, 301) and the rate of total acid excretion increases (91, 313, 314) during the nighttime hours when glomerular filtration (and hence glomerular HCO_3^- filtration) declines, it is reasonable to surmise that decreased GFR is probably to some extent responsible for the usual circadian rhythm of renal acid-base excretion, especially in view of the striking correlation between the nocturnal decline in acid excretion and decreased urinary excretion of Na^+, K^+, and Cl^- (651, 652). On the other hand, because the rate of secretion of acid into the nephrons is automatically adjusted to the rate of glomerular sodium filtration (II-7F), fluctuation of urinary acid-base excretion with GFR is in general minimized, and other factors undoubtedly contribute to nocturnal urinary hyperacidity.

As the postabsorptive period is prolonged, the body's own tissues are increasingly drawn on as the source of energy for vital needs; the acidity of the load increases when body protein and fat are burned, compared with oxidation of diets containing considerable carbohydrate or alkali metals or both, and so an increased acid load may contribute to increased nocturnal acid excretion in some subjects. In addition, sleep can be said to induce mild incipient respiratory acidosis: Comroe states (109, page 95) that alveolar (and therefore arterial) P_{CO_2} rises from 40 to 46 mm Hg or more during sleep, owing to decreased sensitivity of the respiratory mechanism to the stimulus of circulating CO_2. Endres (153) was the first to invoke CO_2 retention to explain nocturnal urinary hyperacidity, and according to Stanbury and Thomson (591) "there can be no doubt that the cycle of sleep and wakefulness is normally associated with an alternation of hypoventilation and relative hyperventilation."

Postural effects on renal function (653, 154, 522, 357, 212, 215, 604, 605) are complex in their causation. Orthostasis, through its direct hydrostatic effects on the vasculature and possibly for other reasons as well, causes effective blood volume to decrease in at least certain portions of the circulation in which are located receptors critical for factors controlling certain renal functions. It has generally been agreed that urinary excretion of water and electrolytes is affected in two principal ways by orthostasis: (i) by decreased renal blood flow and GFR (652, page 122) and (ii) by stimulation, especially after several hours of standing, of the renin-angiotensin system, with consequently increased secretion of aldosterone (215), and perhaps also increased vasopressin release (522, 604). Quiet standing has hemodynamic effects similar to those produced by orthostasis effected by tilt-table; in some studies of the effect on kidney function of quiet standing, care has been taken to restrict activity to minimal muscular movements, but ordinary mild activity does not seem to make any important difference in the results. Certainly our own studies (256), in which the subjects, after assumption of the erect posture, made observations while moving about the laboratory or sitting on a high stool, are in general agreement with earlier studies.

Thomas (604, 605) reviewed the literature on postural influences upon urinary flow and electrolyte excretion and reported his own experiments, which have been the most carefully executed and interpreted of the published studies of this subject, particularly in that the effects of orthostasis were compared with control observations on another day at the same time—both morning and afternoon experiments were made—of the same subject. It appears, in brief summary, that the major, prompt effects of assuming erect posture consist of decreased urinary excretion of water, Na^+,

Cl^-, HCO_3^-, and K^+. Decreased GFR appears to be principally responsible. By extending his observations of subjects who had assumed the erect posture for as long as 3 hours, Thomas was able to discern (604) that an increasing portion of the extra Na^+ retained was balanced by extra acid secreted by the nephrons; in addition, small increases in urinary K^+ excretion made their appearance in the 3rd hour of orthostasis in two subjects. These effects were interpreted as reflecting increased secretion, under the influence of the increased aldosterone released within 1 hour in response to orthostasis (215), of acid and K^+ in exchange for additional Na^+ reabsorbed from the tubules.

Urinary pH declines in the standing subject because glomerular HCO_3^- filtration and urinary HCO_3^- excretion are diminished (II-7F). Thomas (604) reported consistent increases in urinary ammonia excretion, which would be expected as a consequence of the declining urinary pH (II-9). Other investigators, however, have not uniformly observed this effect, and Goodyer and Seldin (212) reported decreases. Postural studies in our own laboratory also failed to disclose a uniformly inverse relation between urinary pH and ammonia excretion (Figure II-1.2). The relatively poor correlation between changes in urinary pH and ammonia excretion in response to altered posture appears to us clearly to be due to the simultaneous influence of another posture-related variable upon urinary ammonia excretion. Urinary ammonia excretion is diminished by declining urine flow (II-10), a relation which had not yet become generally appreciated by the time the studies published earlier were carried out. Since the urine flow rate declines when subjects assume the erect posture, the effects of changing flow and pH on ammonia excretion tend to counterbalance each other. In higher-pH urines ($pH > 6.3$), urinary pH exerts relatively little absolute effect upon the rate of ammonia excretion (II-9), and especially when the urine is in this pH range we have observed (Figure II-1.2) predominance of the effect of changing urinary flow over that of urine pH on renal ammonia excretion in response to change of the subject's posture. It is accordingly natural that there should have been some nonuniformity in prior reports of the effect of changing posture upon urinary ammonia excretion.

Other problems of interpretation occur in studies of renal acid-base excretion in intact subjects (healthy volunteers) as a result of nonhomeostatic influences inevitably present to a greater or lesser extent; one such problem is illustrated in Figure II-1.3. The figure shows the urinary bicarbonate and net renal base excreted (II-5) in successive collection periods by a subject who had ingested a large load of alkali before the experiment. The results clearly betray the simultaneous influence of two different factors upon the time-course of events: (i) elimination of much of the alkali which had accumulated in extracellular fluid before the beginning of the experi-

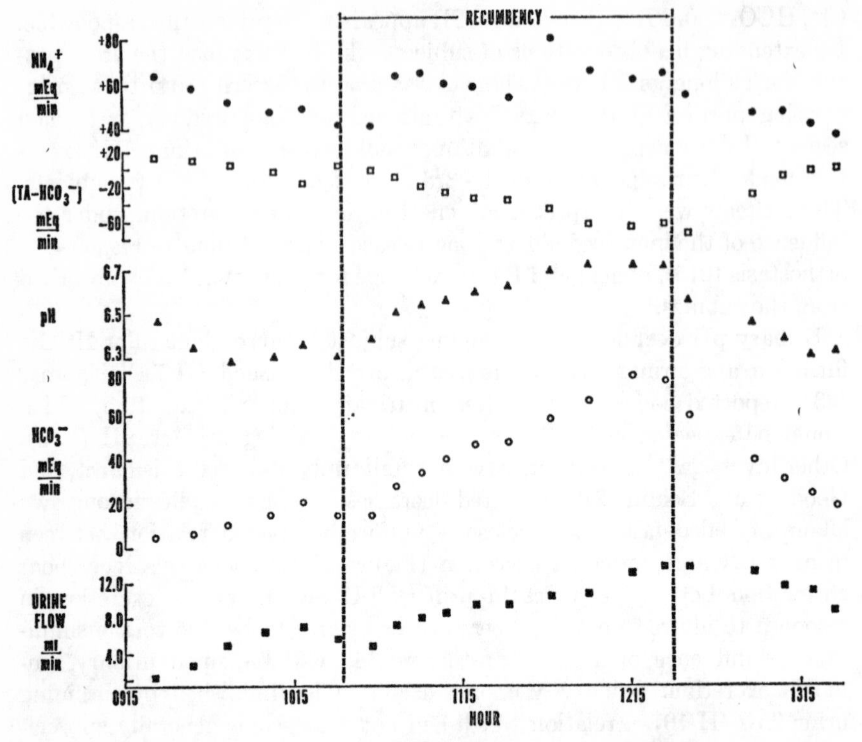

FIG. II-1.2. Effects of recumbency in a volunteer subject (256). When the urinary flow rate is changed by changing water load (or by exogenous vasopressin), urinary HCO_3^- excretion is inversely related to urine flow (II-10); during recumbency urinary HCO_3^- excretion increases, in spite of reduced water reabsorption in distal tubules and collecting ducts, because of increased glomerular filtration and hence glomerular HCO_3^- filtration. Consequently, in this range of urine pH (appropriate to acid-base load near zero) and in higher pH ranges, urinary ammonia excretion is affected, during recumbency, oppositely by (i) an enhancing effect of increased urine flow, and (ii) an inhibiting effect of rising urinary pH.

ment, expressed by a negative base balance throughout the experiment; and (ii) effects of imposed variation of the urine flow rate (II-8) upon the rate of renal net base excretion and of urinary HCO_3^- excretion. In the figure, the two simultaneous influences upon renal base excretion are easily disengaged by the eye. Such an experiment, considered as a study of the effect of water diuresis upon the renal net base excretion, is affected by an unwanted variable (negative base balance involving excretion of excess base at a decreasing rate as a function of time) which could largely, though seldom completely, be eliminated by sufficient care and experience in the experimental design. Provided, however, that change in urine flow rate is imposed independently of elapsed time—in this case by establishing diuretic peaks at the beginning and again at the end of the experiment—the two independent influences on urinary acid-base excretion are readily separable. Similar considerations apply when studying influences (such as water diuresis) on net renal acid excretion (II-10). It is a desirable if not neces-

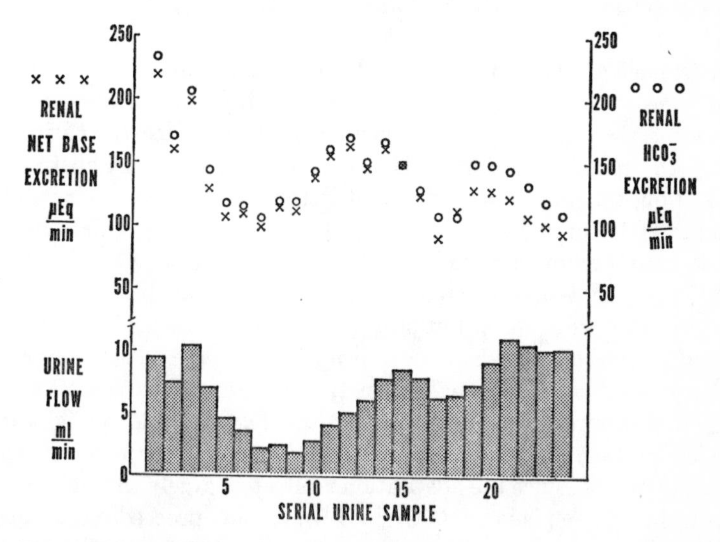

FIG. II-1.3. Urinary HCO_3^- (*circles*) and renal net base excretion (*crosses*) as a function of urine flow and elapsed time in a healthy volunteer subject. Renal net base excretion is reckoned from − (titratable acidity − HCO_3^-) (II-4). Urinary pH range, 6.95 to 7.48; the subject weighed 90.9 kg. Preparation: no breakfast; 75 mEq $NaHCO_3$ at 6 PM and again at 10 PM on preceding day, and 50 mEq 90 minutes before the experiment, which began at 8:30 AM, 2 hours after the subject arose. $NaHCO_3$, 1.44 mEq, administered every 12 minutes during the experiment.

The graph shows (i) that the rates of urinary net base and HCO_3^- excretion increase with rising urine flow and (ii) that there was a decrease with time of the excretion rate of each which would have been continuous except for the effect of changing flow. This decreased base excretion with time represented excretion of some of the large quantity of alkali absorbed just prior to the experiment. Base was being ingested during the experiment at a rate of 120 μEq/minute, but, since some of it must have been neutralized by the acid always metabolically produced during fasting, the base load (II-2B, II-5) was undoubtedly somewhat less. Urinary HCO_3^- excretion generally exceeded urinary net base loss in this experiment by 5 to 25 μEq/minute (cf. II-4; II-7F); urinary net base excretion averaged 133 μEq/minute during the experiment, indicating that the negative base balance to be expected under the circumstances was present throughout the experiment, though the rate of urinary elimination of the accumulated base was notably slowed by antidiuresis. By the end of the experiment, which lasted 4 hours and 20 minutes, renal net base excretion had declined to about 120 μEq/minute, a rate equal to the rate of continuing alkali administration and presumably not greatly in excess of the net base load. Circadian rhythmicity of renal acid-base excretion as tested on another occasion was not prominent in these morning hours in this subject.

sary part of the experimental design of any such study to divorce imposed conditions of interest from any simple relation to elapsed time.

2. Description of Acid-Base Homeostasis

Homeostatic systems can be intercompared in respect to their efficiency in defending the stability of the internal environment; criteria of homeostatic efficiency can conveniently be considered under three headings: speed, precision, and simplicity or ruggedness. Neither in these comparisons nor in the descriptions of the machinery of acid-base regulation which occupy the remainder of Part II can one avoid a certain abstraction or idealization, which, however, does not in the least impugn their validity.

Figure II-1.3 provides data relevant to the speed with which the kidney adjusts to an accumulated acid-base load and to the time-course of events. In the experiment shown, renal net base excretion was brought nearly into balance with a substantial but still physiological base load within about 4 hours after the base load had been abruptly reduced. We may suppose that the time required by normally functioning kidneys to establish fully a steady state appropriate to drastically altered acid-base load might amount to twice or three times the value reflected in this experiment, in general half a day or less. Since the adaptation is effected largely through alteration of the value of the plasma [HCO_3^-] (II-6), it must proceed most rapidly at the start, zero balance being approached asymptomatically, or rather merging into the usual circadian and other fluctuations of renal acid-base excretion. Establishment of a new steady state of acid-base balance in response to a changed acid-base load should not be confounded with renal adaptation, under pathophysiological circumstances, to abruptly altered P_{CO_2} (II-B; III-1); here acid or base are retained, in such a fashion as to minimize deviation of plasma pH, until a steady state of "compensated" respiratory acidosis or alkalosis supervenes (III-1). This latter renal adjustment, which operates through a different mechanism, may require several days to establish a new steady state.

The precision with which acid-base balance and arterial P_{CO_2} are normally regulated in the interest of stabilizing extracellular fluid pH are incomparable in absolute terms. Granted normal arterial P_{CO_2}, the kidney will adjust renal acid-base excretion to load so finely that responsive variation in plasma pH, as the load varies over a normal range, is conventionally taken to be 7.35 to 7.45, corresponding to 44 to 35 nEq (10^{-9} moles) of H_3O^+/liter, an absolute variation of ± 0.000000005 Eq/liter around a mean value of 40 nEq/liter. Speed and precision of responsive alteration of renal acid-base regulation are compared in Table II-1.1 with the two other most important of those aspects of the extracellular fluid which reflect the interrelation of body water and electrolytes—volume and osmolality.

Because of its high degree of automaticity (II-6), the renal system which

TABLE II-1.1. *Approximate comparisons of some renal homeostatic mechanisms in respect to speed and precision**

Aspect of extracellular fluid regulated	Speed: new steady state achieved within	Homeostatic efficiency (extent of normal adaptive variation)
Volume	5 days	± 2+ liters (± 15%)
Osmolality	2 hours	± 4 milliosmoles/liter ± 2 mEq Na^+/liter ± 1%
pH	6 to 12 hours	± 0.000005 mEq/liter (± 10%)

* In each case an abrupt and large, but physiological, change in load is envisaged: respectively (say) a change of 200 mEq daily in the NaCl intake, sudden imbibition of water at a continuing rate of 0.5 liter hour, a change in acid-base load amounting to 150 mEq daily. The most rapid of these responses is that of urinary water excretion, which is effected by variation in plasma osmolality whenever the water balance is altered transiently as by the drinking of water. Speed of adjustment of renal acid-base excretion is considerably less, but much faster than renal adjustment of NaCl excretion to a major change in load.

In percentile terms, osmolality is more efficiently regulated by the kidney, in response to challenge, than are volume or acidity. In absolute terms, regulation of $[H_3O^+]$ is more precise by many orders of magnitude than osmolality regulation.

regulates acid-base balance is simpler and therefore more rugged than the renal systems which stabilize the volume and osmolality of extracellular fluid. In both these latter renal regulatory systems, the renal tubular cells operate continuously under the direction of an elaborate neuroendocrine control system; consequently, improper adjustment of water and salt balance may result not only from structural damage to the kidney or intrinsic functional impairment of the renal cells themselves, but also from malfunction of a variety of endocrine organs or from damage to neural pathways. We need only think of the dramatic and sometimes life-threatening manifestations of diabetes insipidus or panhypoadrenocorticism to secure the most graphic illustration of the extreme vulnerability of these important renal regulatory systems to extrarenal impairment. By contrast, renal acid-base regulation is relatively independent of neural and endocrine influence, and hence much less vulnerable to damage resulting from extrarenal disorders. It is true that nonrespiratory alkalosis and hypokalemia, which may be of endocrine origin, are often associated with renal tubular secretion of acid into the nephrons at a rate inappropriately high in relation to the acid-base load and to the plasma $[HCO_3^-]$ (III-2B); but even here the tubular cell is itself abnormal in its chemical composition. And in all other instances of primary disturbance of acid-base balance, either there is an intrinsic disorder of the kidney itself or of its perfusion, or else the organ is overwhelmed by an abnormally large load of acid or base beyond its capacity to excrete without development of acidosis or alkalosis.

The definition of what constitutes a top normal load of acid or base is inevitably, within limits, an arbitrary one. Acid loads up to 100 mEq daily are not much beyond the acid loads (40 to 80 mEq daily) generally named as conventional in this country; on the other hand, a load of 200 mEq daily is probably never provided by the diet and metabolism of a healthy adult, and suffices to produce acidosis. For the purposes of this book, a load of 150 mEq of acid daily is therefore conventionally taken as the maximal acid load which can be considered physiological for a healthy adult. The same value is assigned as a top physiological base load; in health, base loads are relatively unusual, though not very rare, in this country, but are the rule in some societies. Renal acid-base excretion in the mammal is from the functional standpoint an unbroken continuum, and the renal defenses against base and acid excesses are more or less comparably efficient in stabilizing the pH of extracellular fluid.

two

KIDNEY, LUNG, AND BODY BUFFERS

A. ACID-BASE REGULATION: COLLABORATION OF KIDNEY AND LUNG

It is conventional (I-3) to consider that the pH *in vivo* of extracellular fluid, including blood plasma, is held very close to 7.4 by physiological mechanisms which result in stabilization of the plasma concentration of two substances: HCO_3^- near 24 mmoles/liter and CO_2 near 1.2 mmoles/liter; the normal 20:1 ratio of $[HCO_3^-]:[CO_2]$ in turn defines pH 7.4 at 37° C.[1] It is of course chemically arbitrary to focus upon the concentrations of $[HCO_3^-]$ and $[CO_2]$ as defining the plasma pH. According to the isohydric principle, pH is equally defined by the concentration ratios of the species of any of the other buffer pairs present in the plasma; and any alteration of arterial P_{CO_2}, or any addition of base or nonvolatile acid to extracellular fluid, must affect the position of equilibrium of all the extracellular buffers, not merely that of the HCO_3^- buffer system (I-5). Nevertheless, the contributions of the lung to acid-base homeostasis are made by means of the control exercised by this organ over the value of the arterial P_{CO_2}; and under normal circumstances, including normal arterial P_{CO_2}, $[HCO_3^-]$ in extracellular fluid is set by renal action.

In a broader sense, however, it is the net balance of nonvolatile acid and base which is regulated by the kidney (II-5); in this sense plasma $[HCO_3^-]$ is regulated simply as one among the buffers of extracellular fluid, all of whose positions of equilibrium will necessarily shift as net transfers of protons, to or from the buffers of extracellular fluid, are determined by the acid-base load (II-5) and the renal response to it. Moreover, deviation of arterial plasma P_{CO_2} from normal values causes $[HCO_3^-]$ to vary independently of renal regulation of acid-base balance; it is this fact, inconvenient to the diagnostician though not to the species, which has inspired a great deal of meticulous theoretical investigation and practical developmental work issuing in improved definition, in terms of acid-base analysis of blood, of the renal component of acid-base derangements (III-1B2).

But, though the kidney regulates acid-base balance rather than plasma [HCO_3^-], fluctuation of plasma [HCO_3^-] is indubitably the most important consequence of renal acid-base regulation. There are two principal reasons for the predominance of HCO_3^- among the buffers of extracellular fluid, within the context of the nonrespiratory component of acid-base regulation: one physicochemical, one physiological. Because of the "open system" provided by pulmonary stabilization of arterial P_{CO_2} (I-6B), plus evolutionary selection of 7.4 as the pH of extracellular fluid, the HCO_3^--CO_2 buffer system accommodates more nonvolatile acid accumulating in extracellular fluid, and yields more protons to accumulating base, than all the other buffers present collectively. This is true even in whole blood, in which —though hemoglobin provides exceptionally effective nonvolatile buffering —fluctuation of [HCO_3^-] contributes more than half the capacity to buffer added base or nonvolatile acid (118b). On the other hand, the HCO_3^--CO_2 buffer system cannot of itself moderate the effects exerted by changing arterial plasma P_{CO_2} on the acidic intensity of extracellular fluid; here the nonvolatile buffers of extracellular fluid provide the automatic first line of defense.

The second reason for the preponderant importance of HCO_3^- among the buffers of extracellular fluid is physiological. The system by which the kidney maintains acid-base balance is entirely dependent on the response of the blood HCO_3^- to the changing acid-base load (II-6). Given stable extracellular P_{CO_2}, extracellular [HCO_3^-] will in health[2] move upward from its normal value if, and only if, a net excess of base BOH is added to extracellular fluid:

$$BOH + H_2CO_3 \rightarrow BHCO_3 + H_2O \qquad (1)$$

(or, what is equivalent, if acid is withdrawn). Note that, whenever the resulting base bicarbonate is being removed in the urine, the excess base load is being removed from the body. Extracellular [HCO_3^-] will move downward at stable P_{CO_2} if, and only if, a net excess of nonvolatile acid is added:

$$HA + BHCO_3 \rightarrow BA + H_2CO_3 \qquad (2)$$

(or, what is equivalent, if base is withdrawn). The H_2CO_3 in both reactions is equilibrated with CO_2 gas dissolved in extracellular fluid, and since arterial P_{CO_2} is being held practically constant by the lung, extracellular [CO_2] and [H_2CO_3] are also stabilized. Accordingly, pH rises a little in the first reaction and falls a little in the second, but the only prominent change is alteration of [HCO_3^-]. *This responsive fluctuation of plasma HCO_3^- is the* *principal normal stabilizer of plasma pH as the acid-base load varies because it is responsive change in plasma* HCO_3^- *which adjusts urine pH in such a way that a steady state of zero acid-base balance is always normally approximated* (II-4; II-5; II-6).

Here is accordingly a crucially important difference between the physiological control of the two determinants, $[CO_2]$ and $[HCO_3^-]$, of extracellular pH. Fluctuations of arterial plasma P_{CO_2} and hence of $[CO_2]$ occurring normally in health are, from the standpoint of acid-base balance, incidental phenomena. They are normally related (except at altitude) not to acid-base challenges but to sleep, voluntary or involuntary hypo- or hyperventilation, circadian influences, or other nonhomeostatic phenomena; and they do not in health constitute an essential link in the chain of everyday acid-base homeostatic responses except insofar as arterial plasma P_{CO_2} is self-stabilizing (II-3A). Fluctuations of plasma $[HCO_3^-]$ are by contrast the prime agency by which acid-base balance is regulated (Section B; II-6). It is in fact remarkable how successfully arterial P_{CO_2} is held by the lungs near 40 mm Hg in view of the very substantial changes which normally occur in the rate of metabolic CO_2 production, notably the abrupt and very large augmentation effected by sudden violent exercise. Of course, when the lung is for any reason unable to prevent rise in arterial P_{CO_2}, or in the presence of pathological hyperventilation, renal acid-base regulation is affected by the influence of abnormal arterial P_{CO_2} upon renal acid secretion; but in such cases we are no longer dealing with normal regulation of acid-base balance.

B. ACID-BASE BALANCE

The previous section drew attention to the central role played by fluctuations of plasma $[HCO_3^-]$ in regulating the acidic intensity of the urine. Adjustment of the acid-base composition of the urine is the quintessential renal operation in the regulation of acid-base balance (II-6), and this adjustment can be effected automatically by responses of plasma $[HCO_3^-]$ to changing acid-base load only because of the functional organization of the kidney. Just as the HCO_3^- buffer system plays the leading role among the blood buffers in the stabilization of the pH of extracellular fluid, in the glomerular filtrate it is also this buffer system which is of prime importance to renal regulation of urine pH. The renal tubular cells continuously secrete acid in massive quantities into the nephrons, and base at a corresponding rate into the peritubular blood; the effect would be massive alkalinization of extracellular fluid were it not for the fact that base is continuously being lost from extracellular fluid as HCO_3^- filtered at the glomerulus. When the rate of regeneration of plasma HCO_3^- by the nephrons exactly equals the rate at which HCO_3^- is being filtered, extracellular base is being effectively conserved in spite of the continuous loss of filtered HCO_3^-. Not much of the

HCO_3^- being lost from the extracellular fluid in the glomerular filtrate, however, appears in the urine under these circumstances, because most of the acid being secreted reacts in the nephrons with the filtered HCO_3^- and converts it to CO_2, which is passively reabsorbed.

Continuing regeneration of plasma HCO_3^- in the peritubular blood, combined with decreased loss of HCO_3^- into glomerular filtrate, is the kidney's fundamental method of countering the effect of acid loads upon plasma pH. When nonvolatile acid begins to be added at a rapid rate to extracellular fluid, all its buffers are being acidified, and plasma [HCO_3^-] declines; the predominance of HCO_3^-, among the buffers countering the acidifying effect of accumulating nonvolatile acid in extracellular fluid, impelled Van Slyke and Cullen (627) to refer to blood HCO_3^- as the "alkaline reserve." In the presence of advanced nonrespiratory acidosis, where plasma [HCO_3^-] has been grossly reduced below minimal normal values, loss of extracellular HCO_3^- in the glomerular filtrate is automatically greatly reduced, though of course not arrested; when the acidosis is stabilized, regeneration of HCO_3^- in the peritubular blood, though also reduced, equals the sum of the HCO_3^- being lost into the glomerular filtrate and that disappearing by reaction with the invading acid. Both under normal conditions, i.e., in response to physiological increases of the net acid load (II-5B), and in the presence of acid loads large enough to deplete the "alkaline reserve" and produce acidosis, the fundamental renal defense against excess of nonvolatile acid lies in the relatively well preserved rate of peritubular regeneration of plasma HCO_3^- while the rate of loss of extracellular HCO_3^- in glomerular filtrate declines. Conversely, as plasma [HCO_3^-] rises due to an increasing base load, the fundamental defense against alkali excess resides in the increase of HCO_3^- filtration that automatically results from the increasing plasma [HCO_3^-]. Because filtered HCO_3^- under these conditions progressively exceeds the rate of secretion of acid into the nephrons, all the filtered HCO_3^- cannot react in the nephrons with the secreted acid, and so the excess is excreted as urinary HCO_3^- (II-6C).

Because of the rather limited range over which the urinary pH can be varied, transport into the urine of adequate amounts of buffer for the excesses of acid or base which must be excreted is also an essential part of the renal mechanism that regulates acid-base balance (II-6). The principal urinary buffer for excess base is the CO_2-HCO_3^- system; the buffering takes place in extracellular fluid by reaction of base with H_2CO_3, but the resulting HCO_3^- is transported into the urine by glomerular filtration. Base is usually excreted entirely as HCO_3^-, but when the urinary pH exceeds plasma pH, nonvolatile buffer systems do contribute to renal base excretion. Even when they do not, however, base excretion is not synonymous with urinary HCO_3^- (II-4C). One can manipulate urinary [HCO_3^-], and hence the calculated rate of HCO_3^- excretion, simply by equilibrating the urine at various values for P_{CO_2}; but this maneuver does not affect the calculated renal net excretion of acid or base (II-4C). Raising urinary P_{CO_2}, for example, generates urinary HCO_3^-, but the extra urinary base so generated is exactly equal to the additional urinary buffered acid that is generated by protons simultaneously transferred from H_2CO_3 to the other urinary buffers to yield the HCO_3^-.

Renal acid-base regulation is of course most visibly reflected in the acid-base composition of the urine. Renal net base excretion is principally accounted for by losses of urinary HCO_3^-; it is in bicarbonate-poor, lower-pH urines that the urinary buffers for excess acid, particularly ammonia, play an essential role in renal maintenance of acid-base balance (II-6). Urinary buffered acid cannot attain high concentrations until urine pH falls below 6.0, that is, until most of the filtered HCO_3^- has previously been removed from the luminal fluid in the nephrons by reaction with secreted acid. Urinary buffers for excess acid include a number of nonvolatile conjugate pairs and the NH_3-NH_4^+ buffer system, ammonia being transported into the urine as NH_3 by diffusion by an ingenious system which makes more NH_3 available as the need for it increases (II-9C). The central physiological importance of the urinary buffers for excess acid is that they permit acid secretion into the nephrons, and hence peritubular regeneration of HCO_3^-, to be sustained under conditions where acid secretion exceeds HCO_3^- filtration, luminal pH declines, and acid must be secreted against a pH gradient which, however, is reduced by the presence of these buffers (II-8A; II-8D).

Summarizing, urinary excretion of HCO_3^- (or loss of HCO_3^- by any route) represents from the physiological standpoint loss of base which has been or is being added to extracellular fluid; accumulation of extracellular base automatically increases plasma [HCO_3^-] and hence also glomerular HCO_3^- filtration and urinary HCO_3^- excretion. The kidneys stabilize the plasma [HCO_3^-] at a variety of levels within the range of normal, each appropriate to some particular acid-base load, because the plasma [HCO_3^-] automatically moves to such a level that net excesses of either acid or base being added to the body will be lost in the urine at the same rate. Fluctuation of plasma [CO_2] is not an integral part of the mechanism of normal adaptation to changing acid-base loads, though it is of great importance in the superstructure of renal acid-base regulation (II-3B) and in pulmonary compensation for acidemia or alkalemia of nonrespiratory origin (Table III-1.1).

The close analogy of "acid-base balance," defined physiologically as the net balance of base or nonvolatile acid, to other physiological balances with the maintenance of which the kidney is charged, such as that of N or Na or K, consists in the fact that in preserving balance the kidney brings urinary excretion into equilibrium with *net load* (II-5B) by a negative feedback system. In all these renal homeostatic systems, accumulation—absolute or relative to other constituents—in extracellular fluid of the particular material of interest incites a chain of responses the upshot of which is that the kidney prevents further accumulation of that constituent by increasing its excretion rate. Conversely, absolute or relative loss of the constituent from extracellular fluid incites its renal conservation through reduction of its urinary excretion rate. Special physiological features of acid-base balance, among these renal systems, are extreme simplicity (absence of elaborate hormonal and neurological control elements), high precision, and the rapidity with which normal adjustments to changing acid-base loads are made (II-1B2). A chemical feature unique to regulation of acid-base balance is the reciprocal relation between acid and base, such that acid gain in extracellular fluid is equivalent to loss of base while base gain is equivalent to loss of acid. The balance technic can be applied to quantitative estimation of acid-base balance, as it can be applied to any other balance for maintenance of which the kidney is charged, with certain special limitations to be discussed (II-5C; Appendix G).

C. ROLE OF THE LUNG: STABILIZATION OF ARTERIAL P_{CO_2}

Even under basal conditions the rate of formation of the principal end product of mammalian metabolism is large (about 2 pounds daily in adult man). CO_2, or at least its hydrate H_2CO_3, is indubitably an acid, and it is clear that elimination of CO_2 at the same rate at which it is being added to the body fluids is essential to the preservation of normal pH of extracellular fluid. Except for the tiny proportion of the total eliminated CO_2 present in the urine as "total urinary CO_2" (plus any lost in other fluid loss and in theory a minute amount evolving from the surfaces of the body), all the CO_2 produced must be excreted via the lungs; retention of even a tiny proportion of the CO_2 normally exhaled would produce severe respiratory acidosis. In this sense it is valid to draw attention to the size of the normal total "CO_2 turnover" of around 24,000 mEq daily, as compared with the much smaller quantity (around 50 mEq daily) of acid eliminated by the kidneys of subjects who are on usual protein-containing diets, or "total metabolic hydrogen turnover" (165).

On the other hand, it should be realized that CO_2 exhalation and net renal acid excretion are not commensurable types of "acid excretion." CO_2 exhalation can be quantified as the product of the volume of air expired per unit time and the difference of $[CO_2]$ between the breath and the environmental air. If we chose to call CO_2 "respiratory acid," then the rate of elimination of CO_2 via the lungs can be termed the rate of pulmonary acid excretion. It is not possible to express renal net acid excretion in an analogous manner as a chemical quantity definable independently of the extracellular fluid (II-4C). In addition, we do not in practice attempt to evaluate pulmonary acid-base regulation by measuring the difference between CO_2 production less extrapulmonary CO_2 losses and exhaled CO_2; the quantities are so large relative to the difference between them, which is where our interest lies, that such an approach would not be operationally feasible. Fortunately, we have available a far more satisfactory means of determining the manner in which the lungs are contributing to acid-base regulation: measurement of the arterial P_{CO_2}. Estimation of this parameter, now generally and very satisfactorily accomplished with the aid of suitable electrodes (II-3), furnishes a direct appraisal of pulmonary acid-base regulation at the moment of measurement; no comparably simple estimate of the renal component of acid-base regulation is available under pathophysiological circumstances (III-1).

When in this book, therefore, we consider *renal acid-base regulation as synonymous with regulation of acid-base balance*, and *pulmonary acid-base regulation as synonymous with control of arterial* P_{CO_2}, we categorize the subject in a manner which is theoretically sound, one which corresponds to the operational approach which we will take, as physicians and physiologists, to evaluate individually the function of the two organs upon which acid-base regulation depends.

D. NONVOLATILE BUFFERS OF BODY FLUIDS

The buffers other than the HCO_3^--CO_2 system of extracellular fluid are essentially the nonvolatile buffers. The conjugate pair NH_4^+-NH_3, vitally important as a urinary buffer, can make no significant contribution to buffering in extracellular water because of the extremely low concentration of total ammonia in all the body fluids.

The ratio of each conjugate pair of nonvolatile buffers of extracellular fluid must vary as its pH varies, with changing acid-base loads, over its normal range of variation, and their presence reduces the deviation of extracellular pH which will attend net accumulation in extracellular fluid of any given quantity of base or nonvolatile acid. Nevertheless, these buffers cannot influence the extent to which plasma pH must be altered in order to establish a new acid-base steady state appropriate to an altered acid-base load. For example, if we start to add excess base to extracellular fluid at a substantial rate, as by a constant intravenous infusion, it will be necessary for the arterial [HCO_3^-] to rise to some particular value in order for an appropriate new steady state to be established by the kidney through increase in the rate of urinary HCO_3^- excretion. The value of the plasma [HCO_3^-] necessary for this purpose is affected not at all by the presence of the nonvolatile buffers; all that these buffers do under such circumstances is to retard somewhat the rate of rise in plasma pH and [HCO_3^-] by accommodating some of the extra base accumulating in the body during the transitional period. When acid begins to be added to body fluids at an increased rate, the nonvolatile buffers similarly accommodate some of it, but their presence does not affect the new values for arterial plasma [HCO_3^-] and for pH which must be reached to effect a new state of balance between the altered net acid load and renal net acid excretion.

The nonvolatile buffers of extracellular fluid do complement collaborative acidity regulation by lung and kidney by stabilizing the pH of blood passing through the capillaries to the veins. Deviation of systemic venous pH (other than the pH of renal vein and vena cava blood plasma) from the arterial value is not directly accessible to control by either lung or kidney. If we assume that circulatory carriage of CO_2 is normal in all respects save for the total absence of any buffers for CO_2 in the blood—HCO_3^- is not a buffer for CO_2—we can calculate (Table II-2.1) that the pH of the mixed venous blood will be around 7.0, nearly half a unit below arterial plasma pH.

In chronic derangements of acid-base homeostasis it still remains true that pulmonary action in regulating arterial P_{CO_2} and renal function in maintaining acid-base balance are the essentials for survival. Though P_{CO_2} may remain abnormal, or an accumulation of nonvolatile acid or base may persist in extracellular fluid, life can continue if, and only if, the resulting derangement of the internal environment enables the failing lung or kidney, respectively, to hold arterial P_{CO_2} steady at an abnormal but tolerable level, and to approximate acid-base balance even though the accumulated base or acid cannot be gotten rid of. Grossly elevated arterial P_{CO_2}, for example, can be tolerated for a very long while, but death must ensue within minutes unless the ailing lung, assisted by the abnormally raised alveolar P_{CO_2}, succeeds in bringing the rate of CO_2 exhalation into balance

TABLE II-2.1. *Hypothetical effect of CO_2 added to venous blood if blood were unbuffered**

CO_2 production	17.0 mmoles/minute†
Cardiac output	8.0 liters/minute
CO_2 added to blood in capillaries	2.0 mmoles/liter
Mixed venous [CO_2]	3.2 mmoles/liter
Venous pH if [HCO_3^-] were 24 mmoles/liter	$pH = 6.1 + \log (24/3.2)$ $= 6.1 + \log 7.5$ $\cong 6.1 + .9$ $\cong 7.0$

* Arterial [HCO_3^-] is assumed to be 25 mmoles/liter. The addition of 2.0 mmoles/liter of metabolic CO_2 in the capillaries would generate a negligible amount of HCO_3^- in the absence of buffer.

† Assuming CO_2 output of a normal active adult to be 24 moles/day (165, page 64).

with CO_2 production, i.e., stabilizes arterial P_{CO_2}. Similarly, the diseased kidney may achieve acid-base balance after allowing the plasma HCO_3^- to fall to abnormally low values and the plasma pH to decline by 0.2 or 0.3, but death will soon ensue unless the net rate of excretion of nonvolatile acid can then be brought into balance with the net load in a state of stabilized acidosis.[3]

The nonvolatile buffers are indeed of the greatest importance in defending plasma pH in acute pulmonary dysfunction. Sudden hypercapnia can occur very abruptly in disorders affecting pulmonary function; the only immediate defense against the serious and possibly lethal deviation of plasma pH which would otherwise result is that provided by these buffers. Renal compensation, though it begins at once, and ultimately contributes in an important way to the defense of the internal environment under these circumstances, cannot establish an effective alteration, compensatory to abruptly altered arterial P_{CO_2}, in extracellular fluid composition for hours to days. In sudden total or near-total incompetence of the kidney, the nonvolatile buffers assist pulmonary compensation to retard the development of acidemia. During transitional periods in acid base disorders when extracellular pH is being altered, the nonvolatile buffers make an important contribution to the defense of arterial plasma pH.

The nonvolatile buffers of the body fluids are not confined to the extracellular compartment (III-1A). Yoshimura and his colleagues (676), studying the disposition in dogs of acutely administered, sublethal quantities of mineral acid (HCl), found that about 40% of the acid had been accommodated by buffers of cell water at the end of a 3-hour infusion. After 24 hours, about a quarter of the acid had been eliminated by the kidney and the extracellular pH was in the normal range, so that close to 75% of the acid had been accommodated by the cellular buffers. This sequestered acid was then gradually released into extracellular fluid, and excreted in the urine, over about a week. Cells also contribute to the accommodation of large loads of mineral acid, and to accumulation of extra CO_2, by metabolic responses as well as physicochemical buffering (II-3A; III-1A). Cellular defenses against accumulation of acid or base in extracellular fluid can hardly play much role in acid-base homeostasis when lungs and kidneys are normal and acid-base loads not excessive, but they become increasingly important as acidosis or alkalosis develops.

NOTES

1. The P_{CO_2} of interstitial fluid, since it reflects local cellular catabolic activity, must be a little higher than that of arterial blood plasma as well as (like local venous plasma P_{CO_2}) less uniform. Blood or blood plasma samples are the materials routinely available which reflect the acid-base equilibria of extracellular fluid. Arterial samples, which directly reflect pulmonary control, are preferred; available mixed venous (pulmonary arterial) samples are more reliable guides than are local ones to acid-base regulation in the organism as a whole.

For convenience of exposition, whenever this involves illustrative calculations applying to the normal state at rest, certain conventional values are assumed, in spite of the small errors involved, to apply indifferently to extracellular fluid and arterial blood plasma, particularly these:

a. Midpoint of $[HCO_3^-]$: 24 mmoles/liter; range, 20 to 28.

b. Midpoint of pH: 7.4; range, 7.33 to 7.46.

c. P_{CO_2}: 40 mm Hg; $[CO_2]$, 1.20 mmoles/liter.

d. Solubility of CO_2 in plasma under physiological conditions: 0.03 mmole liter^{-1}/mm Hg. This value is defined by the values listed above for P_{CO_2} and $[CO_2]$.

It may be that in fact extracellular fluid P_{CO_2} and pH are closer to venous values, which might be 45 mm Hg and 7.37, respectively, in pulmonary arterial samples taken from a resting subject. $[HCO_3^-]$ will in actual fact also be less in plasma than in plasma water and interstitial fluid because of the volume occupied by plasma colloids. More precise values for blood are compiled elsewhere (418). It is recommended (418) for very precise work that pK be redetermined in each laboratory.

Where experimental data relating to urine composition are presented, the best available values for physical constants have in general been used in this book rather than these conventional values.

2. Loss of extracellular water can increase $[HCO_3^-]$ (contraction alkalosis) and gain of extracellular water can reduce it (dilution acidosis) (III-1).

3. Skeletal base may be drawn upon in the chronic acidosis of renal failure to prevent progressive fall in extracellular pH when some acid is being retained daily (II-5; Appendix G).

three

ARTERIAL P_{CO_2}

A. STABILIZATION OF ARTERIAL P_{CO_2}

From the standpoint of acid-base homeostasis, arterial P_{CO_2} can be regarded as stable at 40 mm Hg in health. However, circadian and other everyday fluctuations of the value occur which are not of primary homeostatic significance, though they exert transient effects upon acid-base balance (II-1B). The principal alteration of arterial P_{CO_2} regularly occurring over the 24-hour period is related to the depression of the respiration effected by sleep (II-1B); the hypoventilation results from decreased sensitivity of the respiratory center to the stimuli to which it is normally responsive. The effect of sleep on respiration, like that of neurodepressor drugs, is essentially a by-product of diminished irritability of the nervous system: unlike the renal contribution to acid-base regulation, pulmonary regulation of arterial P_{CO_2} is continuously directed by the brain. It is therefore vulnerable to derangement by functional or structural disorders of the central nervous system as well as of the lung itself.

Normally the homeostatic control of respiration appears to be vested in the arterial P_{CO_2} and functions to hold this parameter within narrow limits (271, 340a, 109, 165, 37). However, respiration serves not only to remove waste CO_2, but also to provide the organism with O_2. Under some pathological circumstances hypoxia provides a powerful stimulus to the respiration (456, 271, 109). If the hypoxia is due to compromised ventilation of the alveoli, it may be accompanied by hypercapnia, but under other circumstances (e.g., high altitude or cyanide intoxication) the hyperventilation caused by anoxia may cause the arterial P_{CO_2} to fall to abnormally low levels. Nevertheless, at sea level, except when vigorous exercise is undertaken, alveolar ventilation in liters per minute is controlled in health fundamentally by the arterial P_{CO_2} (271, 109, 340a); any fluctuation of the rate of metabolic CO_2 production that tends to alter the steady-state arterial P_{CO_2} immediately affects the respiration in such a way as to restore the original value. Altitude, which stimulates the ventilation because arterial P_{O_2} is reduced along with ambient P_{O_2}, tends to produce chronic hypocapnia, so that the parameters of steady-state acid-base regulation are affected, producing respiratory alkalosis.

The regulatory role of CO_2 tension, rather than the total CO_2 of blood, was recognized early as a result of the work of J. S. Haldane and his colleagues. In recent years the employment of suitable electrodes for estimating the P_{CO_2} of blood and of other physiological fluids has become a part of clinical medicine. In awake, healthy persons at sea level, the precision with which arterial P_{CO_2} is stabilized is very remarkable. Aside from the circadian variations previously considered, and incidental changes produced by cortical influences such as voluntary breath-holding, pain, emotion, talking, singing, and so forth, we may consider normal variation in arterial P_{CO_2} to be no more than ± 2 mm Hg around a conventional mean of 40 mm. The exquisite sensitivity of the respiratory center to small changes in arterial P_{CO_2} was first noted by Haldane and Priestley (222), who reckoned that a doubling of the ventilation resulted from an increase of 0.02 atmosphere P_{CO_2}. Astonishing also is the dispatch and efficiency with which the lung is able to respond to the manyfold increase in the rate of metabolic CO_2 production that accompanies sudden physical exercise with an increase of minute ventilation sufficient to hold arterial P_{CO_2} (and O_2) nearly steady. This adjustment, which occurs too rapidly to be ascribable to reflex self-regulation of arterial P_{CO_2}, is effected initially by neural stimuli apparently supplemented promptly by hormonal and other humoral factors, and subsequently by some degree of acidemia and acidification of the cerebrospinal fluid (CSF) pH effected by accumulation of lactic acid (271).

It is the volume of air moved in and out of the alveoli per unit time which must be adjusted to the CO_2 production for homeostatic purposes; this function in turn is mechanically the resultant of the rate and the depth of respiration, whose integration is effected in the respiratory center. By this latter term (surrounded by quotation marks by some writers) are designated the highly organized but poorly localized collections of neurons, present in the reticular substance of the medulla oblongata, which emit impulses passing to the respiratory muscles via their motor innervation. The rhythms of normal respiration reflect the activity of the center, which is influenced by stimuli from various sources. The pulmonary reflexes affect the depth of respiration (and hence influence the relation between depth and rate of respiration at any value for minute ventilation); cortical stimuli may change the rhythm, rate, and depth of respiration; but the arterial P_{CO_2} controls the minute respiratory volume in such a way that deviations of its own value produced by the cortical stimuli are only transient. This control, and the stimulation of the respiration by hypoxia under pathophysiological circumstances, are both mediated by neural impulses transmitted to the respiratory center from chemosensitive structures located at a distance from it. These structures are classifiable into two groups: (i) those of the central chemosensitive area in the medulla and (ii) the peripheral receptors, located in man mainly in the carotid body and in other species also at several locations in the region of the aortic arch and subclavian artery.

Discovered by Heymans and his associates in 1927 (250), the peripheral receptors are now believed to play only a minor role in the control of respiration under normal physiological circumstances. But they are powerfully stimulated by severe hypoxemia, the stimulus being potentiated in the presence of hypercapnia; after interruption of their innervation, hypoxia often depresses rather than stimulates the respiration. They respond briskly to sufficiently severe acidemia with increased impulses to the respiratory center, but under usual clinical circumstances they appear to account for not more than a third of the total ventilatory response to acidemia.

The primacy of hypercapnia over hypoxia as the normal physiological stimulus to respiration was originally demonstrated in 1892 in experiments of elegant simplicity by J. S. Haldane and Lorrain Smith in which it was shown that the respiration of a subject in an enclosed chamber remained normal much longer if accumulation of CO_2 in the chamber was prevented by an absorber of the gas (222a). Leusen's observation (341, 341a) in 1954 of a marked influence on P_{CO_2} of the pH of CSF passing through the cysterna magna led to recognition and characterization of the central chemoreceptor area, located in the ventrolateral substance of the medulla (340a). It lies sufficiently close to the cavity of the fourth ventricle that topical application of nicotine or acetylcholine stimulates respiration, while application of procaine or stripping of the pial membrane inhibits it. The immediate response of the chemoreceptor cells to change in pH of the CSF, whether effected by change in its P_{CO_2} or its content of nonvolatile acid and base, suggests ready communication and probably some degree of continuity between CSF and the extracellular fluid (ECF) surrounding the responsive neurons. At least in the steady state the pH and [HCO_3^-] of the two compartments appear to be very similar (567a).

It seems, as originally proposed in 1911 by Winterstein (664, 665), that the final common pathway by which changes in the acid-base composition of blood and CSF affect the activity of the central chemoreceptors is probably the pH of the local ECF (355a). However, because the blood-brain barrier retards diffusion of charged particles between the blood and the CSF, acute nonrespiratory changes in blood pH exert a lesser immediate effect upon the respiration than acute hypercapnia or hypocapnia. As first discovered by Walter in 1877 (II-4A), nonrespiratory acidemia powerfully stimulates the respiration. However, the resulting compensatory increase of ventilation immediately lowers CSF as well as blood P_{CO_2}, a change tending to counteract the effect of acidemia on ECF pH in the area of the central chemoreceptors. Conversely, nonrespiratory alkalemia, because it depresses the respiration, causes CSF P_{CO_2} to rise, partially negating the direct depressant effect of the alkalemia upon the respiratory center.

The stabilization of the pH of the CSF itself by variations in local lactic acid production and by active movement of ions in and out of the fluid, and the relation of this ion transport to potential differences between CSF and blood, have been the subject of extensive investigation in recent years, particularly by H. H. Loeschcke, J. W. Severinghaus, and B. K. Siesjö and their associates. In spite of the immediate effect of a substantial change in arterial P_{CO_2} upon the P_{CO_2} of the CSF, the pH of the fluid is found to be surprisingly little affected within less than a day after an imposed change (555, 340a, 355a). Within 24 hours after removal of a person to high altitude, for example, the pH of his CSF is essentially normal, though its P_{CO_2} is reduced and alkalemia and hypocapnia are present. The ion transport responsible for stabilization of CSF pH plays a role in acclimatization, since inhibition of hyperventilation by an alkaline CSF is eliminated. Nonrespiratory acidosis (III-3A2) and alkalosis also affect the pH of the CSF remarkably little (567a), CSF [HCO_3^-] being altered by active ionic transport in such a way as to compensate for the altered CSF P_{CO_2} which maintains near-equilibration with the altered P_{CO_2} of arterial blood. The regulation of the composition of the CSF in relation to respiration has recently been reviewed by Leusen (340a). It should be borne in mind that rapid correction of established nonrespiratory acidosis by aggressive alkali therapy may temporarily *lower* the pH of the CSF as plasma and CSF P_{CO_2} rise (431); this phenomenon doubtless contributes to the continued hyperventilation, producing alkalemia with normal blood [HCO_3^-], often seen when the acidosis is too aggressively treated (III-3A2).

B. SOMATIC AND RENAL RESPONSES TO ALTERED ARTERIAL P_{CO_2}

Acute alteration of arterial P_{CO_2} effects immediate parallel changes in extracellular [HCO_3^-] for purely physicochemical reasons, independently of any renal regulatory activity (III-1B2); there is also an immediate striking effect upon renal acid secretion which within several days brings about additional and generally larger changes in [HCO_3^-] (Table III-1.1).

Yandell Henderson and H. W. Haggard (221, 245), working with dogs, appear to have been the first to observe that hypocapnia evokes a lowering of the blood [HCO_3^-] sufficient to moderate the alkalemia, and hypercapnia an elevation of blood [HCO_3^-] sufficient to moderate the acidemia. They thought at first that only shifts of alkali between blood and tissue were involved, but Leathes (333), and J. S. Haldane and J. B. S. Haldane and their associates (221b, 120, 221a) showed that in man renal responses come promptly into play. The urinary response to hypocapnia produced by forced ventilation was later studied in some detail in human subjects by McCance and Widdowson (388) and by Stanbury and Thomson (592); it consists of markedly increased excretion of HCO_3^-, Na^+, and K^+, which results from the respiratory alkalosis and not from the ventilatory effort (388). In the presence of either salt depletion or nonrespiratory acidosis, the urinary base loss is diminished and there is little or no natruresis.

Longson and Mills (357) reported that volunteers breathing 5 to 6½% CO_2 did not manifest any greater increase in urinary Na^+ and K^+ excretion than was observed in control subjects breathing air when observations were made during midmorning, a time when there is normally a considerable increase in the rate of urinary excretion of Na^+, K^+, and HCO_3^-; they also found that the human kidney does not correct respiratory acidemia within 3 hours. In acute hypercapnia and acute hypocapnia, the initial defense against gross deviation of the pH of extracellular fluid is in fact almost entirely provided by extracellular buffering and by ionic exchanges between extracellular fluid and cell water (147, 160a), but the bulk of the available evidence continues to indicate that renal responses come promptly into play under most circumstances (33, 147). The effect of this renal response upon the acid-base composition of extracellular fluid becomes increasingly prominent, over a period of several days, in the defense of plasma pH as the abnormality of arterial P_{CO_2} continues; it constitutes the principal active defense against deviation of the plasma pH in chronic acidosis or alkalosis of respiratory origin (Table III-1.1). Later, however, the urinary acid-base composition naturally becomes appropriate to the acid-base load in spite of continuing abnormality of arterial P_{CO_2} (147). Elkinton, Singer, Barker, and Clark, who studied with particular thoroughness both tissue buffering (147) and the renal response (33) to acutely altered arterial P_{CO_2} in man, referred to the temporary acidification of the urine in acute hypercapnia, and its temporary alkalinization in hypocapnia, as the "displacement phase" (33). Basing their calculations on measurements of the chloride space, these authors inferred that tissue buffering in acute respiratory acid-base abnormalities involved movement of protons into and out of cells, approximately half the proton transfer being matched, in terms of equivalents, by movement of Na^+ in the opposite direction. Significant exchange of cell K^+ for protons was not observed in their experiments, but probably occurs in the more drastic experiments which can be carried out in the dog (197).

Increased metabolic production of lactic acid contributes to the defense against acute hypocapnia (9a, 67a, 77); its cause remains uncertain. Gamble and his associates (187) found no interaction between the buffering of mineral acid and that of blood carbonic acid in experiments on severe acute acidosis in dogs. Over a 6-hour period the plasma [HCO_3^-] rose by 1.0

to 1.3 mmoles/liter per 10 mm Hg increase in arterial P_{CO_2} whether or not additional change was produced by simultaneous administration of mineral acid. Renal adjustments to chronic hypercapnia and hypocapnia are summarized later (Table III-1.1).

Stanbury and Thomson (591) suggested that the effect of hypocapnia upon the acid-base composition of the urine was probably explicable in terms of a stimulating effect of the CO_2 tension of the plasma on the exchange by which Na^+ is reabsorbed from the glomerular filtrate in exchange for acid secreted into the nephrons. That renal acid secretion does vary directly with the arterial P_{CO_2} has been many times confirmed by studies of renal HCO_3^- clearance (68a, 203, 501b, 135, 552). Because acute hypercapnia increases plasma [HCO_3^-] and hence also filtered HCO_3^-, some stimulation of acid secretion into the nephrons might be accounted for by the lesser blood-to-lumen pH gradient against which acid is then secreted; but acid secretion is very sensitive to small changes in arterial P_{CO_2} and its responsiveness is not believed to be explicable entirely in terms of the electrochemical gradient. The subject is further discussed elsewhere (II-8B).

four

RENAL EXCRETION OF ACID AND BASE

A. EXCRETION OF (NONVOLATILE) ACID

Excretion of acid via the kidney has consistently attracted more attention among investigators than base elimination. This partiality is doubtless mainly ascribable to the facts (i) that mammalian urines, including those of the species to which the investigators belong, are very commonly acidic, a finding early taken to mean—correctly in most though not all instances (II-6)—that acid is being eliminated in the urine; (ii) that some acid is always produced metabolically from neutral precursors, and that diets containing ample protein do result in a net load of excess acid which requires elimination in the urine; (iii) that all starving mammals produce an excess of acid to be eliminated; and (iv) that certain clinical disorders of metabolism result in gross metabolic overproduction of nonvolatile organic acids in quantity sufficient to result in fatal acidosis because the renal defense against excess acid is overwhelmed. Large base loads do not result from abnormal metabolism of organic compounds, and large intakes of excess alkali are clinically relatively uncommon, generally being iatrogenic or drug-induced through self-medication. Nevertheless, severe nonrespiratory alkalosis is a common condition which usually occurs clinically as the result of abnormal loss of nonvolatile acid, especially by vomiting of gastric HCl, but sometimes also through inappropriate renal regulation of the pH of extracellular fluid (Table III-2.1).

The overthrow of the phlogiston theory by Lavoisier, inaugurating the revolution in chemistry which two centuries ago placed the subject on a solid scientific basis, also laid the firm foundations of a chemical physiology by making clear that the production of animal heat and energy through chemical transformation of the organic materials of the diet is a kind of procrastinated combustion—less violent than that produced by burning or roasting but equally resulting almost entirely from oxidation of carbon and hydrogen. The way was now open to scientific investigation of animal metabolism and nutrition and of the chemical composition of the tissues. Fundamental aspects of acid-base balance were first systematically studied in the laboratory of Justus Liebig in Giessen as one aspect of widely ranging investigations of the nutritional and metabolic roles of mineral as well as organic substances.

Indicators of vegetable origin had earlier sufficed to prove that mammalian urine might give either an acidic or an alkaline reaction, and it had already been established by the early 1800's that the urine of herbivores is characteristically alkaline, while that of carnivores is acidic (349, 350). Gilbert Blanc had further discovered that human urine, normally acid, could readily be rendered alkaline by ingestion of sodium or potassium salts of organic acids, the cation being recoverable in the urine (349, page 189); and Friedrich Wöhler demonstrated that a fruit diet alkalinized the urine and increased its cation content (350). Liebig and his pupils, who first noted the predominance of Na as extracellular cation and of K among the cations of the tissues, systematically examined the ash of various foodstuffs, of the tissues, and of the excreta, that is, the incombustible material remaining after tissues, foodstuffs, urine, or stool are incinerated in a kiln. These studies soon showed that the animal organism maintains characteristic tissue levels of the several mineral elements irrespective of variation of dietary minerals because the minerals assimilated are excreted in urine and stool. Systematic studies of the metabolism of various species of carnivorous and herbivorous animals showed clearly that the reaction of the urine did not normally depend upon species-specific metabolic peculiarities, but rather upon the dietary ash (349, pages 189 and 190). Whereas combustion of meat gives rise to a net excess of acid because the P and S it contains are largely oxidized to phosphoric and sulfuric acids, the combustible portion of vegetable foods is largely organic anions ingested as salts of the alkali earths and metals Na, K, Ca, and Mg; the effect, after combustion of the organic anions, is as if strong alkali had been added to extracellular fluid (I-3B; II-5B1; III-1B1).

Fruitful further development of these studies over the following century included (i) the foundation of nutritional studies, including investigation of dietary mineral requirements in adult and immature animals and the metabolic role of the tissue minerals and other accessory food factors in animal metabolism (350, 367, 388a); and (ii) the origins of our knowledge of acid-base balance (II-5). Liebig had presciently pointed out that the much greater quantities of ammonia present in the acid urine of carnivores, as compared with herbivores, indicated that ammonia was providing base for partial neutralization of the excess acid derived from the metabolism of meat (II-9A). He did not, however, examine the quantitative effect of measured loads of excess mineral acid or base upon the acid-base composition of the urine.

The first recorded experiment of this kind appears to be that of Miquel (405), who in 1850 observed that about 30 mEq/day of a strong mineral acid (H_2SO_4) orally administered to a dog caused increased excretion of Na and K in the urine, indicating that the organism was drawing upon its alkali reserves (I-3B) to neutralize the extra acid. An extensive series of experiments on the effect of administration of excess mineral acid was shortly thereafter carried out in both man and dog at the University of Dorpat by Clare, Eylandt, Wilde, Frei, Kurtz, and Gäthgens (100, 156, 655, 190, 191). As a result of these observations, four important facts were established: (i) the reaction of the blood was found always to be alkaline; (ii) substantial loads of administered mineral acid were excreted in the urine, as manifested by an increased quantity of alkali required to titrate the urine to neutrality when an acid load was being given; (iii) acid loads, provided they were large enough, caused extra urinary excretion of the Na and K of the blood and tissues, but not nearly in quantities equivalent to the acid being lost; and (iv) the human and canine subjects studied tolerated large administered quantities of strong mineral acid well, with little or no depletion of the blood's "soda" and without disappearance of its alkaline reaction. Gäthgens (190, 191) placed particular emphasis upon the conservation of the body's alkali stores in blood and tissue during acid loading. He realized that the operation of some physiological mechanism for conserving the body's supplies of mineral base was in evidence, but he had at first no explanation for the findings.

Investigation of the disposition, by other species, of acid loads resulting either from feeding acid-ash diets (albumin, taurine) or from administration of mineral acid later disclosed important species-specific differences in the ability to tolerate large acid loads (267, 536, 328, 537). The experiments of Salkowski (536, 537) and of Lassar (328) demonstrated that the rabbit could not conserve its blood alkali, after administration of a large acid load, as effectively as dogs and cats, and that rabbits succumbed relatively readily to fatal acidosis after administration of large acid loads given in the form of H_2SO_4 or of the sulfur-containing compound taurine which yields H_2SO_4 when metabolized. Salkowski revived Liebig's earlier suggestion that the difference in the tolerance of herbivores and carnivores for an acid load probably resided in a superior ability of the carnivore to neutralize the acids with urinary ammonia, allowing conservation of the mineral bases regarded as constituting the alkaline stores of the body. Walter (645) first demonstrated that the dog does in fact respond to a continued large acid load with a large increase in urinary ammonia (II-10B), and confirmed the relative vulnerability of the rabbit, which normally excretes an almost ammonia-free urine and does not respond as promptly to acid loading with large increases in the rate of urinary ammonia excretion, to large acid loads. Walter also found that very large acid loads always killed his animals before he could render the reaction of their blood acid to litmus paper. He reported the marked stimulation of the respiratory center produced by acute acidosis, and showed that apparently agonal acutely acidotic animals, whose respiration had slowed and then stopped, could be revived promptly by administration of $NaHCO_3$ intravenously.

Although the protocols of these seminal observations disclose a prompt effect on urinary ammonia excretion of alkalinizing the urine of a dog (Salkowski), and a delayed and larger progressive effect of continued acid loading in the dog (Walter), the fact that administered acid can affect urinary ammonia excretion in two different ways was overlooked for several generations; it was finally unequivocally demonstrated in individual experiments only when Ferguson (161) used inhibitors of carbonic anhydrase to alkalinize the urine promptly while gradually acidifying the body fluids. Of the two mechanisms by which acid excess affects urinary ammonia excretion, only the increase in urinary ammonia production during acidosis was clearly recognized and studied over the ensuing half-century. There is always a temptation in physiological investigations to consider that normal physiological regulation is reflected when unphysiological conditions are imposed in order to induce large and easily discernible effects. In addition, interest in the pathophysiology of acid-base balance was powerfully stimulated just at this time by major advances in the study of the disturbed intermediary metabolism associated with diabetic ketosis and lactic acidosis. By 1884 Minkowski (404) and Stadelmann (585, 586) had shown unequivocally that diabetic acidosis results from an enormous overproduction of the organic acids (hydroxybutyric, diacetic) which Magnus-Levy (371) showed to be the products of incomplete oxidation of fat; Hallervorden (224) and others (586) demonstrated that this type of clinically occurring acid overload also produces a great increase in urinary ammonia excretion. And before 1900 Araki (10–12) had shown that anoxia leads to increased metabolic production of lactic acid, lacticemia, and lacticiuria.

Because it was understandably hoped that alkali therapy might be lifesaving in the treatment of diabetic ketoacidosis, these discoveries also led to the first systematic studies of the effects of administration of alkalinizing organic salts of Na (586). Apart from showing that the mammalian organism can defend itself effectively against large excesses of base, these experiments also abundantly confirmed the reduction of urinary ammonia excretion brought about by administration of alkalinizing salts.

Henderson and Palmer (239, 243, 244) first clearly quantified total renal excretion of nonvolatile acid, which we may symbolize as TE_A, as:

$$TE_A = TA + NH_4^+ \tag{1}$$

where TA is the urinary titratable acidity and NH_4^+ is the urinary ammonium ion, or, practically speaking, urinary (total) ammonia. The expression is exact provided TA is taken to signify the rate of renal transfer of protons to the nonvolatile urinary buffers, not the titration by which this physiological process is subsequently evaluated in the laboratory (Section C; Appendix F). Similarly, if differences in ionic strength between glomerular filtrate and urine are corrected for, every urinary NH_4^+ contains a proton which was detached from combination with some base in extracellular fluid; and since virtually all the urinary ammonia is present as NH_4^+ (I-6B), urinary total ammonia excretion represents, to a very near approximation, protons transferred out of extracellular fluid to the only volatile urinary buffer available to accept them. The remainder of the buffered acid present in acid urine must evidently have been accepted by the urinary nonvolatile buffers, for urinary H_3O^+ is always negligible; even at pH 4.0, the lowest value ever reported and one considerably less than any urine pH normally attainable (II-6A), $[H_3O^+]$ would amount by definition to only 10^{-4} Eq or 0.1 mEq/liter, a negligible quantity compared with the 40 to 80 mEq or more of acid excreted daily by meat-eaters. Consequently, without needing to know the chemical identity of the various urinary nonvolatile buffer compounds, we can evaluate how many protons per milliliter of urine were collectively accepted by them as the urine was being acidified in the nephrons, by titrating CO_2-free urine from the pH of the collected urine to the pH of glomerular filtrate (Section C; Appendix F).

Though dissolved CO_2, or rather its hydrate H_2CO_3, is undoubtedly an acid, urinary CO_2 gas has never been regarded as a constitutent of urinary acid excretion, which is not a chemical quantity definable independently of the extracellular fluid (255; Section C). Insofar as arterial P_{CO_2} is stable, all the CO_2 being lost from the body—the relatively tiny amount excreted in the urine as well as the greatly preponderant loss via the lungs—simply flows through the extracellular fluid without affecting its acidic intensity. Any CO_2 produced within the body and excreted as such in the urine has caused no net transfer of protons into or out of extracellular fluid. When, on the other hand, arterial P_{CO_2} deviates from normal values under pathological circumstances, the crucial effect of depletion or augmentation of $[CO_2]$ in the extracellular fluid upon its pH is the result of pulmonary dysfunction; it has nothing to do with renal excretion of CO_2 gas. In this book the traditional practice is followed of excluding the volatile urinary "acid" CO_2 when considering net renal excretion of acid and base; renal acid excretion is by definition elimination of nonvolatile acid. Similar considerations apply to losses of NH_3 gas, whether via the expired air or in the urine; urinary ammonia gas (which is of course in any case negligibly small) does not even theoretically contribute to urinary base loss, which is by definition elimination of nonvolatile base. NH_3 added to extracellular fluid simply passes through it, in the steady state, like CO_2, insofar as urinary NH_3 does not accept a proton.

B. EXCRETION OF (NONVOLATILE) BASE

The disposition of net excesses of base has always attracted less attention than the disposition of loads of nonvolatile acid. After potentiometric methods had disclosed in the first decade of our century how very near to neutrality the reaction of the blood is (II-6A), it was ascertained that blood can accept large amounts of alkali, as well as acid, with surprisingly little alteration of its acidic intensity (584, 177). This remarkable buffer strength of blood, especially *in vivo*, provides protection against the perturbations of arterial plasma pH resulting from rapid accumulation of large base loads; but defense against continuing base excess depends rather upon establishment of steady states such that renal net base excretion is brought into balance with the load than upon the blood's capacity to buffer excess base.

The more acid urines contain practically no HCO_3^-, but as urinary pH rises above 6.1 there is appreciable and increasing excretion of urinary $[HCO_3^-]$ (Figure I-6.3); and as Gamble first recognized (181), excretion of HCO_3^- represents physiologically loss of base from the body. This is so because any base added to extracellular fluid accepts protons from the buffers of extracellular fluid including H_2CO_3; and when the resulting HCO_3^- is excreted in the urine, the base is being eliminated, even if the pH of the urine is lower than that of the plasma. For example:

$$BOH + H_2CO_3 \rightarrow B^+ + HCO_3^- + H_2O \tag{2}$$

where B is the cation accompanying the hydroxyl ion as it enters the extracellular fluid and is excreted matched (I-3B) against HCO_3^- in the urine. Any nonvolatile base, not only a hydroxide, detaches protons from H_2CO_3, leaving HCO_3^- to be excreted in the urine:

$$B_2CO_3 + H_2CO_3 \rightarrow 2\ B^+ + 2\ HCO_3^- \tag{3}$$

Accordingly, *net* renal acid excretion is obtained by deducting urinary HCO_3^- excretion from renal total acid excretion. In urine whose HCO_3^- content is negligible (pH $<$ 6.0), net renal acid excretion E_A is equivalent to renal total acid excretion because base loss in the urine is negligible. But as the urine pH rises above 6.1, urinary $[HCO_3^-]$ rapidly becomes a significant part of renal acid-base excretion, and must be deducted from renal total acid excretion to obtain renal net acid excretion:

$$E_A = TA + NH_4^+ - HCO_3^- \quad (4)$$

When urinary pH rises sufficiently high for the rate of HCO_3^- excretion to exceed that of $TA + NH_4^+$, there is *net renal base excretion* E_B:

$$E_B \ (= -E_A) = HCO_3^- - TA - NH_4^+ \quad (5)$$

Rarely, the sum of the terms on the right is zero; such urine, in which there is no net excretion of urinary acid or base, is appropriate to a state of acid-base balance at zero acid-base load.

That E_A and E_B are really two halves of a continuous function will be apparent from Equation 5; the point is made graphically by Figure II-6.2. Just as $E_B = -E_A$, so $E_A = -E_B$; these equalities simply reflect the fact that acidity is the negation of basicity and basicity the negation of acidity. That is why we obtain "net" acid and base excretion by Equations 4 and 5, respectively, by means of a subtraction not involving identical entities, like subtracting 3 apples from 5 apples to obtain a net of 2 apples. But there is no mystery here: a positive value for E_A means physiologically net transfer of protons from the buffers of extracellular fluid, and a positive value for E_B signifies net proton transfer in the reverse direction. The overall process by which HCO_3^- reaches the urine is that some of the CO_2 plus H_2CO_3 produced metabolically (plus any that is ingested, as in soda water) does not all flow out through the lungs and in the urine, as is the case in very acid samples, but instead some H_2CO_3 transfers a proton to the buffers of extracellular fluid, and the HCO_3^- generated as a result is excreted in the urine matched against the cation of some basic compound (like NaOH) with which the H_2CO_3 reacted (II-8B). If the urinary nonvolatile buffers and ammonia are accepting protons from extracellular fluid at a greater rate than that at which H_2CO_3 is donating them to extracellular fluid, then there is net acid excretion, E_A; if at a lesser rate, then there is net base excretion E_B; and very occasionally the net proton transfer between urine and extracellular fluid is zero.

In describing the renal excretion of excess acid, Henderson (239) wrote in 1911: "The acid end products of metabolism, without appreciably changing the actual alkaline reaction, constantly take up alkali from blood and protoplasm. In this manner there is a tendency to disturb the normal protective equilibrium between bases and acids. This tendency is held in check by the kidney, which in the process of urine formation reverses the reaction of neutralization of acid and restores that alkali which has served as a carrier of acid." Contemporary acid-base theory makes it particularly easy for us to visualize renal acid excretion as inevitably accompanied by equivalent secretion of base into the renal venous effluent, thus regenerating base in extracellular fluid, and to visualize renal base excretion as inevitably accompanied by secretion of acid into the renal venous effluent, thus regenerating acid in extracellular fluid.

The urinary nonvolatile buffers usually are urinary proton acceptors, hence the term "titratable acidity" of urine; but occasionally these urinary buffers serve as proton donors to extracellular fluid, or in other words contribute to renal base excretion. The pH of high-flow urine is never as high as plasma pH (cf. Figures II-9.1 to II-9.4), so that during water diuresis the nonvolatile urinary buffers of the glomerular filtrate have always accepted some protons by the time the end urine has been formed. There is never, in other words, any "titratable alkalinity" in high-flow samples, and base excretion in such urines is entirely effected by elimination of HCO_3^-. But urinary pH can rise as high as 8.0 during antidiuresis when there is a large load of base to be excreted; under these circumstances the nonvolatile buffers of glomerular filtrate have transferred protons to the buffers of extracellular fluid, and TA is a negative value.

C. EVALUATION OF NET RENAL ACID OR BASE EXCRETION

Equations 4 and 5 are expressions of a perfectly general definition of the excretion term of acid-base balance (255). Since TA is defined as the overall physiological process by which protons are transferred between the buffers of extracellular fluid and the nonvolatile urinary buffers of the end urine, differences in ionic strength μ, and hence in the value of pK_2' for phosphate, between glomerular filtrate and the diluted urine sample in the titration beaker must be corrected for. These differences affect urinary pH for purely physicochemical reasons, independently of any exchange of protons between extracellular fluid and end urine. This problem is considered in detail in Appendix F. Schwartz, Bank, and Cutler (545) have recommended that for accurate results the value of TA be calculated, and other procedures for minimizing this type of error have been described by Nutbourne (426).

E. K. Marshall (384) first drew attention to the fact that substantial alterations of the pH and P_{CO_2} of urine often result from spontaneous evolution of CO_2 during or after the urine collection and before the measurements are made. When CO_2 diffuses off, the HCO_3^--CO_2 equilibrium is displaced, and HCO_3^- reacts with protons donated by the other urinary buffers (H^+, at the left) to restore equilibrium:

$$H^+ + HCO_3^- \rightarrow H_2CO_3 \rightarrow H_2O + CO_2 \rightarrow \text{diffuses off} \qquad (6)$$

The large rapid losses of CO_2 from high-pH urines, as compared with lower-pH specimens, result only in part from the initially increased CO_2 tensions that occur with increasing urinary pH; much more important is the fact that the urinary P_{CO_2} stays high much longer as CO_2

diffuses off higher-pH urines, permitting continued rapid diffusion of CO_2 down a CO_2 pressure gradient from urine to air, in contrast to urines containing little HCO_3^-. The explanation of the failure of CO_2 tension of high-pH urines to decline rapidly, as CO_2 diffuses away, is found in the extensive transfer of protons by the nonvolatile buffers to HCO_3^-; the magnitude of this conversion of HCO_3^- to CO_2 is in turn dictated by the high urinary HCO_3^--to-CO_2 ratio (I-6) which ensures that absolute change in $[HCO_3^-]$ is large relative to change in pH (and in P_{CO_2}). Unless adequate measures are taken to reduce to a minimum loss of CO_2 during collection of the sample, P_{CO_2} of high-pH urines will be markedly underestimated and their pH will be markedly overestimated; earlier reports of urinary pH exceeding 8.0 were undoubtedly artifactual and resulted from failure to prevent evolution of CO_2. There is indeed some appreciable loss of CO_2 from low-flow, highest-pH urines as they traverse the renal pelvis, ureters, and urinary bladder (261).

As CO_2 diffuses out of HCO_3^--containing urines, each bicarbonate ion converted to CO_2 absconds with a proton donated by the other urinary buffers. When there is net renal base excretion, so little ammonia is present in the urine that the donated H^+ indicated on the left (Example 6) is practically all furnished by the urinary nonvolatile buffers, and there is virtual chemical equivalence between the HCO_3^- disappearing and the decrease in titratable acidity produced. Evolution of urinary CO_2 therefore results in a decrease in the quantity of alkali which will be required during titration to restore the position of equilibrium of the nonvolatile urinary buffers to what it was in glomerular filtrate, and the decrease in measured urinary TA must be chemically equivalent to the loss of HCO_3^-. Since $HCO_3^- - TA$ is virtually unaffected by loss of CO_2 from the sample, the estimate of E_B according to Equation 5 is also unaffected. Even when the acid-base load is close to zero, in urines of around pH 6.3 to 6.5, conversion of urinary NH_4^+ to NH_3 hardly competes, as CO_2 diffuses off, with the nonvolatile buffers as the source of the protons required for the conversion of HCO_3^- to CO_2. Below pH 6.3, where NH_4^+ comes to equal and then rapidly to exceed acid buffered by the nonvolatile urinary compounds (II-6), too little HCO_3^- is present for $TA - HCO_3^-$ to differ greatly from TA. Estimate of either E_B or E_A accordingly remains (for practical purposes) unaffected by evolution of CO_2 from the urine prior to analysis of the urine sample, provided, of course that no CO_2 is permitted to evolve between the measurements of HCO_3^- and TA.

Albright and his associates (5, 6) popularized the expression "titratable-acidity-minus-CO_2," which will be symbolized as $(TA - HCO_3^-)$, to characterize the variable obtained by a procedure in which $TA - HCO_3^-$ is estimated as a single combined value. $(TA - HCO_3^-)$ has also been termed "residual titratable acidity." The procedure involves addition of sufficient measured mineral acid to a suitably diluted aliquot of a timed urine sample to convert all HCO_3^- present to CO_2; practically, this is accomplished by adjusting urine pH to around 5.0. After this, the sample is aerated to remove the CO_2 (so that none can be titrated to HCO_3^-); the quantity of alkali required to bring the diluted sample to plasma pH is then ascertained by titration, the previously added acid is deducted from the titration figure, and the TA is expressed as an excretion rate. The method seems to have been devised in Gamble's laboratory (181); the excretion rate of ($TA - HCO_3^-$) (it may be a positive or a negative value) may be substituted in Equations 4 and 5 for TA and HCO_3^- separately determined:

$$E_A = (TA - HCO_3^-) + NH_4^+ \quad (7)$$

$$E_B = -(TA - HCO_3^-) - NH_4^+ \quad (8)$$

Albright and co-workers (5, 6) added an extra term, namely 49% of the urinary calcium, to the right-hand side of Equation 7 to obtain a quantity which they called the "sum of the base-sparing mechanisms" (II-5), a quantity with which we shall not be concerned in this book (Appendix G).

Because of the attractive analytical simplicity of the procedure by which it is obtained, as well as the fact that no precautions against CO_2 loss from the urine sample are necessary, evaluation of ($TA - HCO_3^-$) is universally preferred to independent estimation of TA and HCO_3^- when E_A is to be measured. Corrections for any discrepancy between the ionic strength μ of the glomerular filtrate and that of the diluted sample in the beaker at the end of titration can be applied exactly as in urines containing essentially no HCO_3^- (Appendix F). In studying renal net base excretion, it may be desired also to evaluate the urinary pH or P_{CO_2}; in such cases, where gas loss from the urine must be prevented in any case, E_B can alternatively be obtained from independent estimates of $[HCO_3^-]$ and $[TA]$, the latter being determined from the titration curve of the CO_2-free diluted urine between the urinary pH originally measured and the blood pH. The percentile error of estimate of E_B arising from neglect of the effect of sample dilution on phosphate pK_2' is here small, because when any substantial net base load is being eliminated, $[TA]$ is very small relative to $[HCO_3^-]$ (Appendix F). As long as the paired values for urinary pH and P_{CO_2} at any P_{CO_2} are made the basis of this technic, the estimate of renal net base excretion remains unaffeçted by any prior alteration of the urinary P_{CO_2}; accordingly, evaluation of renal net base excretion from $-(TA - HCO_3^-)$ is simply the limiting condition, at $P_{CO_2} = o$, at which this entity can be evaluated from the paired values of urinary P_{CO_2} and pH. The satisfactory congruence, previously reported

(255), of estimates of E_B by the two different methods available is visible in a graph reproduced elsewhere in this book (Figure II-8.1).

"Renal net excretion of nonvolatile acid" does not mean excretion of acid chemically defined; rather it expresses the rate of transfer of protons from the buffers of extracellular fluid to those of urine, a *physiological* function (Section B). This important point, which is also touched on in the ensuing chapter and in Appendix G, can conveniently be illustrated here by considering, as an example, the urinary phosphate excretion. $H_2PO_4^-$ can be both a base and an acid in the Brønsted-Lowry system; its functional significance in the urine, in terms of the physiology of renal acid-base regulation, depends entirely upon the effect of its excretion upon the acidic intensity of extracellular fluid. This effect can be ascertained only by comparing the relative rates of excretion of the two species of the physiological buffer pair $HPO_4^=$-$H_2PO_4^-$ to their ratios in extracellular fluid. If the ratio of the rates of excretion of the two species in the urine is identical with their ratio in extracellular fluid, i.e., 4:1, their removal has had no effect on the position of equilibrium of the extracellular phosphate buffer system, nor, *a fortiori*, on the position of equilibrium of any of the other buffers of extracellular fluid. It is true, but entirely irrelevant to physiology, that a 4:1 ratio of $HPO_4^=$:$H_2PO_4^-$ in the urine represents in terms of chemical theory the urinary excretion of three net excess particles of the conjugate base. In the more acid urines, in which the urinary ratio $HPO_4^=$:$H_2PO_4^-$ has declined below that of plasma, we can quantify the contribution of phosphate buffer to the rate of urinary acid excretion as the product of the increase in the proportion of $H_2PO_4^-$ in urine over extracellular fluid and the total urinary phosphate excreted over a given period. Conversely, in urine of pH 8.0, where most of the phosphate is $HPO_4^=$, urinary base has obviously been generated, during the conversion of glomerular filtrate to urine, and protons derived from the conversion of filtered $H_2PO_4^-$ to $HPO_4^=$ have been transferred to extracellular fluid; the rate of base excretion as phosphate buffer can be reckoned similarly as total phosphate excreted per unit time multiplied by the increase in the proportion of $HPO_4^=$ in urine over extracellular fluid.[1] The contribution of the excreted phosphate to renal acid-base excretion can be defined only relative to extracellular fluid.

Unlike the nonvolatile urinary buffers, the rate of donation of protons by the urinary CO_2-HCO_3 buffer system to the buffers of extracellular fluid is not expressed by differences between the plasma and urine ratios HCO_3^-:CO_2, but by the rate of urinary excretion of HCO_3^-. Essentially, this is because all the urinary conjugate base of this buffer system represents base loss, which is not true of the urinary nonvolatile buffers: every milliequivalent of urinary HCO_3^- has been derived from a milliequivalent of H_2CO_3. It was the donation of a proton by this H_2CO_3 to the buffers of the extracellular fluid which left behind the HCO_3^- which (in the net) was excreted in the urine. Moreover, change in the ratio of species of the volatile urinary buffers as compared with plasma does not necessarily signify any exchange of protons between urinary and plasma buffers; the ratio can be changed simply by altering the urinary gas pressures. When urinary P_{CO_2} is raised or lowered—irrespective of whether one manipulates the P_{CO_2} of the urine sample after collection or whether the change from 40 mm Hg represents equilibration at the papilla at a different P_{CO_2} or changes occurring beyond the papilla or in the urinary tract—urinary $[HCO_3^-]$ increases or decreases with the P_{CO_2} for purely physicochemical reasons. However, since every microequivalent of HCO_3^- appearing when P_{CO_2} is raised has increased TA by 1 μEq, and every microequivalent disappearing when P_{CO_2} is lowered has decreased TA by 1 μEq, the only effect of altering urinary P_{CO_2} is to cause exchange of protons between the urinary buffers. The partitioning of the total renal base excretion between urinary HCO_3^- and $-(TA)$ is therefore operationally arbitrary, and the variable of physiological significance is always $TA - HCO_3^-$.

NOTES

1. The significance, for acid-base excretion, of elevation of the urinary phosphate ratios over the ratios in plasma is clear, but the actual sequence of events in the nephrons is somewhat complex. The pH of urine can be raised above that of plasma only when urinary HCO_3^- excretion is high in low-flow urines (II-10). During the dehydration of the bicarbonate-rich luminal fluid in the distal nephron, protons are transferred from the nonvolatile buffers to HCO_3^-, which is thereby converted to CO_2 and largely reabsorbed. Renal base excretion as HCO_3^- is being reduced, by these proton transfers between two classes of urinary buffers, to exactly the extent that base excretion as $(-TA)$ is being increased; the actual net transfer of protons from urine to extracellular fluid had already occurred when HCO_3^- derived from extracellular H_2CO_3 was filtered at the glomerulus.

five

ACID-BASE LOAD (L) AND BALANCE (b)

A. ACID-BASE BALANCE DEFINED

A balance b is of interest in physiology as a means of determining whether the bodily content of a constitutent is increasing or decreasing. The first precise, convincing demonstration of a physiological balance was carried out in a cat by Bidder and Schmidt in 1852; they showed near-equality of fecal plus urinary N with the dietary N intake (388a, page 104). The term "balance" (*Bilanz*) was introduced in conjunction with studies of N metabolism by Carl Voit of Munich. To this investigator and his associates and pupils, including Pettenkofer and Rubner, we owe the development of the metabolic balance technic as a systematic means of examining quantitative aspects of the metabolism of intact experimental animals and humans (367).

For nonmetabolizable materials like Na^+, K^+, or atomic N, which can neither disappear in the body nor arise *de novo:*

$$b = I - O \tag{1}$$

where I is intake into the body by all routes and O is outgo from the body by all channels. Where Equation 1 is applicable, estimation of b over a period of time quantifies any gain or loss of the bodily constituent of interest; b gives the absolute change in the body content of the material incurred over the period of observation, without furnishing any information about the total quantity present at any time.

The concept of balance as an expression of the change in the organism's content of a constituent has been extended to include both water balance and acid-base balance, but in neither instance is Equation 1 applicable. Water is continuously formed within the body as the principal end product of dietary H, so that change in content of body water over any period is given only if an additional term representing metabolic water production is included on the right-hand side of the equation. An additional term is similarly required in the case of acid-base balance, since animal metabolism always results in production of certain nonvolatile acidic compounds from neutral precursors; also acid and base can be *generated* in the body.

In examining the renal mechanisms which underly steady states of balance of bodily constituents, it is convenient and conventional to introduce the concept of *load*, L, a term popularized if not introduced by Gamble (182). This quantity is the appropriate object of attention in any study of renal regulation, because it is the load which defines the task which the kidney must carry out in discharging its homeostatic function. When the kidney regulates the concentration of a substance in extracellular fluid, the magnitude of the load presented to the kidney for excretion in the steady state is not often given simply as intake plus the net metabolic production, as is nearly true of urea. On the contrary, the rate at which the substance must be excreted by the kidney, if balance is to be preserved, is influenced just as essentially by all obligatory extrarenal losses. These include, in the case of water, all normal or abnormal extrarenal loss of liquid plus "insensible loss" of water vapor by several routes. A further complication, in the case of acid-base balance, is the reciprocal relation between acid and base in any solution (I-5); from the acid-base standpoint, though not from the standpoint of osmolality, net loss of base from extracellular fluid is equivalent to net gain of acid and net loss of acid to net gain of base. Furthermore, in order for the load to be characterized generally as a positive quantity, the kidney must be considered (255) to be charged alternatively with the duty of maintaining acid balance b_A, or base balance b_B, in the face of net loads (respectively) of acid (L_A) or base (L_B):

$$b_A = L_A - E_A \tag{2}$$

$$b_B = L_B - E_B \tag{3}$$

Equations 2 and 3 focus attention upon the renal excretion term E by consolidating into L the net acid-base influence of all processes contributing to L_A or L_B: intake of acid and base as such, net metabolic production of acid from neutral precursors, and net extrarenal losses of acid and base (Figure II-5.1). Evaluation of E_A and E_B, which are physiological processes, not chemical entities, has been considered previously (II-4).

B. THE ACID-BASE LOAD L

The word "acid" is commonly used in combination, in organic chemistry and metabolism, in a sense different from the meaning of "acid" in con-

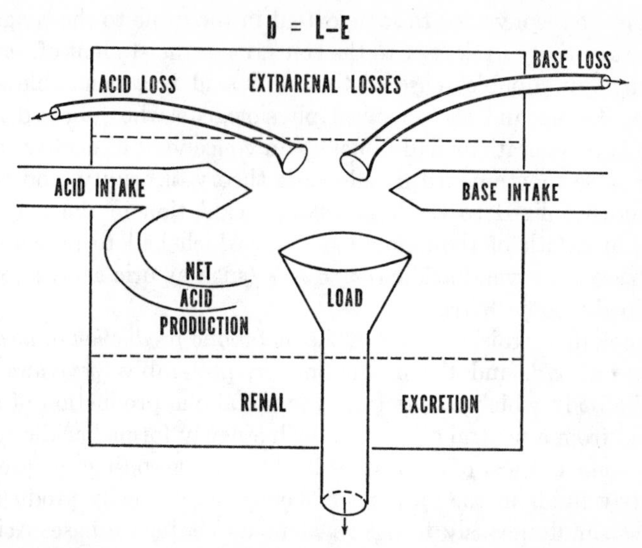

FIG. II-5.1. Constituents of the acid-base balance b. The extracellular fluid is represented schematically as located between the *dotted lines*; net renal excretion of acid or base E is represented by the funnel below the *bottom dotted line*; contributions to the load L are shown above it. 1. The *arrows* comprise (i) nonmetabolizable compounds which upon absorption will instantly transfer protons to, or accept them from, the extracellular fluid, and (ii) partially metabolizable salts which, after their metabolism is completed, will have brought about similar proton transfers. 2. Net production of acidic compounds from neutral organic dietary precursors and from oxidation of organically bound dietary S and P is represented as an endogenous contribution to the previously mentioned *arrow* signifying acid intake. 3. Acid is also generated in extracellular fluid whenever base is lost in gastrointestinal secretions, and base is generated when gastric HCl is lost.

Obviously the acid load L_A is in general less than the total rate of addition of acid to extracellular fluid, and the base load L_B is always considerably less than the total rate of addition of base to extracellular fluid. It is the *net* load of acid or base which must be excreted unceasingly and quantitatively by the kidney as E_A or E_B. "Bien loin . . . que l'animal élevé soit indifférent au monde extérieur, il est au contraire dans une étroite et savante relation avec lui, de telle façon que son équilibre résulte d'une continuelle et délicate compensation établie comme par la plus sensible des balances" (59, page 113).

temporary acid-base chemistry, one which embraces *both the Brønsted acid and its conjugate base.* A particular compound referred to as an acid, such as "uric acid," is said to be metabolically produced even though it never exists to any appreciable extent as a Brønsted acid in the body fluids. From the tolerant standpoint of the organic chemist, production of uric "acid" really means production of the conjugate pair uric acid-urate, irrespective of their position of equilibrium. The facts (i) that "uric acid" exists essentially as urate ion in extracellular fluid and (ii) that it may or may not be reconverted to some extent in the urine to the congener acid are both essentially irrelevant to the concerns of the student of, say, purine biogenesis. This double sense of the word acid is a venerable source of confusion. Liebig and the chemical physiologists who followed him over the next two generations had difficulty in conceiving of acidity in the abstract because they enjoyed no adequate theory of acidity, and they were consequently obliged to think of acidity regulation in the very unsatisfactory framework of the amount of base which had to be conserved to "neutralize" the several acidic *compounds* (such as uric acid) known to be present in the body fluids.

This book distinguishes between the metabolic *production* of a *compound*, including *an acid*, and the production or, preferably, *generation* of *acid* within the body which results from the metabolic production of an acidic compound from a neutral precursor. Such a newly formed acidic compound transfers some or most of its protons to the various conjugate bases present in the body fluids in the very act of being metabolically produced, while it is itself simultaneously transformed to its conjugate base. Acid can of course also be *added* as such to the body (e.g., by ingestion), and it can be *generated* within the extracellular fluid in ways other than by production of an acid: loss of base via the kidney or by any other route causes proton transfer to the buffers of extracellular fluid and so generates acid in extracellular fluid. *Base generation* is applicable in an analogous fashion to proton transfers by the buffers of the extracellular fluid, mostly to acid excreta; organic bases are also produced metabolically in minor amounts.

Constituents of the acid-base load L are essentially dietary mineral acid and alkali, present as such in the diet or yielded by disappearance of the metabolizable portion of mineral salts (Section B-1); organic acids metabolically produced; and acid and base generated in extracellular fluid by extrarenal losses of (respectively) base or acid (Section B-2). Oxidation of the body's own tissues results in net production of an acid load consisting of both mineral and organic acid which must be excreted by the kidney, but we shall confine our attention almost entirely in this chapter to analysis of acid-base balance under steady-state conditions in which caloric balance is being maintained by means of an adequate diet. Figure II-5.1 summarizes contributions to L, including extrarenal losses which, though small in completely healthy persons, may be very substantial in patients who can still maintain normal acid-base balance.

1. Intake and Metabolism

Strong mineral acids or bases are occasionally taken into the body as such, and naturally contribute their full equivalence to the load. As previously explained (I-3B), the net mineral contribution to L depends largely upon the balance of dietary acid-forming and base-forming elements, since metabolism of the anions or cations with which these elements may be associated in the food results in generation of strong acid or base within the body, while on the other hand extensive neutralization ensures that only the net excess of mineral acid or base so generated contributes to *L*. Though most of the organic material of the diet is oxidized to CO_2 and H_2O and so does not contribute to *L*, net metabolic production of some organic acids from neutral dietary precursors contributes to the acidity of *L*. Any organic acid taken into the body and not metabolized would also contribute.

Phosphate salts are the only dietary minerals which are nonvolatile buffers in extracellular fluids, and it depends both upon the immediate effect of their absorption and upon the manner of their subsequent elimination whether absorption into the body contributes (for example) to any overall acidification. Consider the immediate effect of intravenous administration of a neutral phosphate buffer solution (of ionic strength equal to plasma). Evidently, if this infused phosphate is chemically neutral—by definition, an aqueous solution in which $[H_3O^+] = [OH^-]$ is neutral—, it will contribute to the acidity of *L*, under these circumstances, immediately upon entry into the body, because it will immediately transfer protons to extracellular fluid; administered phosphate buffer must be at pH 7.4 if it is to accept no protons, and yield none, as it is infused. Moreover, from the overall standpoint, the effect of any extrarenal elimination of the neutral phosphate upon the load must also be taken into account; for example, if fecal phosphate excretion results in transfer of protons to extracellular fluid, it will contribute to the acidity of *L*. As a matter of fact, some of the infused phosphate would ordinarily be eliminated in the feces in more alkaline form than it was taken in, a process which would generate additional acid in extracellular fluid and so make a further contribution to L_A. The contribution of extrarenal losses to *L* is further considered in Section B-2.

a. Dietary Ash

Investigations of the mineral ash of the diet pioneered in Liebig's laboratory were progressively refined over a period lasting nearly a century. For a long while, despite continuing improvement, the analyses did not suffice for accurate quantitative evaluation of the mineral balances, though they documented abundantly the acid-forming effect of dietary Cl, P, and S and the base-forming effect of dietary Na, K, Ca, and Mg (I-3B). (Other minerals are present in the diet only in trace amounts and do not figure in the acid-base balance.) Cl and the four base-forming elements are readily recovered in full in the ash, but S and P, particularly the former, give rise on heating to volatile oxides which escape analysis when simple kiln combustion is carried out. By 1900, total alkaline ash of vegetables was being routinely examined in agriculture in order to determine the rate of removal of mineral cations from the soil in which the crops were grown; but the *net* ash of foodstuffs could be ascertained only after methods had been devised, based upon separate processing of multiple aliquots of a given food sample for specific elements, which incorporated adequate precautions against loss of S and P during the analysis (176, 22, 299, 80, 564, 563, 566, 562, 90, 209, 89). Sherman and his associates (564, 563) carried out over a period of some years perhaps the first entirely satisfactory determinations of the net dietary ash of foods of vegetable as well as animal origin.

The difference, in equivalents, between the sum of the base-forming and the sum of the acid-forming elements of the diet gives a quantitative value for the net ash of the diet. But though the net ash offers a theoretically exact estimate of the contributions of Cl and the base-forming elements to L, and one which is not often very inexact for P, it leaves out of account the fact that not all dietary S will be fully oxidized to the strong acid H_2SO_4 in the course of metabolism, as it is in ashing. In fact, an appreciable and variable proportion of the dietary S is excreted as urinary neutral S (536, 537, 670, 563, 567, 293). Unless a correction is made by deducting this urinary neutral S from the ash S of the diet, L_A will to this extent be overestimated (or L_B underestimated).

b. Organic Acids

The contribution of intake and metabolism to the acid-base load is by no means determined exclusively by the corrected balance of the minerals of the diet, even in health; excretory end products of normal mammalian metabolism characteristically include certain nonvolatile organic acids derived from neutral precursors (628, 424, 204, 531, 162), and net production of such organic acids naturally contributes to the acidity of the load. In health the net rate of organic acid production is relatively small, but in certain diseases enormous quantities of such acids may be produced, whether by incomplete oxidation of carbohydrate or fat (as in lactic acidosis or diabetic ketoacidosis, respectively) or by ingestion of compounds metabolized to strong organic acid (as when methanol is oxidized to formic acid) (III-3A).

Most if not all the organic acids metabolically produced have pK values low enough (uric acid about 5.6, most of the others 4.0 or less at 37° C) that their metabolic production represents generation of strong acid; i.e., all their protons are contributed to extracellular fluid. It is true that production of uric acid, since this compound can be excreted almost entirely as the conjugate acid in very acid urine, does not always contribute, overall, to acidification of extracellular fluid. However, since when the urine is maximally acid, uric acid contributes its full equivalence to the buffered urinary acid which we measure as titratable acidity (II-4C), we must also assign the compound its full equivalent contribution to L when estimating acid-base balance.

Only the *net* amount of organic acids produced which must be excreted in the urine and stool in the steady state concerns us in computing their contribution to *L*. For example, the steady-state concentration of the lactate ion (lac$^-$) in extracellular fluid is normally considerable; but a stable, sizable concentration of this ion does not necessarily nor usually signify a sizable contribution of lactic acid production to *L*. Virtually no lactate is normally excreted in the urine; only to the (normally negligible) extent that fecal or other extrarenal losses of lactate occur is metabolic synthesis of this organic acid contributing to L_A under normal circumstances. But whenever net metabolic production of lactic acid is providing, at an appreciable rate, lactate which is being lost to the body via urine or any other channel, the lactic acid produced is a constituent of L_A.

Net metabolic production of lactate from ingested glucose or other carbohydrate is a metabolic alternative to complete oxidation of the carbohydrate to CO_2 (Figure II-5.2). How rapidly lactic acid is contributed to the extracellular fluid, as part of the acid component of the load requiring renal excretion, is determined, not by any factors related to acid-base regulation, but essentially by factors determining the intermediary metabolism. Here we need only notice, using lactic acidosis as an illustrative example, that the net extent to which strong organic acid is produced from neutral dietary precursors can be influenced profoundly by disturbances of intermediary metabolism.

As indicated in Figure II-5.2, the plasma lactate concentration in health is principally determined by two factors: (i) the rate of delivery of lactic acid to the circulation by muscle and other tissues, and (ii) a dynamic equilibrium between the lactate in extracellular fluid and lactate and pyruvate within the hepatic cells, pyruvate being an end product of glycolysis and a precursor of the Krebs cycle compounds. Lactate, contributed to the circulation at variable rates by the muscles and other tissues, is taken up by the liver and oxidized to CO_2 via the Krebs cycle (or incorporated into glucose or glycogen); if the blood lactate level rises owing to increased contribution by the muscles, then the rate of entry of lactate into the hepatic Krebs cycle also increases. Because the renal threshold for lactate is high enough in health to prevent renal lactate loss at normal plasma lactate concentrations, there is normally little net contribution of lactate by the body cells as a whole to extracellular fluid, though about 1 to 2 mEq of lactate derived from dietary carbohydrate is generally found as lactate salts in the feces (162). Net hepatic uptake of any circulating lactate ion (no matter what its source), together with its metabolism to CO_2 within the hepatic cell, must be accompanied by uptake of a proton from extracellular fluid by the liver, just as the net production of lactic acid from glucose by the muscles releases the lactate ion plus a proton into extracellular fluid.

Accordingly, lactic acid synthesized by the muscles or administered,

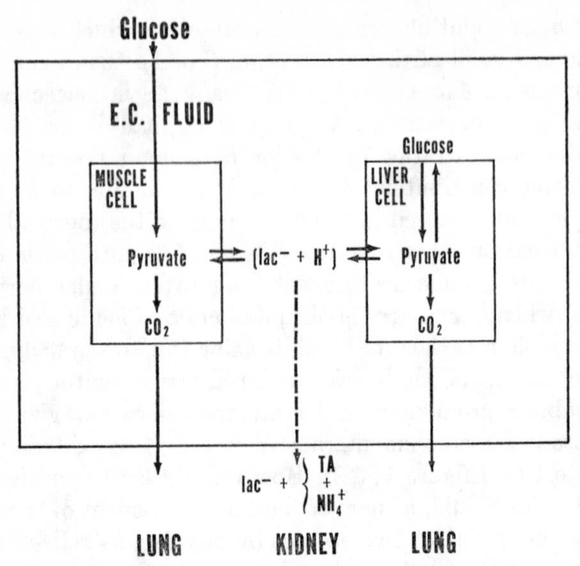

FIG. II-5.2. Metabolic production of lactic acid. Ordinarily the rate of delivery to the circulation of lactic acid, mainly by the muscles, is equalled by hepatic uptake. Plasma lactate (*lac*⁻) fluctuates, rising with increased contribution of lactate from the muscles during exercise and declining during rest, but these fluctuations of plasma lactate occur over a concentration range lying well below the renal threshold. Therefore, except for a very small fecal loss of lactate, there is normally no net production of this compound. The very small loss of lac⁻ in the alkaline feces does entail generation of 1 mEq of acid for each lac⁻ lost as a salt in the stool.

Significant net production of lactic acid begins only when an imbalance develops between contributions of lactic acid to the circulation and the hepatic capacity to remove it; this happens in lactic acidosis. Under these pathological circumstances (*dashed line*), plasma lac⁻ level must rise until the renal threshold for lac⁻ is exceeded; this anion is now lost in the urine at a substantial rate which (in the steady state) is balanced by net production of lactic acid by the organism. Every molecule of lactic acid being released under these conditions into the body fluids instantly dissociates into lac⁻ and a proton; this proton is accepted by the buffers of the body fluids. Evidently under these conditions there is continuous *net generation of acid* in extracellular fluid owing to *metabolic production of a strongly acidic compound from neutral precursors* (carbohydrate). The kidney will automatically waste lac⁻, since it can now no longer reabsorb the anion as rapidly as it is filtered; but if acid-base balance is to be maintained, the lac⁻ must be accompanied in the urine by additional buffered acid. This objective is attained as long as the urinary lac⁻ is electrically matched (I-3B) by additional urinary NH_4^+ plus titratable acidity (*TA*). If acid excretion cannot be increased, then it will not be possible for the kidney to conserve the filtered Na^+ normally, and some of it will be lost in the urine as a matching ion with lac⁻.

In health, increased delivery of lac⁻ to the circulation displaces the hepatic equilibrium, probably by mass action, so that more lac⁻, accompanied by a proton, reenters the cells, enters the Krebs cycle, and is metabolized. This is true whether it is exercising muscle which is delivering additional lac⁻ plus protons to the circulation, or the physician is administering lac⁻ plus Na^+ as a therapeutic measure. In the latter case, the effect is normally alkalinizing, since Na^+ is being added to extracellular fluid and H^+ is being removed from it.

provided that it will enter the Krebs cycle within the hepatic cells, will have no overall effect on the acid-base load L; but administration of sodium lactate (with the same proviso) will contribute (overall) an equivalent quantity of base to extracellular fluid, because (overall) the administered Na^+ will have replaced the proton captured by the lac^- in its conversion within the liver to CO_2. For this reason sodium lactate was formerly extensively used (before stable, sterile sodium bicarbonate solutions were available) as an alkalinizing solution in the treatment of nonrespiratory acidosis; this therapeutic use of lactate was of course predicated upon the expectation that administered lactic acid would indeed be taken up by the normal liver and converted to CO_2. Therapeutic use of sodium lactate as an alkalinizing solution would, of course, be disastrously ineffective in lactic acidosis, a condition in which lactic acid is already being added to extracellular fluid by the muscles and other tissues at a rate well in excess of the liver's capacity to take it up (Part III). Because, as a result, plasma $[lac^-]$ has risen in this condition to a level that exceeds the renal threshold, lactic acid is being both added to extracellular fluid by the skeletal muscles and other tissues and excreted at a rapid rate in the urine (*dashed line*, Figure II-5.2). From the standpoint of acid-base balance, net addition of lactic acid synthesized in the tissues to extracellar fluid in lactic acidosis is indistinguishable from addition of any other acid to extracellular fluid: *net* metabolic production of lactic acid makes its full contribution to the acidity of L. Sodium lactate given therapeutically to a patient with lactic acidosis would simply be lost from the body as additional lactate matched in the urine against the Na^+ which was administered with it; there would be no alkalinizing effect in extracellular fluid.

The principles governing the relation between acid load and metabolic production or consumption of lactic acid are directly applicable to net production of other organic acids, including not only acid compounds abnormally overproduced like the ketoacids in uncontrolled, severe diabetes mellitus,[1] but also a variety of organic acids normally produced and ordinarily excreted in both urine and stool. Neutral dietary precursors are converted metabolically into a variety of compounds other than CO_2, H_2O, NH_3, and urea; a number of these other compounds are strongly acidic. If, in the net, these compounds are being added to extracellular fluid, they add to the acidity of the load because they donate all their protons to the buffers of extracellular fluid; these protons must be transferred (other things being equal) to the urinary buffers if acid-base balance is to be preserved. But insofar as the acidic intermediates of fat or carbohydrate metabolism are, in the net, being oxidized to CO_2 and H_2O rather than being added to extracellular fluid, they make no contribution to L_A even though their (stable) plasma level may be quite appreciable.

Sherman and Gettler (563) were the first to carry out a metabolic acid-base balance study in the course of which satisfactory ashing technics were

TABLE II-5.1. *Increased acidity of net dietary ash, when rice is substituted for potato, compared with increased renal total acid excretion (= titratable acidity plus ammonia)**

	Potato	Rice	Δ
		mEq acid/day	
A. Mineral contribution to L_A	21.9	52.0	+30.1
B. Mineral contribution to L_A, corrected	9.4	42.1	+32.7
Urinary acid			
Ammonia	23.8	34.5	
TA	21.4	34.1	
C. Total	45.2	68.6	+23.4
C − A	23.3	16.6	
C − B	35.8	26.5	

* The data are abstracted from Table 4 of Sherman and Gettler (563). The volunteer, a healthy male, was studied under nearly uniform conditions; the only material change, during the 10-day study, was alteration of part of the diet on Days 5 through 8. During this period, an approximately isocaloric quantity of rice, an acid-forming food, was substituted for potato, a base-forming one.

applied to the collection of metabolic balance data in a human subject. Table II-5.1 summarizes data from one of their studies which compared the change in renal total acid excretion, ΔTE_A (II-4A) with the change in load, ΔL_A, calculated from dietary ash, imposed by substituting rice for potato in the otherwise constant diet of a healthy volunteer. Note that correction of the contribution of sulfur ash was made for urinary neutral sulfur excretion; Lemann and Relman (336) are assailing a straw man in stating that "there has never been any proof of the . . . quantitative relationship between sulfur metabolism and the endogenous generation of acid" which they erroneously hold to be "implied by classical calculations." It was on the contrary known by 1860 that appreciable quantities of S are excreted in the urine in neutral form, and the proportions of the S of fed cystine and of other organic sulfur compounds recoverable as urinary neutral sulfur and as sulfate were recorded in many animal species including man by Salkowski, Goldmann, Wohlgemuth, Sherman and Gettler, and others (536, 537, 670, 563, 567, 293) from 1873 onward. Shohl and Sato (567) also deducted nonsulfate urinary sulfur in reckoning the contribution of dietary S to the acidity of the ash.

The correction in Table II-5.1 of the analytical values for dietary ash on each regimen represents deduction from the net acid-producing value of the dietary minerals of that portion of dietary S which appeared as neutral S in the urine. Correction did not much affect the calculated increase of the net acid-producing value of the dietary minerals when rice (an acid-forming food) was substituted, but it did substantially decrease the calculated acid-producing value of both diets. Renal total acid excretion consistently exceeded the metabolic acid production reckoned from the dietary minerals, especially when the latter term was reduced by deducting dietary S excreted as neutral S.

The discrepancies reflected in all values for "C − A" and "C − B" at the bottom of the table are to be expected, for three reasons: (i) allowance was not made for the contribution of L_A of organic acids metabolically produced from neutral dietary precursors (628, 424, 204, 531, 162); (ii) base loss in the stool (90, 209, 89), which is always alkaline (272, 7, 90, 209, 89), was not taken into account (see also Section A-2); and (iii) renal net acid excretion (E_A) was overestimated insofar as urinary HCO_3^- was not deducted from renal acid excretion to obtain E_A. Only 10 years later would Gamble (II-4B) call attention to the necessity of deducting urinary HCO_3^- excretion from total acid excretion to obtain net renal acid excretion, and only 8 years later did Van Slyke and Palmer publish their method for determining urinary organic acids (628).

"C − A" and "C − B" therefore do not indicate a negative acid balance (base retention). The study was designed not as a quantitative measurement of the absolute acid-base balance b—hardly a practical enterprise in 1912—but as an evaluation of the change in acid-base balance, Δb, occurring as a result of a change in the ratio of acid-forming to base-forming minerals in the diet. In theory (after a brief period of adjustment) Δb should be 0, since the true acid-base balance ought to have been 0 on both diets. In other words, the effect of altering the diet upon either C − A or C − B ought in theory to have been 0. In fact, (C − B) fell by 9.3 mEq/day when rice was substituted for potato, and C − A fell by 6.7 mEq/day, a change which might be taken to indicate that a positive acid balance (acid retention) had developed on the rice diet. Sherman and Gettler thought that this apparent increased acid retention of 9.3 mEq/day might be accounted for in terms of lesser excretion (not taken into account) of fecal base when the subject was on the rice diet, thought they recognized that their evidence on this point was not very persuasive.

c. *Extrarenal Loss of Acid and Base*

Leaving out of account menstrual flow, sweat, and seminal fluid and other trace losses of miscellaneous fluids, the only extrarenal net loss of nonvolatile acid or base that requires consideration in completely normal circumstances is base excreted in the stool. Nevertheless, physicians deal very frequently with continuing or repetitive loss of fluids from the gastrointestinal tract; whether or not these losses are extensive enough to result in nonrespiratory acidosis or alkalosis, they are often enough major contributors to the net acid-base load. Loss of HCl through vomiting or withdrawal of gastric contents is of course tantamount to regeneration of base in extracellular fluid; it is the most common cause of nonrespiratory alkalosis. The secretions of the lower gastrointestinal tract are predominantly alkaline (pancreatic juice, bile, and succus entericus), and loss of these bicarbonate-rich fluids, whether as ileostomy fluid or through diarrhea, suction, or fistulas, is tantamount to generation of acid in extracellular fluid. Naturally when all these gastrointestinal secretions are reabsorbed, there is no net generation of acid or base in extracellular fluid; it is only their loss from the body which must be counted toward the acid-base load L.

C. QUANTITATIVE EVALUATION OF ACID-BASE BALANCE

The theory, utility, and degree of precision of evaluation of the absolute acid-base balance b with the aid of the metabolic balance technic have recently become matters of interest in consequence of publication of studies of this kind from several laboratories. Contemporary investigators are able to approach theoretical soundness more closely in evaluating b_A than could earlier investigators (563, 566) because the renal excretion term can now be evaluated accurately and because the contribution of organic acid production to L_A can be estimated, even if in a manner not wholly satisfactory technically or theoretically. On the other hand, the accuracy with which dietary and stool ash can be determined, and the unavoidable vagaries inherent in the balance technic proper, probably remain essentially unchanged since the study of Sherman and Gettler (Table II-5.1).

The recent literature on evaluation of acid-base balance by the metabolic balance technic has concerned itself almost entirely with elimination of an acid load L_A and has been characterized by controversy; the running battle over the past decade between the two principal schools of investigation active in this area is reviewed in Appendix G. But there is now increasing evidence (Appendix G) of the attainment or imminence of a broad consensus that the dietary and metabolic sources of nonvolatile acid which the kidney must eliminate if balance is to be preserved are: (i) dietary Cl $-$(Na + K + Ca + Mg); (ii) that portion of dietary S which in the net is oxidized metabolically to acid; (iii) that portion of the dietary P which is fully oxidized in the diet or metabolically (this generally includes the bulk of the dietary P which gives rise to phosphoric acid either by hydrolysis of diester linkages or by combustion of organic cations with which dietary phosphate is associated); and (iv) net metabolic formation of organic acid from neutral precursors less any net metabolic formation of nonvolatile base. Naturally the acid load L_A must also take into account any generation of acid by base loss, while generation of base by acid loss, or addition of preformed nonmetabolizable base to the body fluids, must be deducted from L_A. It is a clinical and physiological commonplace that a readily calculable extra intake of acid or base, as by administration of

$NaHCO_3$ or NH_4Cl (in effect a substantial and readily absorbable additional contribution to the ash in the diet) will be reflected faithfully in the deviation of rate of net acid excretion E_A from prior control values (cf., for example, III-3B). There is no doubt that acid-base balance is in fact preserved by the kidneys of healthy subjects, in accordance with the doctrine of homeostasis, nor that in principle metabolic balance data obtained in healthy adults should serve, if adequate analytical methods are deployed, to confirm the fact experimentally. However, complications (Section A) connected with the quantitative evaluation of acid-base balance data, as compared with other metabolic balances, should always be borne in mind in interpreting published studies.

The most troublesome of these, from the theoretical standpoint, centers about the contribution of endogenously generated acid, especially production of the organic acids, to L_A. The practice of Relman and Lennon and their associates has been to evaluate the contribution to the acidity of the load of the organic acids synthesized in the body by recovering these compounds in the excreta.[2] As Camien, Simmons, and Gonick (89) have pointed out, however, the necessity of determining the production rate of these acids by the fact of their excretion introduces a unique element of circularity into the calculation of acid-base balance. Recalling that a positive value of b is supposed to tell us that the constituent of interest is accumulating in the body, we see that the contribution of the endogenously produced organic acids to L_A can only be reckoned correctly in this way if in fact acid-base balance is 0. The error of estimate will be maximal in an anuric and stool-less person who is becoming acidotic; that portion of the acid accumulating in extracellular fluid which is ascribable to metabolic production of organic acid will be invisible since it is not being excreted. This difficulty attaches also to estimations of b_A in any subject who is in fact in positive acid balance, i.e., who is retaining some of the acid load; that portion of the retained acid accountable to organic acid accumulation will not appear in the balance calculation, and b_A will be to that extent underestimated. Evidently also, accurate evaluation of b, even in subjects who are in fact in acid-base balance, is dependent upon complete recovery in the excreta of organic acids produced in the body and added to extracellular fluid. This objective has certainly not been accomplished in balance studies using methods available until now, and the magnitude of the error involved in the assumption of complete recovery is not well established.

Quantitative balance data (other than acid-base balances) are often used in conjunction with sampling of extracellular fluid or with tissue analysis to investigate shifts of constituents between compartments of body water. In view of the theoretical difficulty that evaluation of the absolute acid-base balance b is inherently imprecise, comparisons of this kind seem likely to prove somewhat treacherous in acid-base physiology. Thus Lennon and his colleagues, reckoning from balance data that acid is constantly being retained by subjects in stable, chronic nonrespiratory acidosis, have concluded from the failure of extracellular $[HCO_3^-]$ to decline that base is being made available by dissolution of skeletal calcium salts to neutralize retained acid. But neither concomitant demonstration of a negative calcium balance nor data on the precision of estimate of b provides independent evaluation of the reliability of the estimated absolute value of b (Appendix G).

Albright and his associates (5–7) measured what they called "S.B.S." ("the sum of the base-sparing mechanisms"), defined as E_A plus 49% of the loss of calcium from the body. Like Lennon and his associates, these authors considered that when alkaline calcium salts of the skeleton are yielded to the extracellular fluid of patients with chronic metabolic acidosis, as indicated by a negative calcium balance, the extracellular fluid is drawing on a reserve of base which is available in the skeleton and in this way preventing further the further acidification of extracellular fluid which must otherwise occur if E_A is exceeded by L_A. Evidently, "base-sparing" must be understood in this context to refer to the extracellular base reserves, not those of the whole body.

D. DIETARY ASH AND THERAPEUTICS

It was mooted as late as 1952 whether the degree of acidity or basicity of L might be a matter of practical concern in the nutrition of normal persons. Realization of the astonishing effectiveness of the normal kidney in maintaining near-constancy of the acidic intensity of extracellular fluid over the entire spectrum of L, represented by practically all self-selected diets anywhere in the world, has removed this question from further consideration in public health dietetics. A tendency is also discernible in current books on nutrition and in diet manuals (406, 427) to reduce to a minimum practical counseling about the prescription of dietary ash. This tendency has apparently been reinforced by doubts expressed, quite without warrant, as to the importance of the influence of the dietary ash upon net renal acid excretion (279, 502). This trend is to be deplored, for there is a continuing need for convenient, varied, and palatable prescriptions of alkaline-ash and ash-acid diets for the convenience of physicians who wish to control the urinary pH of their patients in part or entirely by dietary means (Part III). Gonick, Goldberg, and Mulcare (209) have ascertained that the ordinary American diet is nearly neutral in ash; an acidic load and net renal excretion of acid result when such a diet is taken because of the metabolic formation of organic acids from neutral precursors and because of substantial loss of the base-forming dietary cations Ca and Mg in the stools.

Acid-ash diets may be wanted to assist with therapeutic acidification of the urine in conjunction with the use of antimicrobial agents, or in the management of all types of calcium urolithiasis except those secondary to renal tubular acidosis (III-4). Alkaline-ash diets are sometimes advantageous for control of acidosis in high-grade glomerular insufficiency (III-3B3); they are also useful in the management, and in some cases highly effective in the prevention, of uric acid urolithiasis (III-4). They may also be used as an adjunct to alkali salts in the treatment of renal tubular acidosis (III-3B2). Alkaline-ash diets decrease the quantity of medicinal alkali salts which must be taken as dietary supplements, and sometimes obviate the necessity for them, both in renal acidosis and when alkalinization of the urine is sought for other reasons; they may in fact be indispensable to optimal management of such patients when a high sodium intake is for any reason undesirable, or for patients who are rendered uncomfortable by a large intake of sodium bicarbonate.

NOTES

1. In the case of ketoacids, it is the liver which yields the compounds to the extracellular fluid and the muscles which take them up. The blood level of ketoacids rises above the renal threshold because muscle uptake cannot keep pace with hepatic release; the roles of liver and muscle as supplier and consumer are here reversed as compared with their roles in lactic acidosis. From the point of view of renal regulation of acid-base balance, the question as to which organs and tissues are suppliers of organic acids within the body, and which are consumers, is an irrelevant detail; *net* contribution by the tissues, generally to extracellular fluid, of acids which must be eliminated in urine or extrarenally defines the contribution of these acids to L_A.
2. These compounds are excreted in the stool entirely as salts of the alkaline metals and earths. Except for uric acid, they are excreted also to the extent of 90% or more in the urine as (dissociated) salts (including ammonium salts) since their pK lies below the pH of the most acidic urines. Uric acid can be excreted mostly as such (in low-pH urine) or almost wholly as the urate ion (at urine pH exceeding 7.1) (cf. III-4).

six

THE FUNDAMENTAL SYSTEM OF RENAL ACID-BASE REGULATION

Renal maintenance of acid-base balance (II-5) can be analyzed into two necessary functions: (i) alteration of urinary pH in response to fluctuations of the acid-base load, and (ii) transport into the urine of buffers of suitable types and in suitable quantities, again at variable rates in response to changing acid-base load. The first is the quintessential function, in that renal tubular secretory processes capable of acidifying or alkalinizing the glomerular filtrate could in theory operate without the aid of urinary buffers if the amount of energy required for transporting protons were of no importance. But, in fact, it is clear that the evolution of the renal system of acid-base regulation was decisively shaped by the economy achievable by the kidney in its acid-base regulatory activity through its employment of CO_2 and NH_3 as uniquely effective urinary buffers (254). The large and appropriate variation in the rate of transport of these buffer gases into the urine as an automatic accompaniment of changing urinary pH is an amazingly effective and economical acid-base regulatory device.

In the next three chapters the intrarenal mechanisms concerned with the acidification of glomerular filtrate and the transport of buffer compounds into the urine are examined in detail. As a preliminary, it will be desirable to consider in this chapter the changing acid-base composition of urine, in relation to the acid-base load, largely from the descriptive standpoint, i.e., to examine the normal range of variation of urine pH (Section A) and the kind and quantity of urinary buffers found as the pH fluctuates (Sections B and C). Changes occurring spontaneously in the pH, P_{CO_2}, and P_{NH_3} of urine after it emerges from the papilla into the gross passages of the urinary tract can alter markedly the acid-base composition of the urine; these postpapillary changes in urine composition tend to conceal the fact that the effective urinary pressures of the buffer gases at the papilla are stabilized through equilibration of urinary P_{CO_2} and P_{NH_3} with the papillary tissue pressures. In this chapter the *fundamental system of renal acid-base regulation* will be described in terms of the acid-base composition of papillary urine (Section C). The postpapillary changes in the acid-base composition, especially, of higher-pH urines reflect one of several renal acid-base regulatory devices which are superimposed upon the fundamental system and increase its already striking efficiency in preserving acid-base homeostasis; these superimposed mechanisms are considered later (II-10).

A. URINARY pH

The earliest electrometric measurements of the acidic intensity of the urine, carried out by Höber (265, pages 248 to 251; 266) and von Rhorer (506), revealed the considerably greater normal variability of $[H_3O^+]$ of urine as compared with blood. Both authors also made experimental comparisons of urinary acidic intensity and the quantity of acid present as judged by titration methods, as did von Skramlik (576), who re-investigated the already well known difference in rates of urinary acid excretion among persons consuming different diets (meat, vegetable, milk). He followed the electrometrically determined urinary acidity of such persons as well as their total urinary acidity as determined by titration; the meat diet was found to cause the urine to become most acid by both tests. Henderson, in a communication (238) describing the application of indicators to determinate $[H_3O^+]$ in urine, was the first to state explicitly the crucial significance for acid-base homeostasis of the acid-base composition of the urine.[1] Henderson (239), Ringer (509), and Henderson and Palmer (242) presented the first estimations of the range of normal variation of the acidic intensity of urine; Henderson and Palmer, who were the first to report their results in terms of Sørensen's then new pH unitage, found that the pH of 100 normal samples, collected without taking prior steps to influence the acid-base load, ranged between 4.82 and 7.46.

Various limits have subsequently been assigned to the pH of normal human urine. Smith (577, page 377) listed limiting values of 4.8 to 8.2; Pitts (467, page 184) identified the normal lower limit as 4.4; and Pitts, Ayer, and Schiess (472) found values ranging between 4.4 and 7.8 in their study. Values given in the earlier literature are not always directly applicable to physiological analysis, because spontaneously CO_2 loss from higher-pH urines was often not prevented (II-4C), and the urinary pH was customarily estimated at room temperature. Owing to the effect of temperature upon the pK values of the urinary buffers, estimates of urinary pH at 25° C may exceed values at 37° C by 0.2 to 0.3. Experimental conditions which will increase markedly the distal transtubular electrochemical gradient (hormonal stimulation of Na^+ reabsorption during Na^+ depletion) (549) can lower urine pH below the lowest values normally found; the lowest recorded values produced in this way were three samples of pH 4.4, 4.3, and 4.0 (549). Reports of urine pH exceeding 8.0 continue to appear from time to time in the clinical literature; these are almost certainly artifactual, ascribable to evolution of CO_2 (II-4C).

Both the highest and lowest values for urinary pH are found exclusively in low-flow urines; during brisk water diuresis the normal range of urinary pH never exceeds the limits of 5.0 and 7.3 at 37° C (259). The expansion of the urinary pH range effected by antidiuresis is the result of increased reabsorption of the buffer gases CO_2 and NH_3 from the distal tubules and especially the collecting ducts secondary to increased water reabsorption (II-10C). In bicarbonate-rich urines, because of the predominance of urinary HCO_3^-, the principal acid-base effect of reabsorption of the two buffer gases is conversion of HCO_3^- to CO_2; the nonvolatile buffers yield protons, and urinary pH rises. In acid urines containing little HCO_3^- but relatively large amounts of ammonia, the principal effect is acidification, as NH_4^+ transfers protons to the nonvolatile buffers to replace some of the NH_3 diffusing off. Acid urines consequently become more acidic as urine flow falls, while the pH of high-pH urines rises.

On the basis of his own experience with 107 urine samples of flow rates between 0.5 and 3.0 ml/minute, systematically collected from healthy trained volunteers, with stringent precautions against gas loss and representative of a wide range of normal acid loads (cf. Figures II-9.1 to II-9.4), the author believes the normal limits of variation (II-1) of urinary pH in healthy humans to be 4.5 and 8.0 at 37° C. These figures represent equilibrium pH values; urine pH is often lower at the papilla than in the bladder, because as the H_2CO_3 present in papillary urine at concentrations above equilibrium is spontaneously dehydrated in the urinary tract, urinary pH rises as HCO_3^- receives protons from the other urinary buffers to replace some of the decomposing H_2CO_3 (I-6C; II-10A).

B. TRANSPORT OF THE BUFFER GASES AND THEIR IONS INTO THE URINE

Table II-6.1 brings out total dependence of the organism, for acid-base regulation, upon the urinary buffers, which increase the capacity of the urine to accept protons from the extracellular fluid, or to yield them to the extracellular fluid. Protons which could be transferred to a liter of unbuffered urine, and to unbuffered urine secreted per 24 hours, are reckoned directly from the maximal H_3O^+ concentration possible in high-flow urine (top row) and low-flow urine (bottom row); only a fraction of a milliequivalent of acid could be eliminated per 24 hours without the aid of the urinary buffers, a negligible proportion of average acid loads of 40 to 80 mEq

TABLE II-6.1. *Maximal rates of proton transfer to hypothetical urine containing no buffer**

Urine pH	A	B	Proton Transfer	
	Urine [H_3O^+]	Plasma [H_3O^+]	Per liter (A − B)	Per 24 hours
	mEq/liter		*mEq*	
5.1	0.00794	0.00004	0.00790	0.079
4.5	0.03162	0.00004	0.03158	0.016

* In unbuffered urine all protons would be present as H_3O^+, and the difference between urinary [H_3O^+] (Column A) and plasma [H_3O^+] (Column B) in milliequivalents per liter would express the proton transfer from blood plasma to 1 liter of urine. During water diuresis minimal urine pH is about 5.1; 1 liter of unbuffered urine at pH 5.1 would have accepted 0.0079 mEq, and so, allowing 10 liters of urine daily in extreme water diuresis, 0.079 mEq of acid could be eliminated/24 hours. During antidiuresis, the maximal excretion rate of acid would be less, in spite of the higher [H_3O^+] attainable at minimal low-flow urine pH of 4.5, because of the low 24-hour urine volume of around 500 ml. Since highest-pH urine is only 0.6 pH unit higher than blood plasma, still less base could be excreted in unbuffered urine.

daily. Still less base could be excreted in unbuffered urine at the maximum possible urinary pH of 8.0. Conveyance of the proper kind of buffer into the urine at rates appropriate to the acid-base load is therefore essential.

1. Nonionic Diffusion

The NH_4^+ (as well as the NH_3) present in plasma undoubtedly passes through the glomerulus with the filtrate, but for reasons considered in detail later (II-9C) the ammonia filtered at the glomerulus is of no significance for urinary ammonia excretion. Rather, ammonia gains access to the urine essentially as NH_3 by diffusion down an ammonia pressure gradient from renal tissue, movement of NH_4^+ being restrained by cellular plasma membranes and other membranes. This process is termed "nonionic diffusion." Both CO_2 and HCO_3^- pass freely through the glomerulus in the ultrafiltrate of plasma water which constitutes the glomerular filtrate; both members of this buffer pair are filtered at much greater rates than they are excreted, and HCO_3^- filtration is a process essential for urinary HCO_3^- excretion. However, filtrate HCO_3^- is largely converted to CO_2 in the nephrons as a result of acidification and dehydration of glomerular filtrate; most of this HCO_3^-, as well as most of the filtered CO_2, diffuses back into the blood that perfuses the kidney by simple transtubular diffusion down a CO_2 pressure gradient (II-7C; II-8C). As glomerular filtrate is progressively acidified within the nephrons, converting luminal NH_3 to NH_4^+, P_{NH_3} of luminal fluid is sustained by diffusion of NH_3 into the nephrons, both processes proceeding more rapidly when acid urine is being secreted (II-9C). CO_2 diffuses out of the nephron, not into it like NH_3, but the P_{CO_2} of urine at the papilla, like its P_{NH_3}, is held relatively steady, as urinary pH and $[HCO_3^-]$ vary, by equilibration of the urinary gas pressure at the papilla with the tissue gas pressure. How these papillary tissue gas pressures are stabilized is considered later: P_{CO_2} in Chapters II-7 and II-8, P_{NH_3} in Chapter II-9.

The glomerular membrane differs from most other membranes of the animal body in being the site of continuous bulk transudation of water propelled by hydrostatic pressure. This membrane's principal function is to restrain the blood's formed elements and macromolecules; its selectivity is accordingly far below that of most other membranes of the body, and all plasma crystalloids and smaller ions traverse it freely and rapidly with the water in which they are dissolved. The membranes of cells have to be freely permeable to the flow of respiratory gases moving independently of hydraulic flow, but are relatively efficient, compared with the glomerular filter, in obstructing passage of ions and of most uncharged molecules except the smallest ones—H_2O, O_2, CO_2, and NH_3. Urea diffuses through the membranes of the body generally, but not extremely rapidly like H_2O and the blood gases.

That highly diffusible gases rapidly traverse most animal membranes under the influence of small transmembrane gas pressure gradients, and that ionic congeners of these gases are restrained, are facts which have been familiar for generations. In the case of CO_2, the concept is very nearly implicit in the realization that this end product of carbon metabolism is transported by diffusion from the cells into the blood, and from the blood into the respiratory passages. The idea of nonionic diffusion in this sense goes back over a century to the pioneering studies of the young Lothar Meyer and the systematic investigation of CO_2 conveyance in the blood by Pflüger and his pupils in Strassburg (II-3); these led to the realization that the CO_2 continually being produced in the cells diffuses into the capillary blood and is carried in the venous blood to the lungs, where it diffuses down a very small pressure gradient into the expired air (II-3). A half-century ago Merkel Jacobs, observing the color changes of indicator dyes within cells as the external P_{CO_2} was altered, obtained evidence that CO_2 acidifies cell fluid by diffusing preferentially down pressure gradients into cells, even out of alkaline media, where the ratio $HCO_3^-:CO_2$ is high, indicating that HCO_3^- is restrained by the cell membrane (283); evidence of free ingress of NH_3 into cells in preference to NH_4^+ was demonstrated by the same technic (284). Jacobs continued to study the exchange of ionic and nonionic materials between several types of cells and their surrounding media over many years (285–287).

The term "nonionic diffusion" is not generally taken to include transport of nondissociable compounds like urea by diffusion. Nonionic diffusion was first recognized as the basis of an important class of renal excretory mechanisms when it was realized that restraint by the wall of the nephron upon movement of the ionized congener of a buffer pair present in peritubular blood and luminal fluid, while the uncharged molecule is much freer to diffuse into or out of the nephron, could account for striking dependence of the urinary concentrations and excretion rates of compounds of this kind upon the urinary pH (Figures I-6.2 and I-6.3). In 1939, Kempton (303) discovered that excretion of neutral red by the kidney was greatly accelerated by acidification of the urine, and inferred from his data that transtubular diffusion plays a role in the excretion of this dye. The following year Travell (614) observed that urinary acidity affected the rate of renal excretion of nicotine, an observation soon confirmed (220); and over the next few years dependence of the rate of urinary excretion on pH was demonstrated for a variety of antimalarial agents (220a, 152, 14, 612, 289). It was the British Army Malaria Research Unit at Oxford (14) which first connected the dependency of the excretion of such organic compounds on urinary pH with the already familiar effect of the acid-base balance on urinary ammonia excretion (II-9). Trager and Hutchinson (612) were inclined to ascribe these effects to a specific influence of urine acidification on the transport of the ammonium ion into the nephrons, whereas Jailer, Rosenfeld, and Shannon (289) proposed that the concentration of total ammonia in the distal tubules was determined by transtubular diffusion of the uncharged base, in the same fashion as the concentration of total drug in the case of weak organic bases such as the antimalarial drugs. "The data are in keeping," they wrote, "with the assumption that the free base is more freely diffusible across the cells of the distal segment. . . . Such being the case there may be expected to be net gain or loss of drug depending upon the relation between the hydrogen ion concentration of the tubular urine and that of the peritubular fluid."

The concept that the relation between urinary ammonia excretion and pH depends on nonionic diffusion of ammonia in the kidney was carried further by Pitts and his colleagues (461, 27, 600, 179, 595, 469), Orloff and Berliner (434), and Milne, Scribner, and Crawford (403). Equations have been developed (403, 125) which relate urinary concentrations and excretion rates to the blood concentrations of substances entering tubular urine by nonionic diffusion, but not diffusing to equilibrium. We shall not consider these here, as there is good evidence that the NH_3 is in diffusion equilibrium, under all ordinary conditions, between urine and renal tissue at the papilla (Section B-2). Whereas antimalarial drugs and other weak organic bases undoubtedly are enabled to diffuse through the lipid constituents of the walls of the nephrons by their solubility in fats (125), it is probable that the much more highly diffusible NH_3 owes its rapid equilibration between luminal fluid and tissue water not to high solubility in the lipids of the membranes and cells, but rather to passage through aqueous pores of the membranes. NH_3, unlike CO_2, is not very soluble in lipid solvents (430).

Evidence that transport of any compound is entirely passive must be cumulative, and can never be conclusive beyond peradventure. Active transport of CO_2 has never been proposed, while the evidence that non-ionic diffusion of NH_3 can always account for observed ammonia transport is so extensive and consistent that active transport of NH_4^+ (to say nothing of NH_3) anywhere in the body must be considered unlikely (II-9D1). In the author's opinion, attempts to evaluate the P_{NH_3} of tissues by the Henderson-Hasselbalch equation from measurements of tissue total ammonia and pH are at present ill-advised; it is certain that errors, of unknown magnitude but probably very sizable, will be created owing to the high ionic strength of most cell water and the uncertainty as to NH_3 solubility in the aqueous and lipid phases of cell and extracellular water in tissue, even if overestimation of [ammonia] due to rapid artifactual ammonia release can be entirely eliminated. Because of the extremely high rate of passive transport of CO_2 and NH_3 gas across cell membranes, it is difficult to believe that the effectiveness of a pump for HCO_3^- or NH_4^+ would not be sabotaged by leakage of the congener gases in the reverse direction.

In every situation so far investigated, ammonia has been found to move freely through renal tissue down any identifiable NH_3 pressure gradient, and immediately to move at a more rapid rate when the gradient is increased. Such situations include the following.

1. Urinary ammonia concentration increases nearly exponentially as urine pH is progressively reduced (Figures II-9.1 and II-9.3).

2. The rate of urinary ammonia excretion is always accelerated when the urinary flow rate is increased (Appendix H), a new steady state appropriate to the higher rate of urine flow being established within a few minutes at most. (Both reduction of urinary pH and increase of urine flow rate necessarily increase the [undetectably small] NH_3 pressure gradient from renal tissue to urine.)

3. Sampling of the luminal fluid of the nephrons by micropuncture technics, beginning with observations in the laboratory of A. N. Richards at the University of Pennsylvania, have consistently shown that ammonia is always secreted into the luminal fluid in strict topographic parallelism with progressive acidification. Significant acidification of glomerular filtrate and secretion of ammonia were found by Montgomery and Pierce (409) and by Walker (640) to commence in the distal tubules. In the mammal, significant acidification begins already in the proximal tubules (II-7E), but, in both rat (203a, 430) and dog (98), ammonia secretion appears to proceed hand in hand with acidification as glomerular filtrate passes down the proximal and then down the distal convolutions.

4. Ammonia injected during acidosis into the canine renal artery exhibits both precession and postcession (27) in comparison with a simultaneously injected "glomerular substance" like creatinine (one excreted essentially by glomerular filtration without tubular reabsorption). Earlier appearance of the injected ammonia in the urine (precession) indicates short circuiting of ammonia in the nephron, i.e., diffusion of NH_3 from the descending limb of Henle's loop to the collecting duct or from proximal to distal tubule. Transient rise in renal tissue P_{NH_3}, due to accumulation of ammonia in the tissues above control values persisting until after the creatinine has passed through the nephrons, results in postcession: ammonia is excreted in temporary excess of control values after the administered creatinine has all been eliminated.

5. When acidifying or alkalinizing substances are injected into the renal artery, they affect renal venous blood pH more rapidly than urine pH; under these circumstances some ammonia is initially diverted from urine to renal venous effluent, in accordance with the thesis of passive ammonia transport (600).

6. Rapid increase in the urinary flow rate causes *exaltation*, and rapid decrease *abatement* (606, 494, 258), of the urinary ammonia excretion (Figure II-6.1). The same phenomena occur in the case of urinary urea excretion, where they were first observed and named by Shannon (560) and Schmidt-Nielsen (543). The mechanism is the same in both cases (315, 258); after urine flow increases abruptly, ammonia excretion increases transiently above values under steady-state conditions at that flow rate, because the outflow velocity constant (II-9) has increased without instantaneous change in renal tissue P_{NH_3}. However, since ammonia is now being removed more efficiently in the urine, the P_{NH_3} of renal tissue declines to a new steady-state value characteristic of the higher flow. Conversely, an abrupt decline in urine flow (producible most rapidly by vasopressin) means that the product of urinary [ammonia] and urine flow, i.e., the rate of urinary ammonia excretion, has been abruptly reduced to values below those expected at that flow rate in the steady state. The expected rate of urinary ammonia excretion will be attained only after renal tissue P_{NH_3} has risen to the point where a stable new steady state of the system is reached.

2. Diffusion Equilibrium

Acetazolamide and related inhibitors of carbonic anhydrase, when given by intravenous injection, very rapidly raise urine pH and lower slightly the pH of renal venous blood; under these circumstances there is immediate and massive diversion of some of the renal ammonia supply previously flowing out through the urine into the renal venous effluent (478, 443). The magnitude of the observed effect is compatible, as an approximation, with the explanation that the rate of renal ammonia

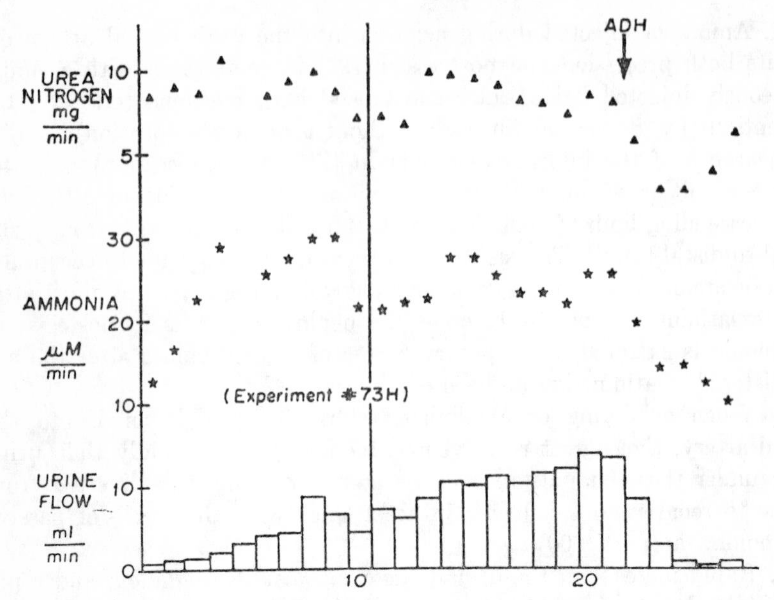

FIG. II-6.1. Exaltation and abatement of urinary ammonia and urea-N excretion Reproduced from Hills and Reid (258). The urine flow rate of a healthy male volunteer was caused to rise at first, and again after urine Sample 10, by two suitably spaced prior drinks of water; diuresis was interrupted rapidly upon collection of Sample 21 by intravenous antidiuretic hormone (*ADH*, vasopressin). Mean duration of collection periods was 6 minutes. *Exaltation* is seen in Samples 3, 4, and 8; these values during rising flow are as high as, or higher than, values at stable higher flow, such as those of Samples 16 to 19. *Abatement* is seen in Samples 10 and 22; compare the higher excretion rates at fairly stable lower flow rates in Samples 6 and 7.

production remains unchanged and that the ammonia supply is passively distributed between the two available outflow channels.

Denis, Preuss, and Pitts (126) were able to show in acidotic dogs that renal venous P_{NH_3} is essentially equilibrated with mean cortical cellular P_{NH_3}: this they did by raising renal arterial P_{NH_3} by means of infusion of ammonia salts into the renal artery. Renal venous P_{NH_3} under control conditions scarcely differed from the value found after renal arterial P_{NH_3} had been artificially raised to equal renal venous P_{NH_3}—in other words, when all phases of cortical water must have been in diffusion equilibrium. Oelert, Uhlich, and Hills (430) studied rats given carbonic anhydrase intravenously to prevent the disequilibrium pH normally present in the distal tubules (II-8C). They showed with the aid of the micropuncture technic that free-flow samples of proximal and distal tubular fluid do not detectably differ in P_{NH_3} from each other or from renal venous effluent. Micro-perfusion with an ammonia-free phosphate buffer solution at pH 6.3 disclosed that perfusate P_{NH_3} attained a constant value, similar in proximal and distal convolutions, within 0.2 second. The permeability π of the cortical nephrons to NH_3 was calculated from the data to be extremely high ($\pi > .01$ cm/second).

These data provide satisfactory evidence of equilibration of NH_3 throughout all phases of cortical water under normal conditions in the mammal. Stone and Pitts (596) have shown furthermore, with the aid of ^{15}N-ammonia injected into a renal artery, that the renal ammonia pool from which the ammonia flows out of the kidney via urine and renal venous blood is a homogeneous one. Injected labeled ammonia is equilibrated with the renal ammonia pool within less time than is required for blood to pass through the kidney: measured specific activity of the urine is thereafter equal to calculated specific activity of the intrarenal ammonia pool. Hempling (235) has estimated from the data that complete equilibration between the renal tissue ammonia pool and the urine is attained within 14 seconds while equilibration between pool and the blood perfusing the kidney is still more rapid.

Though the data exclude consequential barriers to free passage of NH_3 between the various compartments of renal tissue water, it cannot be inferred that there is no NH_3 pressure gradient in the kidney. In fact, a considerable gradient axially directed from papilla to cortex is demonstrable in the carbonic anhydrase-infused dog under all conditions except when high-flow, low-pH urine is being secreted (261; II-10A). The effect of varying urine flow upon the rate of urinary ammonia excretion, which is particularly marked in the concentrated urines (II-10), indicates that urinary P_{NH_3}, as well as osmolality, remains equilibrated at all times in the medullary collecting duct. The P_{NH_3} of high-pH urines, however, is normally higher (during antidiuresis very much higher) than values observed during infusion of carbonic anhydrase (261); the difference must therefore be due to rise in P_{NH_3} beyond the last point where osmolality, P_{CO_2} and P_{NH_3} are still equilibrated with the tissue values, namely at the papilla.

C. THE FUNDAMENTAL SYSTEM

The acid-base composition of the urine appropriate to any acid-base load is profoundly affected by the urine flow rate because of the effect of terminal water reabsorption on transport of the buffer gases CO_2 and NH_3 into the urine (Section A). Terminal dehydration of the urine in the distal tubules and collecting ducts to yield hypertonic urine, a process also responsible for the most prominent of the physicochemical alterations of urinary acid-base composition occurring beyond the papilla, is not only topographically the last of the major events transforming glomerular filtrate into urine; it also reflects the concluding chapter of the evolutionary history of the mammalian kidney (467). When attention is confined to acid-base regulation during water diuresis, we examine not only the workings of a vertebrate kidney in which differences from the kidneys of adult amphibians and their ancestors have been minimized; we also discern more clearly the essential elements of *the fundamental system* of mammalian (and vertebrate) acid-base regulation than when the effects of formation of hypertonic urine have been superimposed (254).

For this reason it will be advisable, in this survey chapter, to confine our analysis of the variation of urinary acid-base composition, in relation to varying acid-base load, to conditions of water diuresis. One of the most striking differences in this respect between high-flow and low-flow urine is the much lesser variation, with changing acid-base load and urine pH, of urinary P_{CO_2} and P_{NH_3} during diuresis; indeed, urinary P_{CO_2} is very well stabilized over the widest possible range of variation of the pH of high-flow urine, scarcely exceeding the limits 50 to 60 mm Hg. Gamble's recognition (181) of the fact and central importance of the stabilization of the urinary P_{CO_2} for renal acid-base regulation is an excellent example of the kind of penetrating insight into simplicities of nature whose significance remains undimmed by subsequent experimental demonstrations of small but significant variation.[2]

1. Stabilization of Effective Urinary P_{CO_2} and P_{NH_3}

It is very striking to what extent the principal features of renal acid-base regulation can be reproduced, in quite reasonable quantitative approximation, by a simple physicochemical construct representing the acid-base composition of high-flow urine as calculated simply from the urinary pH, over its normal range of variation, and from constant values assigned to the effective urinary concentrations of the buffer gases CO_2 and NH_3 (Figure II-6.2). Nothing could more vividly convey the importance to renal acid-base regulation of conveyance of total CO_2 and NH_3 into the urine at widely varying rates appropriate to the acid-base load, and the figure also speaks quite eloquently to the role of *stabilization* of the effective pressures of the buffer gases in the urine. The values chosen for $[CO_2]$ and $[NH_3]$ in the construct are derived from the means of what are here called the *effective* gas pressures, those with which the urine is last equilibrated in leaving the kidney. Evidence presented in detail later (II-9; II-10) makes clear that these "effective pressures" are the papillary tissue gas pressures. To the extent that effective urinary gas pressures—or "quasi-pressure" in the case of CO_2 ($[H_2CO_3]$)—do vary somewhat with the urine pH, the relations presented in this physicochemical construct deviate from those which would be given by a plot of the experimentally determined relationships; the discrepancies are relatively small, however. Nonvolatile urinary buffer excretion as a function of urinary pH, as represented in the graph, is based on empirical data.

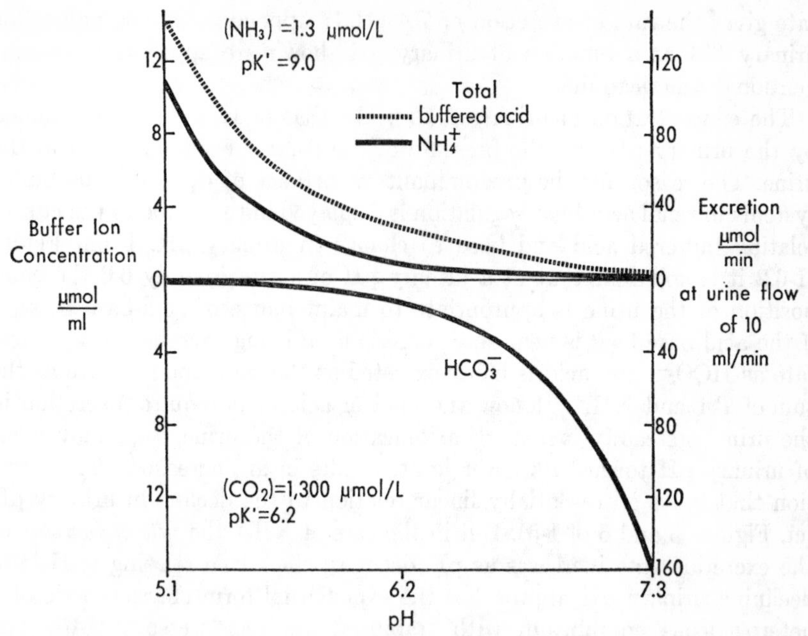

FIG. II-6.2. Theoretical urinary excretion of HCO_3^- and NH_4^+ as urinary pH varies. Slightly modified from Hills (254). The figure is a physicochemical construct which assumes invariant effective (papillary) P_{CO_2} and P_{NH_3} as the urinary pH varies. See text for details.

At high rates of urine flow, values of urine pH observable under normal circumstances extend from about 5.1 to 7.3, as indicated on the abscissa of the figure, whereas during extreme antidiuresis the pH range is expanded at both ends by about 0.7 (II-10C; cf. Figures II-9.1 to II-9.4). Concentrations of the buffer ions HCO_3^- and NH_4^+, as well as of titratable acid (*TA*), are represented on the left-hand ordinate, whereas the right-hand ordinate presents the corresponding excretion rates reckoned on the assumption that the urine flow rate is 10 ml/minute. Calculated as continuous functions of the urinary pH are (*below* the *zero line*) urinary bicarbonate, which is necessarily negligible below pH 6.0 in any solution of the specified CO_2 pressure, and (*above* the *zero line*) ammonium ion (*solid line*) and titratable acid (*dotted line*), the latter being the sum of the protons which have been accepted from extracellular fluid by transfer to the urinary nonvolatile buffers. If the increase in the proportion of the acidic congener in urine over extracellular fluid is multiplied for each nonvolatile buffer pair present by the concentration of that buffer pair in the urine, and these products are summed, we arrive (ignoring small corrections for a different pK′ in urine and extracellular fluid) at the concentration of *TA* in the urine sample; and multiplication by the flow rate gives the rate of excretion of *TA* (II-4; Appendix F). The values for urinary *TA* as a function of urinary pH shown are appropriate to conventional American diets.

The concentration of urinary NH_4^+, like that of HCO_3^-, is determined by the urinary pH and the pressure of the congener gas dissolved in the urine. The reason for the predominant importance of these volatile buffer systems in renal acid-base regulation is displayed in the shape of the curves relating buffered acid and base to change in urinary pH. From Figure II-6.2 it is apparent that at a urinary pH of approximately 6.3 the composition of the urine is appropriate to maintenance of acid-base balance if the acid-base load is near zero: base is then being excreted at a modest rate as HCO_3^-, and acid is being excreted at the same modest rate as the sum of *TA* and NH_4^+. If now the load of acid that requires excretion in the urine is steadily increased, acidification of the urine, with movement of urinary pH toward its lower limit, results in an increase in *TA* excretion that bears an essentially linear relation to the decline in urinary pH (cf. Figures 5 and 6 of I-5). But in the case of NH_4^+ the *rate of increase* of the excretion rate increases as pH declines; the curve relating $[NH_4^+]$ to declining urinary pH approaches the exponential form characteristic of a heterogeneous equilibrium with stabilized low gas pressure (I-6). The graph brings out the great suitability of this volatile buffer system for the accommodation of a large excess of buffered acid in maximally acid urine, even though such urine can attain no great acidic intensity. Since secretory work must be done to lower the pH of the glomerular filtrate, and since there is a stringent limit on the capacity of the tubular cells to acidify the urine (II-8D), it is of the utmost importance that a renal transport mechanism should exist by which the buffer strength of the urine will be markedly increased, automatically and economically, as the lower limit of pH is approached. These objectives are met by the very inexpensive process of simple diffusion of NH_3 down the NH_3 concentration gradient created by urine acidification from renal tissue to nephron lumen.

The relation of urinary HCO_3^- concentration to the urinary pH is similarly represented by an essentially exponential curve. Urinary $[HCO_3^-]$, negligible below pH 6.0, is provided increasingly and at an increasing rate as urine pH rises from 6.1 to 7.3, a relation exhibiting once again the unique fitness of these volatile buffer systems to serve the kidney in its task of acid-base regulation. In a sense, however, the kidney is doing nothing at all, at pH 7.3 during water diuresis, to eliminate base. Base being added to extracellular fluid is being instantly converted to HCO_3^-, and the rate of addition of base to extracellular fluid is normally balanced (to a near approximation) by the excretion of HCO_3^- at the same rate (II-2; II-4B). From the standpoint of mechanism, however, this excreted HCO_3^- is simply a portion of the filtered HCO_3^- which is being allowed by the kidney to escape in the urine. Even in bicarbonate-rich urines most of the filtered HCO_3^- is being reclaimed by secretion of acid into the nephrons. It is by reclamation in the nephrons of practically all the additional filtered bicarbonate—an essentially automatic response to changing load, as we shall shortly see—that the organism adjusts to a decrease in the net base load to zero or to a load of excess acid.

Figure II-6.2 brings out also that during water diuresis the nonvolatile buffers can never contribute at all to the regulation of base balance, since actually the urine is always at a lower pH than the plasma; they function counterproductively by accepting protons from extracellular fluid until urine pH declines below 6.3 in response to imposition of a net load of acid.

Obviously, Figure II-6.2 is not primarily concerned with the details of the intrarenal events which effect the transformation of glomerular filtrate to urine; rather the figure serves to display the physicochemical principles which permit the large observed fluctuations of the urinary concentration of HCO_3^- and NH_4^+ as the urinary pH is varied over its restricted permissible range. As pointed out earlier (I-6), large variation with varying urinary pH of the urinary concentrations and excretion rates of these two ions, which are from the quantitative standpoint the important urinary buffers, depends upon overall conversion, at varying rates, of CO_2 and NH_3 to their congener urinary ions. Continuous passage of buffer substances into the urine, together with donation of protons to, or acceptance of protons from, the extracellular fluid has been designated *excretory buffering;* the urinary volatile buffering systems represent *dynamic* heterogeneous gas-liquid equilibria (I-6B).

Equilibration of the buffer gas pressures between urine leaving the kidney and the renal tissue gas tensions is fundamental to renal acid-base regulation not only on physicochemical grounds, but also from the evolutionary standpoint. The unique fitness of CO_2 and NH_3 to serve as the principal excretory buffers (whether acid and base are to be eliminated via kidney, gills, skin, or mucous membranes) resides (i) in their very stable molecular configuration, which renders them fore-ordained waste products of C and N metabolism, (ii) in their ability to receive or to donate protons, and (iii) in their diffusivity and volatility (I-6B). Though acid-base regulation by the mammalian kidney has been affected markedly by the development of medullary function and the hydropenia characteristic of the mammal (II-10C), renal acid-base regulation in the mammal during water diuresis differs little from that of the adult amphibian; and we cannot doubt that CO_2 and NH_3 have automatically served as excretory buffers in pre-amphibian excretory organs (gills, mucous membranes, skin), just as they clearly do in tadpoles.

The total urinary concentration of buffer compounds reaches its minimum, in general, when urine composition is appropriate to zero acid-base load, i.e., around pH 6.3 during urinary water diuresis (Figure II-6.2). Volatile buffer is transported into the urine at variable rates determined by the urine pH, whereas it is only the positions of equilibrium of the nonvolatile buffers, not the concentrations of the compounds present, which change as urinary pH varies in health. This is so because the rate of urinary excretion of these compounds is determined in health by considerations unrelated to acid-base regulation; and the range of variation in the rate of excretion of the urinary nonvolatile buffer compounds is small under all normal conditions relative to the large fluctuations, as pH varies, of the urinary concentrations of the buffer ions of CO_2 and NH_3. It follows that total urinary buffer is at a minimum when urinary [ammonia + total CO_2] is least; during water diuresis this occurs at pH near 6.3 (Figure II-6.2). Lowest-pH urines and highest-pH urines accordingly exhibit much less absolute variation of pH in response to small perturbations of the experimental conditions; the difference is referable to differences in total urinary buffer concentration.

It is noted above that one reason for the special fitness of CO_2 and NH_3 to serve acid-base balance consists in their both being end products of animal metabolism, and therefore they are substances exceptionally expendable for excretory buffering. This advantage is absolute in the case of CO_2: the rate at which the body must rid itself of this substance so enormously exceeds the requirement for renal base excretion, even when the latter is maximal and the metabolic rate is minimal, that restriction of the supply of CO_2 can never limit the availability of this compound as a source of urinary HCO_3^-. Healthy adult mammals ingesting ample protein in their diets have an ample supply of waste N because their only need for dietary N is to restore tissue protein degraded by the inexorable minimal tissue protein catabolism (II-9D). This minimal protein requirement is in the neighborhood of 25 g daily in healthy adults, and dietary protein in excess of this quantity yields increments of waste N. Only a small proportion of this ample waste N is normally excreted as ammonia, and even the maximal requirement for urinary ammonia characteristic of severe continuing acidosis can be fully met, without drawing on tissue N supplies, by the waste N available from a conventional protein intake of 70 g daily. Of course, when protein intake is insufficient to preserve N balance, urinary ammonia loss contributes to the negative N balance. Röse demonstrated, in a remarkable experiment conducted over a period of years on himself, that N balance could be maintained indefinitely on a protein intake of about 25 g daily in the face of vigorous activity including mountain-climbing. But manipulation of the urine pH in such a fashion as to alter materially the rate of N loss as urinary ammonia could spell the difference between negative and positive N balance when the adequacy of protein in the diet was marginal (521, 598). A number of experiments in animals (460, 8, 391) also show clearly that conditions leading to increased urinary ammonia excretion contribute to a negative N balance provided that protein intake is inadequate (II-9D2). As far as we know, negative N balance has little or no effect on the renal ammonia production; accordingly, there is automatic allocation of waste N to the urine as ammonia buffer, at rates predictable from urinary pH and flow (II-9B), and renal acid-base regulation takes precedence over N conservation.

2. Automaticity

A striking and unique aspect of renal acid-base regulation, among the homeostatic systems of intact higher organisms, is its high degree of automaticity: this regulatory system is designed in such a manner that its behavior, as the acid-base load varies, is determined predominantly passively by the physicochemical consequences of addition of net excesses of acid or base to extracellular fluid. In contrast to renal adjustment of water balance and of the osmolality of extracellular fluid, and indeed in contrast to pulmonary stabilization of arterial P_{CO_2}, neither endocrine nor neurological control systems enter into the fundamental system by which the kidney regulates acid-base balance. Accordingly, we do not encounter primary disturbances of renal acid-base regulation unless the essential elements of the regulatory system—the integrity of the kidney and its blood supply or the acid-secreting mechanism of the nephrons—have been directly affected.

Figure II-6.3 gives an overview, preliminary to the next three chapters, which emphasizes the automatic, physicochemical character of the renal response to changing acid-base loads. The chain of events produced by a base load (*above*) and by an acid load (*below*) in extracellular fluid and (*inside boxes*) within the lumen of the distal nephrons are indicated by

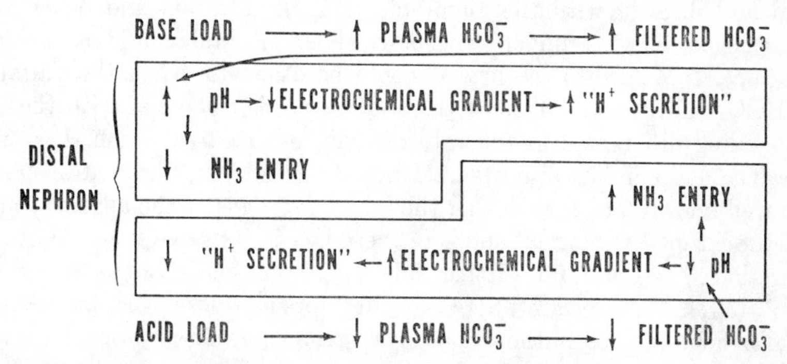

FIG. II-6.3. Automatic effects of changing acid-base load on the acid-base composition of the urine. Reproduced from Hills (255). See text for details.

the *horizontally directed arrows*. Base added to extracellular fluid raises plasma [HCO_3^-] and therefore increases HCO_3^- filtration; more HCO_3^- reaches the distal convolutions so that luminal pH is raised throughout the distal tubules and collecting ducts, and in the end-urine; less NH_3 therefore diffuses into the urine. Acid added to extracellular fluid produces the converse chain of consequences. Plasma [HCO_3^-] and HCO_3^- filtration decline; less filtered HCO_3^- reaches the distal nephron so that the luminal fluid and end urine are more acid; and therefore more NH_3 diffuses into the urine.

A central objective in presenting Figure II-6.3 is to bring out that the reason for the decline in urinary pH in response to an acid load, and for its rise in response to a base load, is not to be sought in a cellular response, specifically in (respectively) an increase or decrease in the rate of secretion of acid into the nephrons. On the contrary, acid secretion decreases when the urine becomes more acidic in response to an acid load, and increases when the urine pH rises in response to a base load (II-8D). In short, the one constituent of the regulatory system which is an energy-consuming, cellular process and not a physicochemical response to imposed conditions is the one constituent which does not respond in an appropriate, regulatory manner to changing acid-base load. Its response is in fact counterproductive; the kidney's response to changing acid-base load would be more efficient than it is were the rate of acid secretion to remain entirely unaffected by changing acid-base load (255). It is the automatic fluctuation of plasma [HCO_3^-] in response to changing acid-base load which is responsible under normal conditions for the *responsive* alterations of the acid-base composition of the urine. The other factors which normally influence urinary pH—the urinary flow rate and the rate of excretion of nonvolatile buffer—do not in the steady state affect the net renal excretion of acid or base.

Most writers on the subject have long been agreed (II-8D) that the rate of acid secretion into the nephrons varies inversely with the acid load and directly with the glomerular HCO_3^- filtration and hence with the plasma HCO_3^-, under physiological circumstances and as acidosis develops. It is hard to see how it could be otherwise when the variation of HCO_3^- filtration, as plasma [HCO_3^-] varies, is compared with the corresponding differences in the rate of renal net acid excretion. Recently, however, Relman has dissented. "I would remind you," he writes, ". . . of the well-known fact that during the initial few days of the administration of an acid load to normal subjects there is a progressive increase of acid excretion . . . despite no further fall in plasma bicarbonate concentration" (501). No data are offered or cited in support of this assertion. It is also held by this author that in a "more normal situation" Lennon, Lemann, and Relman (340) observed increased urinary acid excretion in response to an increased acid load with no detectable change in plasma HCO_3^- or pH.

"Detectable" is a key word if comparisons of estimates of 24-hour net renal acid excretion and plasma [HCO_3^-] are to be urged in evidence against the central role played by plasma [HCO_3^-] in the response to changing acid-base load. In view of the marked circadian fluctuations of plasma [HCO_3^-] and the corresponding circadian fluctuations of renal acid-base excretion (II-1)—fluctuations which moreover are quite variable in pattern and magnitude from subject to subject—, it is quite unrealistic to suppose that estimation of the plasma CO_2 content of one or two samples daily, as carried out by Lennon *et al.*, would necessarily disclose in a statistically significant manner the relation between decreasing plasma [HCO_3^-] and renal net acid excretion per 24 hours. Nevertheless, in one of the two experiments for which the requisite data are presented by Lennon *et al.*, a continuous decline in total serum CO_2 from 28.0 to 25.5 mmoles/liter was recorded over a 5-day period commencing with the beginning of the experimental period in which an acid load was administered. Such a decline in serum total CO_2, followed by a rise to 27.8 and 27.4, constitutes rather impressive evidence of the role of responsive fluctuation of plasma [HCO_3^-] in the renal excretory response to acid administration. Large fluctuations of serum total CO_2 (successively 29.2, 23.5, 24.4 mmoles/liter) listed during the 5-day control period of the same experiment underscore the problem introduced by nonhomeostatic variation of plasma [HCO_3^-].

D. FACTORS INFLUENCING URINARY pH

D. P. Simpson arrives, in the summary of an exhaustive recent review (570) of hydrogen ion homeostasis and renal acidosis, at the conclusion that "regulation of urine pH is a poorly understood process occurring in the collecting duct and depending on the interaction of several variables." The major variables are summarized by this author as (i) the amount of HCO_3^- that escapes reabsorption in the proximal tubule and (ii) the rate of excretion of poorly reabsorable anions, such as sulfate and phosphate. Note is taken that proximal tubular bicarbonate reabsorption (and hence (i) above) is influenced by the P_{CO_2} of the body fluids, carbonic anhydrase activity, and the state of sodium chloride and of potassium balance.

Control of urinary pH is certain to be poorly understood unless the miscellany of factors which can influence it are articulated in such a way as to explain the variable relation between urinary pH and renal net acid or base excretion and to define the fundamental elements of the system that regulates the latter function. Naturally—except for small effects of water and CO_2 reabsorption from renal pelvis, ureter, and bladder—the final acid-base composition of the urine is determined under any particular condition by events in the medullary collecting duct, simply because of topography; but this conclusion is a mere anatomical tautology, telling us nothing of importance about the regulatory system which ensures that renal net excretion of acid or base shall vary in appropriate response to changing acid-base load. As has been emphasized above, this system operates at the level of the whole organism, in which physicochemical responses of the extracellular fluid to the acid-base load play the central role.

The pH of the urine in health is determined fundamentally but by no means entirely by the net rate of renal acid or base excretion, i.e., by the rate (overall) of transfer of protons between the buffers of the urine and of the extracellular fluid. A further important factor influencing urinary pH, in the steady state and at any particular rate of urine flow, is the rate of excretion of volatile and nonvolatile buffers. Not mentioned at all in Simpson's summarization is medullary control of urine osmolality, which is the only operation carried out on collecting duct fluid which in any steady state of acid-base balance affects the urinary pH independent of the most fundamental elements of the normal renal acid-base regulatory system, i.e., HCO_3^- filtration and the total rate of acid secretion into the nephrons (II-10C). Secretion of acid into the collecting duct, if it occurs (cf. II-7E), is important only as a contribution to the total rate of acidification of glomerular filtrate, whereas facultative water reabsorption from the distal nephron—distal tubules and most especially collecting duct—strikingly influences urinary pH at any acid-base load by affecting the relative proportions of nonvolatile buffer and the two volatile buffers. In any steady state of acid-base balance and water balance, net renal excretion of acid or base is independent of urine flow, but urinary pH is markedly affected by the flow. By contrast, any alteration of the total rate of renal acid secretion (including acid secretion into the collecting duct) affects not the urinary pH but the internal environment, once a new steady state of acid-base balance has ensued.

The presence of excessive amounts of $SO_4^=$ in glomerular filtrate—never an important influence on urinary pH under physiological conditions, and not often under any circumstances—is simply one of the many factors which under special circumstances can alter the balance between HCO_3^- filtration and urine acidification. If the rate of acid secretion into the nephrons is, for example, increased, the change influences the urinary pH only transiently; when the plasma HCO_3^- rises in response to the increased acid secretion, as it must do, it will reach a level at which renal net excretion of acid or base (and hence urine pH at any urine flow rate) is again appropriate to load. The changes established after a new steady state has been reached will be (i) lowering of the HCO_3^- clearance and (ii) some alkalinization of the internal environment.

It is of course granted that renal acidosis and alkalosis, by markedly altering the internal environment, will often secondarily affect the acid-base composition of the urine; acidosis for example, will increase ammonia concentration of urine of any given pH and flow, so that the urine pH will be higher than it would otherwise be under the same conditions, in the absence of acidosis. Regulation of urinary pH under entirely physiological conditions, however, depends essentially upon the (i) acid-base load, which determines net renal acid or base excretion in the steady state; (ii) the rate of nonvolatile buffer excretion, which will vary mainly with the dietary phosphorus; and (iii) the urine flow, upon which depends the rate of excretion of the volatile buffers at any acid-base load in the steady state. At whatever stable level the urine flow and the rate of excretion of nonvolatile buffer may be set by the water load and the food intake, respectively, the plasma [HCO_3^-] will automatically adjust itself in such a way that renal net acid or base excretion equals the acid or base load. An analogous case is that of plasma urea, which will adjust itself to that level at which urinary urea excretion equals the urea load; in any such steady state in which urine flow is also stable, the plasma [urea], but not the rate of urinary urea excretion, is influenced by the urine flow rate.

NOTES

1. ". . . it turns out that regulation of the equilibrium between acids and bases in the organism ultimately depends on the variability of this equilibrium in the urine, the case being in this respect exactly analogous to that of the osmotic pressure." (238, author's translation.)

2. The P_{CO_2} of properly collected human urine can vary under normal conditions, in our experience, between about 30 and 100 mm Hg.

seven

ACIDIFICATION (I) AND ALKALINIZATION OF THE URINE

A. INTRODUCTION

It has become customary, in discussing the mechanism of urine acidification, to center the inquiry on whether secretion of "hydrogen ion" (or the redundant "H^+ ion," or even "hydrogen") is complemented by some other process which would generate acid in the luminal fluid of the nephron (487, 200, 659). Secretion, in this parlance, is assumed to consist of movement of hydrogen or hydronium ions into the lumen in exchange for Na^+ reabsorbed; such a cation exchange is now widely held to represent the major if not the only mechanism of urine acidification. But the terms in which this concept is couched are chemically vague, and the conventional formulation of the subject involves certain unproven assumptions; its uncritical acceptance appears to sweep under the carpet some conceptual issues of potential importance relating to the nature of the acidification process (Section D).

No expression in common use refers in a purely descriptive way to the process under consideration, i.e., presupposes or implies nothing about its mechanism. It is in this sense that "acidification" of the urine or glomerular filtrate is to be understood here, though this usage, too, is not entirely satisfactory (Section C). Table II-7.1 lists six types of active (energy-requiring) processes and one passive one, all of which would be capable, simply in terms of chemical theory, of acidifying the urine; their relevance to the function of the mammalian nephron is discussed in Section D.

B. MEMBRANE TRANSPORT AND URINE ACIDIFICATION

Contemporary renal molecular biology has directed much effort to elucidating mechanisms of transport of ions and other substances across the

TABLE II-7.1. *Some processes which could acidify the urine**

I. Active ion transport
   1. Secretion of H^+ in exchange for Na^+ reabsorbed
   1. Secretion of H_3O^+ in exchange for Na^+ reabsorbed
   3. Reabsorption of Na^+ with OH^-
   4. Preferential reabsorption of conjugate bases of weak electrolytes with Na^+
   5. Secretion of H^+ with Cl^- or $SO_4^=$ or with acidic congeners of weak electrolytes.
   6. Reabsorption of HCO_3^- with Na^+

II. Passive acidification
   7. Reabsorption of NH_3 down a NH_3 pressure gradient created by water reabsorption

* All processes which are or may be based on energy-requiring ion transport are included in the "active" category. According to Rector (487), however, it is possible that the transtubular electrochemical gradient created by active Na^+ reabsorption might suffice to allow protons to diffuse into the lumen of the distal tubule at rates sufficient to account for the acidification observed there (II-7A).

wall of the nephron, and to the energetics of these transport processes. Uncharged substances can move passively down concentration gradients existing across a membrane, or, in the case of gases, more exactly down pressure gradients, provided the obstacle offered by the membrane to their diffusion is not too great. Evidently the inherent diffusivity of any particle is a major determinant of the velocity at which transport may occur; for uncharged particles the rate of free diffusion is, as a first approximation ignoring the shape of the molecule, inversely proportional to the square root of the molecular weight (Graham's law). Small uncharged and unhydrated particles like NH_3 and CO_2 are accordingly among the most diffusible of all molecules. Another determinant of the velocity of free diffusion is the magnitude of the locally existing pressure or concentration gradient; and, in the case of charged particles, the sign and magnitude of any net electrical field along the path of transport exerts an effect as fundamental as that of the chemical concentration gradient upon transport energetics of secretion. Another factor is the resistance offered by the membrane to diffusion of the secreted material; particles (including ions) not soluble in the lipids of which much of the membrane is composed can pass only through the small proportion of the surface consisting of water-filled pores. Ions also diffuse much less readily through membranes than do uncharged particles of similar size, because their diffusivity is reduced by hydration, more so in the case of mineral cations than anions (I-2A4) and especially because they are detained by electrostatic attraction to membrane constituents.

Electrical potential differences across amphibian and mammalian nephrons and across membranes of the renal cells have been extensively measured during the past two decades. Transtubular potential differences have been related functionally to transtubular concentration differences; and transtubular transport velocities have been studied not only under more or less normal "free-flow" conditions by the micropuncture technic, but also under highly artificial conditions which can be established by microperfusion of nephrons or their peritubular capillaries, or both, and which facilitate study of the nature of local transport processes (659, 194). Extensive reliance upon voltage measurements across the proximal tubule now believed to have been erroneous because of unrecognized technical problems affecting the validity of the measurements (II-8A), and theoretical complexities affecting model construction, have delayed consensus and lent a somewhat tentative character to current formulations. On the basis of data now regarded as reliable, acidification in the mammalian proximal tubule is considered an energy-requiring process, whereas it is considered possible by some (II-8A) that acidification in the mammalian distal tubule might occur by passive movement of protons down the local transtubular electrochemical gradient.

The renal tubule is of course only one among many mammalian tissues capable of secreting acid; and though it is clear that the process of acid secretion by the stomach, for example, which involves secretion of HCl, must differ fundamentally in its mechanism from renal tubular acid secretion, which does not, it is also true that all transport of ions across secretory membranes must be governed by common principles of some generality at the level of biophysics and molecular biology. Technics applied particularly by Ussing and his associates (623, 624), involving observation of ion transport by surviving tissues *in vitro* in relation to electrical potential differences recorded across the secretory epithelia, permit extensive comparison of the secretory behavior of various types of tissues derived from many species in response to a great variety of imposed experimental conditions. The overall mechanism of transport of the principal ions of glomerular filtrate across the mammalian proximal tubule has widely been assumed to resemble that of transport by tissues extensively studied by these means, such as frog skin and toad bladder, in which active transport of Na^+ is assigned the primacy (Section D); even this elementary formulation, however, is for various reasons not at present universally accepted by workers in the field as directly applicable to the mammalian nephron (435). Pending advances leading to a firmer consensus, the interested reader is referred, for further information concerning *in vitro* studies of ion transport and their relevance to acidification by the mammalian nephron, to reviews of the very extensive literature dealing with this subject (194, 623, 624, 435, 654).

C. ALKALINIZATION (WATER REABSORPTION)

"Urine acidification," though it must *per se* result in decline of the pH of luminal fluid, can (and in the distal nephron under some circumstances does) proceed while the pH of the luminal fluid is stationary or rising in consequence of another, simultaneously occurring process; and it is an unavoidable disadvantage of the term "urine acidification" as used here that the process need not be associated with actual reduction of the urine pH if alkalinization is proceeding simultaneously and independently. Passive CO_2 reabsorption, an alkalinizing process unless acidification is proceeding simultaneously, does in fact occur throughout the nephrons wherever the luminal fluid contains substantial quantities of HCO_3^-. The entire nephron appears to be freely permeable to CO_2; and wherever this is so, as Gottschalk, Lassiter, and Mylle were the first to emphasize (214), reabsorption of filtered water, by concentrating HCO_3^- while CO_2 escapes from the nephron by diffusion, will tend to raise the fluid pH by initiating the following reaction:

$$H^+ + HCO_3^- \rightarrow H_2CO_3 \rightarrow CO_2 + H_2O \qquad (1)$$

Removal of H_2O causes CO_2 to be lost by diffusion down the transtubular CO_2 pressure gradient so created, and the H^+ shown at the left, if it is not furnished by acid secretion, will necessarily be yielded by the other urinary buffers to regenerate much of the H_2CO_3 that decomposes as the reaction chain moves to the right.

If it were not for concurrent massive acidification of glomerular filtrate by acid secretion, reabsorption of much of the filtered water from the nephrons would result in extreme alkalinization of the urine, since all the H^+ shown at the right of Reaction Chain 1 would then have to be supplied by the urinary buffers. But in fact near-equivalence prevails under all normal circumstances between acidification in the nephrons and the alkalinizing effect of water reabsorption. Evidently, when the urine pH is close to 7.4, the effects of water reabsorption and acidification on the position of equilibrium of the urinary HCO_3^- buffer system (and the other urinary buffer systems) are nearly balanced (Figure II-7.1); when the urine pH is low acidification has slightly predominated, and when it exceeds 7.4, water reabsorption has slightly predominated. We do not know of any process other than water reabsorption which raises the pH of the luminal fluid in mammalian renal tubules. Even when very acid urine is being secreted, the acidification process serves mainly, in the quantitative sense, to prevent rise of the pH of luminal fluid secondary to water reabsorption (Table II-7.2). We shall hereafter circumvent the semantic discomfort occasioned by continued reference to "acidification" for a process which may take place with no reduction of pH by referring instead to "acid secretion." But it must be understood that "acid secretion" does not necessarily imply transport of protons from cell interior to the lumen of the nephron (Section D). Expressed as "acid secretion," more than 20 times as much acid is always normally required to supply protons for removal of filtered HCO_3^- (Reaction Chain 1) as serves to reduce the pH of the nonvolatile buffers below that of plasma when acid urine is being elaborated (Table II-7.2).

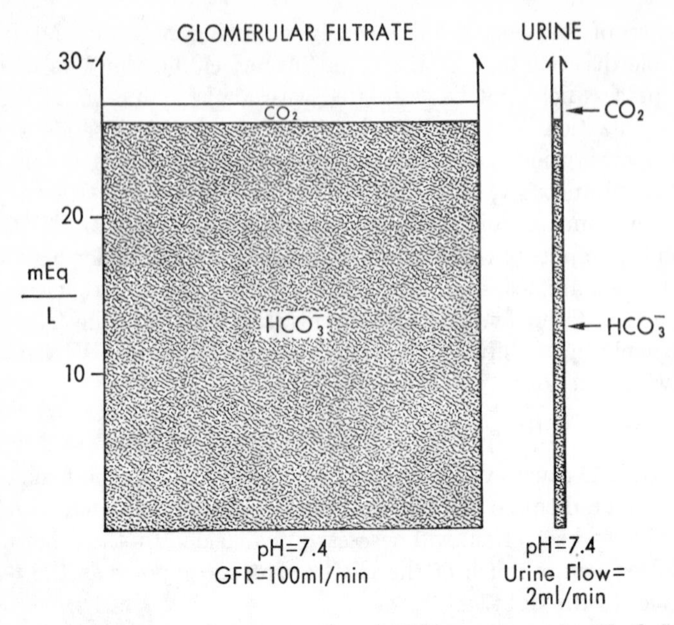

FIG. II-7.1. Rates of filtration of CO_2 and HCO_3^- compared with their rates of urinary excretion. Concentrations of HCO_3^- and CO_2 are represented on the vertical scale; the horizontal dimension represents the volume of fluid (glomerular filtrate and urine, respectively) formed per minute (*GFR*: glomerular filtration rate). The diagram refers to the normal urine composition during a moderate base load, in which HCO_3^- is being excreted in urine of approximately the same pH as plasma; plasma $[HCO_3^-]$ is on the high side of normal when urine pH is 7.4. The higher P_{CO_2} of alkaline urine is simply ignored, as though CO_2 had been allowed to diffuse off until urine P_{CO_2} was close to 40 mm Hg before measuring urine pH; this means that urinary $[CO_2]$ and $[HCO_3^-]$ are almost equal to plasma values.

Only $\frac{1}{50}$ of the CO_2 originally present in glomerular filtrate, and none of that formed in the nephrons from the filtrate HCO_3^- which has vanished from the nephrons, is still present in the urine. The protons required to convert the filtered HCO_3^- to CO_2 all had to be secreted; they could not have been derived from the urinary buffers, since the urinary pH is unchanged from that of glomerular filtrate. Under the stated circumstances, 2.6 mEq of HCO_3^- are being filtered per minute, and $\frac{49}{50}$ of this quantity, about 2.55 mEq, has to be converted to CO_2 per minute. There must therefore be secreted into the nephrons 2.55 mEq *per minute* simply to prevent the rise of urine pH above plasma pH which would otherwise be effected by passive reabsorption of CO_2 down the CO_2 pressure gradient created by water reabsorption. The magnitude of the acid secretion rate required simply to hold the pH of low-flow urine in the vicinity of plasma pH is emphasized when this figure is compared with usual *24-hour* rates of renal net acid excretion in the neighborhood of 50 mEq.

D. POSSIBLE CHEMICAL MECHANISMS OF ACID SECRETION

Only in the terminal nephron segment, and even there only when low-pH low-flow urine is being secreted, may a process other than active ion transport markedly depress the urinary pH: this is loss of NH_3 by diffusion

TABLE II-7.2. *Renal acid excretion and acid secretion compared**

Plasma [HCO_3^-]	21,000 μEq/liter
Glomerular filtration rate	0.1 liter/minute
Filtered HCO_3^-	2,100 μEq/minute
Acid needed to convert all filtered HCO_3^- to CO_2	2,100 μEq/minute
Acid needed for excretion of acid buffered as titratable acid and NH_4^+	100 μEq/minute

* The values are appropriate to a very large acid load sufficient to produce borderline acidosis. Even under these conditions, where acid is being eliminated rapidly (100 μEq/minute corresponds to about 150 mEq/24 hours) over 20 times more acid must be secreted into the nephrons, from which all filtered HCO_3^- is reabsorbed under these conditions, than is excreted. Plasma [HCO_3^-] has declined from a conventional median value of 24,000 μEq/liter in response to the large acid load; the associated decline in the rate at which HCO_3^- is filtered allows more of the secreted acid to donate protons to NH_3 and to the urinary nonvolatile conjugate bases. However, further decrease of HCO_3^- filtration in acidosis is not directly helpful, since the total rate of secretion of acid into the nephrons is limited by the already maximal transtubular pH gradient in the terminal part of the nephron (II-8D); the rate of acid excretion must then be increased principally by increasing the availability of urinary buffer for excess protons, especially NH_3.

down the NH_3 pressure gradient created by water reabsorption from bicarbonate-poor luminal fluid (II-10C). The chemically possible mechanisms of active urine acidification are listed in Table II-7.1 more or less in order of decreasing physiological plausibility; only the first three need be given serious consideration as major means of urine acidification.

There is no persuasive evidence against the first two possibilities—exchange of Na^+ for a proton, hydrated or free, across the wall of the nephron—and there is conclusive experimental evidence (II-8C) that one of these processes, or else reabsorption of OH^- and Na^+, must account for much if not all acidification of glomerular filtrate. The progressive decline, in the presence of an acid load, of the pH of glomerular filtrate as it passes through the proximal tubules, distal tubules, and collecting ducts unfortunately does not make clear what is, in fact, being transported by the epithelial cells to effect this change. The transtubular proton-Na^+ exchange generally assumed to take place in proximal and distal tubules cannot be distinguished, at any rate not by any experimental means so far deployed in investigating the mechanism of urine acidification, from reabsorption of NaOH; and the only theoretical argument which has been advanced against this latter process is that hydroxyl ions would have to be transferred out of a solution (acidified luminal fluid) in which they are present in very low concentration. The argument has some force, but it can hardly be considered decisive; the alternative (proton-Na^+ exchange) must also assume that H^+ or H_3O is pumped out of cells in which protons are present in very low concentration, even if admittedly up to several hundred times greater than [OH^-] in acidified luminal fluid. Moreover, the electrochemical gradient against which transport of either protons of OH^- would have to be effected, which is a principal determinant of the energy required for transport, would be identical under any given circumstances.

The widespread use of the symbol "H^+" in the clinical and physiological literature to represent H_3O^+ is particularly unfortunate in connection with discussions of the mechanism of urine acidification, since it sweeps under the carpet fundamental conceptual discriminations (255). Protons might be transported into the nephrons, in exchange for Na^+ reabsorbed, by transport of the charged molecule H_3O^+; they might conversely move by "hops," in which a proton is passed from one water molecule to another linked to it by a hydrogen bond. This process is represented by Edsall and Wyman (137), in the following manner:

$$\mathrm{H{-}O{-}H} \cdots \mathrm{O(H)(H)} \rightarrow [\mathrm{H{-}O}]^- \cdots \left[\mathrm{H{-}O(H)(H)}\right]^+ \qquad (2)$$

where the dots represent a hydrogen bond. Such hops may occur spontaneously, and in the presence of an applied electrical field they can effect rapid net unidirectional movement of protons; they account for the uniquely high apparent mobility of H_3O^+ and OH^-. There seems to exist no basis for exclusion of the possibility that local electrical fields produced by active Na^+ transport from lumen to peritubular blood might result in net transport of unhydrated protons into the nephron, a process which would be an exceptionally economical means of moving H_3O^+ and OH^-.

The acidification process, whatever its exact nature may be, is to a very great extent coupled functionally with reabsorption of some of the filtered Na^+ (487, 642). The principal factor (normal variation in glomerular filtration rate [GFR]) causing considerable normal fluctuation of the absolute rate of Na^+ reabsorption affects the rate of acid secretion more or less in proportion (Section F); and some pathological states, notably abnormality of extracellular fluid volume, also exert strikingly parallel effects upon the rates of acid secretion and of reclamation of filtered Na^+ (II-8D; Part III). The result is that neither random fluctuations of the filtration rate nor altered extracellular fluid volume much disturbs the normal ratio of about 1 filtered HCO_3^- reabsorbed for every 4 filtered Cl^-.

The homeostatic utility of linking changes in the rate of renal acid secretion with the filtration of Na^+ and Cl^- is clear (182), granted the prominence of these three ions in the electrolyte structure of the plasma and their predominantly proximal reclamation from glomerular filtrate. Their rate of filtration is so high compared with even their maximal excretion rates that most of what is filtered *must* be reclaimed if derangement of the electrolyte structure of extracellular fluid is not to be precipitated simply by fluctuating

GFR. This is true no matter what the whole-body load of Na^+ or Cl^- and no matter what the acid-base load may be: absolute reclamation must approximate filtration, normally around four times greater for Cl^- than for HCO_3^-. In the case of Na^+ about 99% of the filtered material must be reabsorbed in health; about 98% of the filtered chloride must be reclaimed, and 95 to 100% of the filtered HCO_3^-. The bulk of this reabsorptive work must be carried out proximal to the distal nephron, which has comparatively limited transport capacity. When the NaCl load is ample, the healthy kidney succeeds admirably in regulating acid-base balance independently of the balance of NaCl; but under pathological circumstances, acid-base homeostasis can be markedly disturbed by abnormal reclamation of filtered HCO_3^- ascribable to derangements of NaCl balance and homeostasis. For example, the salt-depleted subject without access to adequate dietary Cl^- will generally repair deficits of extracellular fluid volume due to Na^+ depletion through renal establishment of a positive Na^+ balance, even though the process involves production of nonrespiratory alkalosis by a degree of base retention which is inappropriate to the acid-base load (Part III).

It is generally but not universally held (435) that Na^+ reabsorption from the proximal part of the nephron is the principal active electrolyte transport process, in the face of which electroneutrality is preserved in part through coupling of some of the Na^+ transport in some as yet inscrutable manner (487, 642, 622, 342, 375) with acid secretion, but in greater part by passive reabsorption of Cl^- with Na^+ down the electrochemical gradient created by removal of Na^+ from the filtrate. $NaHCO_3$ seems to be reabsorbed more rapidly than NaCl, causing rise of $[Cl^-]$ in proximal tubular fluid; and it has been reported (550a) that in man net proximal Na^+ reabsorption is increased as the ratio HCO_3^-:Cl increases in glomerular filtrate. There are few if any purely electrophysiological grounds for excluding with certainty active urinary acidification as the basis for reabsorption of that portion of the filtered Na^+ which is matched against filtered HCO_3^-, but active Na^+ reabsorption is usually if tentatively assigned the primacy; this is in any event the one active process which could account for reabsorption of filtered Na, Cl, and water as well as urine acidification. But the functional coupling, necessarily present normally, of urine acidification with Na^+ reabsorption renders unlikely *a priori* the fifth mechanism of urine acidification listed in Table II-7.1—secretion of molecular acid. Since practically all the filtered Cl^- must be reabsorbed, any Cl^- secreted with H^+ would necessarily have to be reabsorbed subsequently; and such a mechanism of urine acidification would therefore entail a considerable increase in energy expenditure as compared with the first three mechanisms listed in Table II-7.2, while offering no discernible advantage to the organism. In addition, extensive micropuncture studies have failed to produce any evidence of net proximal Cl^- secretion. The lumen-to-blood Cl^- concentration gradient created by water reabsorption in the proximal tubules and the luminal negativity of the distal tubule are probably sufficient to account for observed rates of reabsorption, but active secretion occurs in Henle's loop (81a). Urinary $SO_4^=$ is also all accounted for as filtered $SO_4^=$; there is nothing to suggest that H_2SO_4 is secreted, nor could this process, nor secretion of the acidic congeners of nonvolatile weak electrolytes like phosphate or urate, account for much of the observed urine acidification; their rates of urinary excretion are all too small to account collectively for much of the total acidification process.

It has been proven (II-8C) that reabsorption of HCO_3^- with Na^+ (Table II-7.1, no. 6) cannot be the mechanism by which mammalian urine is principally acidified, and the weight of present evidence does not assign to this process any important role in urine acidification, though its participation cannot be excluded. As far as luminal acid generation through preferential reabsorption of urinary conjugate bases (Table II-7.1, no. 4) is concerned (e.g., reabsorption of $HPO_4^=$ with two sodium ions), participation of such processes in urine acidification also cannot be excluded rigorously; but clearly they can at most be ancillary, for Pitts and Alexander showed by means of clearance studies in acidotic dogs (471) that acidification takes place at a much greater rate than could be explained even by total reabsorption of all the conjugate base of the filtered nonvolatile substances. Still other possible mechanisms of urine acidification can be excogitated. It has been pointed out, for example (394), that reabsorption of $CO_3^=$ would generate two luminal protons; however, Table II-7.1 leaves out of consideration, as inherently improbable, acidification through transport by the renal tubules of relatively large ions which cannot exist in appreciable quantity at physiological pH.

E. TOPOGRAPHY OF URINE ACIDIFICATION

Fluctuation of plasma [HCO_3^-] in response to changing acid-base load is the principal determinant of urinary pH under normal circumstances (II-6), and always, under any circumstances, an important determinant of the pH. Variation of plasma [HCO_3^-] within the normal range affects the pH of proximal tubular fluid very little, but occasions much larger excursions of urinary pH, and so quite evidently also of the pH of luminal fluid in the terminal segments of the nephron, including the collecting duct—assuming, that is, standardized conditions in which the effects of altered urine flow on the pH in the terminal segments (II-10C) can be left out of account. It follows that the rate of sequential decline of filtrate pH along the whole length of the nephron must increase as the acid load increases and plasma [HCO_3^-] declines at constant urine flow.

Montgomery and Pierce (409) reported that urinary acidification in frogs normally first becomes manifest in the proximal portion of the distal tubule; but Ellinger (149) could show that, when the urine of these animals is rendered strongly acidic by increasing the acid load, the pH of the filtrate already begins to decline in the proximal tubule. The difference reflects the reduced rate of filtration of HCO_3^- secondary to lowered plasma HCO_3^- produced by a large acid load, the rate of proximal acid secretion being relatively unaffected. Micropuncture studies of mammalian nephrons likewise show a greater rate of decline, when plasma [HCO_3^-] is lower, of the pH of glomerular filtrate as it proceeds along the nephron (214, 201, 486, 99, 200, 375). The degree of reduction of luminal pH observed in mammalian proximal convolutions is generally held to exhibit marked species-specific variation (487), though it is in actual fact difficult to be certain from the data provided in the literature that differences in the magnitude of the acid loads imposed relative to the size and metabolic rate of the animal being studied, variations in arterial P_{CO_2}, and other non-uniformities of experimental conditions may not have influenced the interspecific comparisons. Micropuncture experiments in the rat (214, 201, 486, 200, 375) show decline of filtrate pH in the proximal convolutions to as low as 6.8, with [HCO_3^-] as low as 8 mEq/liter; in analogous experiments in the normal dog, proximal tubular [HCO_3^-] was reported to decline only to about 16 mEq/liter (99), though in nonrespiratory acidosis there is a sizable proximal reduction of pH in this animal (97). Substantial decline of pH of the luminal fluid in the proximal convolutions is also reported in the monkey (46).

Table II-7.3 presents estimates of the urine pH and of the pH of luminal fluid in successive segments of the human nephron at very low urine flows in relation to three different, normal acid-base loads; the data are speculative, inasmuch as no direct information is available for man. A principal purpose of the table is to indicate the stabilizing effect, in relation to changing acid-base load, exerted by delayed dehydration of luminal H_2CO_3 upon *in situ* pH in the distal nephron. The low disequilibrium pH produced in this way in the distal nephron is discussed in more detail in Chapter II-8, Section C; it arises from accumulation of luminal H_2CO_3 resulting from reaction of luminal HCO_3^- with secreted acid, and the difference between disequilibrium and equilibrated pH is accordingly greater when higher-pH (bicarbonate-rich) urine is being secreted. According to the table, variation of nearly 2 pH units which would otherwise occur in the mid-distal convolutions, as the acid-base load varies, is probably reduced *in situ*, as the result of the disequilibrium pH, to fluctuation of about 1 pH unit (underscoring). Rector has suggested (487) that such stabilization of the physicochemical environment of the distal tubular cells, as the acid-base load fluctuates, might serve to provide more favorable conditions for the fine regulation of transport processes in the distal nephron where the composition

TABLE II-7.3. *Topography of acidification of glomerular filtrate in the human nephron in relation to acid-base load during antidiuresis**

		Large base load	Zero net acid or base excretion	Large acid load
			pH	
End urine		8.0	6.4	4.5
Nephron				
Proximal convolution (distal portion)		7.45	7.3	7.1
Distal convolution (mid-portion)	1. Disequilibrium	6.9	6.6	X
	2. Equilibrium	7.7	6.9	6.0
Distal convolution (terminal portion)	1. Disequilibrium	6.7	6.1	X
And cortical collecting ducts	2. Equilibrium	7.3	6.4	5.1

* Most of the figures are educated guesses based on data reported in other mammalian species (II-8D1). Left-hand column corresponds to large base load within the normal range, right-hand column corresponds to large acid load, and middle column corresponds to zero urinary net acid or base excretion. The pH of the end urine (top row) can vary over so wide a range only during marked antidiuresis (II-10): note expansion of urinary pH range (upper row) as compared with fluid entering the medullary collecting ducts (bottom row), which is assumed to be similar to high-flow end urine. Magnitude of the disequilibrium pH in the mid-distal convolutions under conditions of base loading is based upon data reported for the rat by Rector, Carter, and Seldin (II-8C); no significant disequilibrium pH can develop or be maintained in bicarbonate-poor fluid whose equilibrated pH is much below 6.4 (right-hand column). A principal purpose of the table is to illustrate the marked stabilizing effect upon the pH of distal tubular contents, as the acid-base load varies, effected by the development of low disequilibrium pH during base loading.

A somewhat anomalous finding is that *in situ* pH values are reported to be about the same in the distal tubule of the acidotic rat as under normal circumstances (374, 375). How representative of the kidney as a whole these values are is uncertain, however, since only superficial distal tubules are accessible.

of the end urine is determined. The low distal disequilibrium pH also suppresses secretion of acid locally by increasing the transtubular electrochemical gradient against which acid must be secreted, and contributes in this way to the efficiency with which excess base can be eliminated at any plasma HCO_3^- concentration (II-8D1).

Facultative additional reabsorption of water from the distal tubules and cortical collecting ducts during antidiuresis can affect the luminal pH in these structures only to a relatively small degree, compared with collecting duct pH, because the *proportional* reduction of fluid volume is relatively small during antidiuresis. Conversely, water reabsorption from the medullary collecting ducts in marked antidiuresis can remove 90% of the fluid entering the ducts; and, though this process exerts little effect on the acid-base composition of urine of around pH 6.4 (Table II-7.3), it is associated with marked rise of the pH of HCO_3^--rich urine (left-hand column, Table II-7.3) and reduction of the pH of urine of high ammonium ion content (right-hand column). The findings are related to reabsorption of the respective congener gases down the pressure gradient created by water reabsorption (II-10C).

On the basis of observations obtained by microcatheters inserted into the atypical large collecting ducts of the hamster (619a), it is generally held that some acid is secreted into the medullary collecting ducts of the mammal (250a, 486, 487, 659, 570). Whether acid generation resulting from reabsorption of NH_3 has been ruled out as a cause of the observed acidification is far from clear, since there is brisk reabsorption of water from these structures even during water diuresis (620, 290), a process quite capable of causing reduction of the urinary pH wherever the (equilibrated) urinary pH is below 6.4. If the ducts are nevertheless assumed to secrete some acid, it would follow that acidification proceeds throughout the mammalian nephron except, in all probability, in a portion of Henle's loop (II-10). Rector (487, page 221) estimates, on the basis of micropuncture data in rats excreting acid urine, reported from several laboratories, that an average of about 85% of the filtered HCO_3^- is reabsorbed in the proximal convolutions and over 10% more before the fluid reaches the distal convolutions, i.e., in the pars recta and loop of Henle. Though the proportion of filtered HCO_3^- removed in the distal tubules is always relatively small, distal acidification is highly effective in lowering luminal pH when the rate of filtration of HCO_3^- is relatively low, because then the absolute rate of delivery of HCO_3^- to the distal segment is very low—say 45 μEq/minute in the archetypal 70-kg adult man (Table II-7.4)—, less, in any case, than the rate of acid secretion into the distal nephron. When the load is alkaline, however, HCO_3^- filtration is relatively high; much larger amounts of HCO_3^- reach the distal nephron, perhaps 250 μEq/minute, and the distal acidification process, during water diuresis, is more nearly balanced by water reabsorption in such a way that the (equilibrated) $HCO_3^-:CO_2$ ratio of the luminal fluid increases relatively little passing from distal convolutions to papilla (Table II-7.4). The pH of the high-flow end urine is therefore relatively close to blood pH.

Though the fraction of the filtered HCO_3^- delivered to the distal segment is always quite small whatever the acid-base load (perhaps 10% with a large base load, less than 5% with an acid load, according to Table II-7.4), the proportional variation of the absolute rate of delivery of HCO_3^- to the distal nephrons must be large in relation to changing acid-base load. According to the rough estimate given in Table II-7.4, HCO_3^- delivered to the distal tubules in man might decline from 250 to 45 μEq/minute as a base load gives way to an acid load. Acid secretion into the distal nephron, even

TABLE II-7.4. *Approximate topography of reabsorption of filtered HCO_3 from the nephrons in an average-sized adult during moderate antidiuresis**

	Base load	Acid load	Water filtration, 100 ml/minute
	HCO_3^- filtration, 2600 μEq/minute	HCO_3^- filtration, 2200 μEq/minute	
Reabsorbed in proximal tubule plus Henle's loop	2350 μEq/minute (90%)	2155 μEq/minute (98%)	85 ml/minute (85%)
Delivered to distal tubule	250 μEq/minute	45 μEq/minute	15 ml/minute
HCO_3^- reabsorption in distal tubule plus collecting ducts	175 μEq/minute	44.9 μEq/minute	14 ml/minute
Excreted in urine	75 μEq/minute (30% of HCO_3^- delivered to distal tubules)	0.1 μEq/minute (<0.3% of HCO_3^- delivered to distal tubules)	1 ml/minute (6% of water delivered to distal tubules)

* A moderately large base load is compared with an acid load of similar magnitude. Conventional values are assumed for glomerular filtration rate and for plasma HCO_3^-; proportional reabsorption of filtered HCO_3^- proximal to the distal tubules is based on figures obtained in micropuncture of kidneys of lower mammals, which exhibit considerable variation (II-8D1). Comparison of those figures with values for urinary [HCO_3^-] in man during moderate antidiuresis yields estimates of reabsorption in distal tubules and collecting ducts.

though the rate of secretion is reduced when acid urine is being elaborated (II-8D), not only clears the urine almost completely of HCO_3^- under these circumstances, but also lowers the urine pH further after practically all the HCO_3^- has been reabsorbed, thereby generating urinary titratable acid and causing more NH_3 to diffuse into the urine; this NH_3 serves as urinary buffer by accepting protons in the lumen to become NH_4^+. The rate of proximal acid secretion is unlikely to be independent of changing acid-base load, for in the proximal tubule, too, the transtubular electrochemical gradient against which acid is secreted must always increase at least slightly as HCO_3^- filtration declines. Still, and in spite of the stabilizing effect of the disequilibrium pH (Table II-7.3) on distal tubular pH, larger variations of the transtubular electrochemical gradient must occur here (II-8D) than in the proximal tubules of man as the acid-base load fluctuates.

F. RENAL HCO_3^- CLEARANCE

The renal clearance of any substance is defined as its rate of excretion divided by its plasma concentration. For some substances always present in extracellular fluid, like urea, the renal clearance is essentially independent of plasma concentration, which is to say that there is always some excretion, and that excretion is in general proportional to the plasma level. Other substances, like HCO_3^-, are "threshold" substances, to use a term introduced in 1877 by Claude Bernard (see 577, page 82): they are normally excreted in the urine in appreciable quantity only when their plasma concentration is raised to some minimum value. The renal clearance of such a substance generally varies directly, above threshold, with its plasma concentration and with the rate of its filtration. Renal HCO_3^- clearance was first carefully studied in relation to varying plasma [HCO_3^-] by Pitts and his colleagues (473, 472). In human volunteers subjected to a considerable prior acid load to reduce plasma [HCO_3^-], urinary HCO_3^- excretion was examined in relation to the rising plasma HCO_3^-, and hence to the increasing rate of glomerular HCO_3^- filtration, produced by gradually increasing the load of alkali by means of sodium bicarbonate infused intravenously at an increasing rate (472). In the course of such a study, commonly if deplorably[1] termed a "titration study," rather striking similarities are observed between the behavior of HCO_3^- and that of certain other threshold substances, archetypally glucose (561, 560, 577).

The glucose "threshold" is the (normally raised) plasma concentration of glucose at which glucose appears in the urine; the rate of glucose excretion in the urine then increases as the plasma concentration is further raised. Since glucose is not transported by the tubules into the urine, the fact that the sugar ultimately appears in the urine means that at least some of the nephrons are no longer able to reabsorb practically all the glucose filtered at the glomerulus. Normally, very little further rise of plasma glucose above threshold then suffices to saturate the total capacity of all the nephrons to reabsorb glucose; at this point the maximal tubular reabsorptive transport capacity (*Tm*) of the kidneys has been reached and all further increments of filtered glucose escape reabsorption. Urinary glucose excretion consequently increases with further increase in plasma glucose concentration, and, if GFR is unchanged, at the same rate as does glucose filtration (Figure II-7.2). The area of "splay" is normally confined to a small range of plasma glucose concentration and is probably due principally to some dishomogeneity of the nephrons.

The conventional HCO_3^- "titration curve" is phenomenologically very similar to that for glucose; in man it reflects a renal threshold (28 mEq/liter in subjects previously subjected to acid loading) and an apparent *Tm* (2.5 to 2.8 mEq/100 ml of glomerular filtrate); the value is similar in the dog (Figure II-7.3). These similarities between HCO_3^- and glucose in their renal handling are nevertheless notably superficial; conspicuous differences between the two processes are summarized in Table II-7.5. The load is not chemically specific in the case of HCO_3^-, as it is in the case of glucose, for any administered base will serve to raise plasma [HCO_3^-] and the rate of glomerular HCO_3^- filtration. Further, the tubular transport

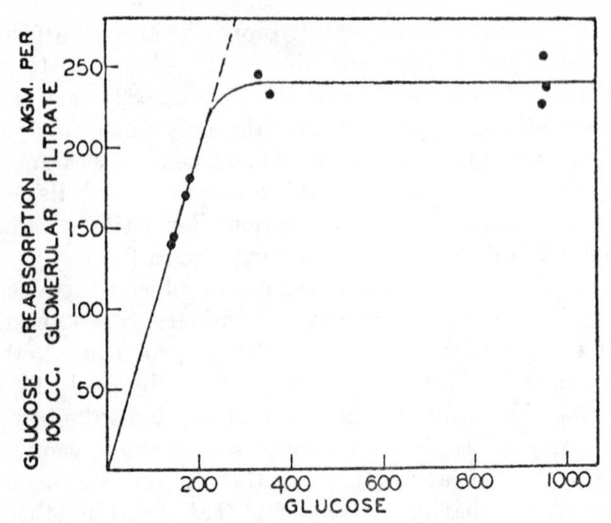

FIG. II-7.2. The maximal reabsorption (*Tm*) of filtered glucose. Data obtained in the dog. Reproduced from Shannon (560), with permission of the publisher. Glucose is not present in appreciable quantity in the urine of normal persons at normal plasma glucose concentrations; accordingly, glucose reabsorption is linearly related to artificially raised plasma [glucose] until some glucose begins to appear in the urine. Over a small area of "splay," some increased reabsorption accompanies increasing excretion of urinary glucose, but soon reabsorption increases a limiting maximal value; this is the *Tm*.

process—which is in this latter case acid secretion rather than true reabsorption of the loading substance, as in the case of glucose—does not, below threshold, proceed at a rate exactly equal to the rate of filtration of the threshold substance. This is because HCO_3^- does not appear in the urine until the urine pH has risen to about 6.1—the exact value specified is somewhat arbitrary since urinary [HCO_3^-] varies essentially exponentially with pH (Figure II-6.1)—; when the urine pH is 6.1 or less, an appreciable proportion of the secreted acid is being excreted as urinary titratable acid and NH_4^+. Whereas glucose is reabsorbed essentially completely from the proximal tubules, both proximal and distal nephron segments are sites of removal of filtered HCO_3^-.

Furthermore, the value of the HCO_3 *Tm* is generally, though perhaps not invariably (607), altered by alteration of the GFR; whereas it is generally though not universally held—the subject has recently been surveyed by Baines (25)—that the glucose *Tm* does not vary physiologically with GFR. In order for the HCO_3^- *Tm* of a person or experimental animal to exhibit a stable value, as GFR varies spontaneously or is caused to vary, *Tm* must therefore ordinarily be divided by GFR, the value being best expressed as HCO_3^- reabsorbed per unit volume of glomerular filtrate (473,

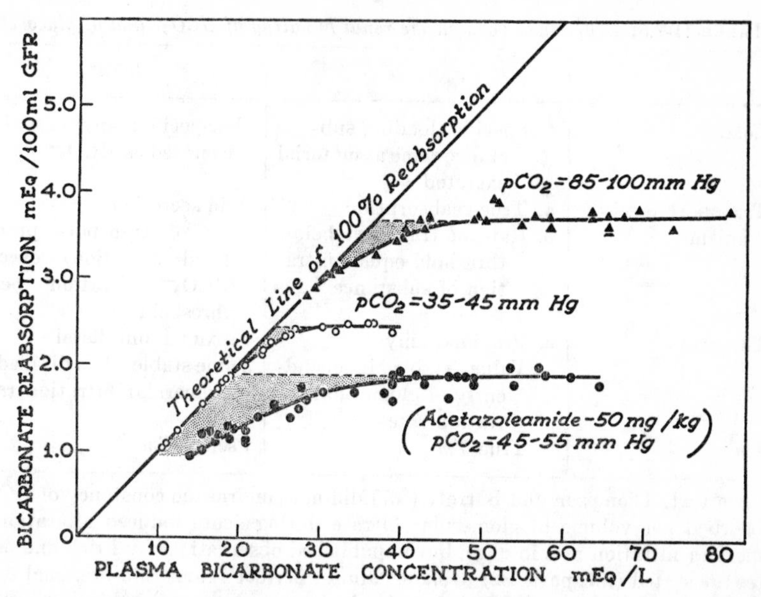

FIG. II-7-3. The apparent bicarbonate *Tm* in the dog and the effect upon it of the arterial P_{CO_2}. Reproduced from Rector, Seldin, Roberts, and Smith (490), with permission of the publisher. The response of the normal kidney to physiological base loads lies within the area of splay (*shaded areas*): as urinary HCO_3^- excretion begins to increase, reabsorption of filtered HCO_3^- from glomerular filtrate also increases, indicating increased secretion of acid into the nephrons.

The effect of variation of arterial P_{CO_2} upon the *Tm* reflects the influence of this variable upon the total rate of secretion of acid into the nephrons, which naturally influences urinary HCO_3^- excretion also at a value for plasma [HCO_3^-] below threshold (Part III).

472, 467). Since most of the filtered HCO_3^- is reabsorbed proximally, the clearance data reflect what has been confirmed by micropuncture studies, that the fractional reabsorption in the proximal convolutions of filtered HCO_3^- remains approximately constant. Constant percentile proximal reabsorption of any filtered material is termed "glomerulotubular balance" and is observed not only in respect to $NaHCO_3$ but also to NaCl and (consequently) to water, as GFR is caused to vary in a variety of ways (642). We may accordingly suppose the dependence of the HCO_3^- *Tm* upon GFR to be mediated, via the functional link between net Na^+ reabsorption and acid secretion previously discussed (Section D), by the mechanism that preserves the glomerulotubular balance of Na^+.[2]

Finally, the apparent HCO_3^- *Tm* observable in experiments such as those of Pitts and associates turns out to be in fact a pseudo-*Tm*, for absolute reclamation of filtered HCO_3^- does increase considerably as plasma [HCO_3^-] rises above the HCO_3^- threshold (Table II-8.2; II-8D). Stabiliza-

TABLE II-7.5. *Differences between the renal handling of HCO_3^- and of glucose**

	Glucose	HCO_3^-
1. Load	Specific: loading substance same as material excreted	Nonspecific: any base load excreted as HCO_3^-
2. Transport mechanism	a. True reabsorption b. Rate of transport below threshold equals filtration of substance	Acid secretion Rate of transport process (acid secretion) exceeds HCO_3^- filtration below threshold
3. Threshold	c. Proximal only Value stable independently of glomerular filtration rate	Proximal and distal Value stable when divided by glomerular filtration rate
4. *Tm*	True *Tm*	Pseudo-*Tm*

* See text. Thompson and Barrett (607) did not confirm the constancy of HCO_3^- reabsorbed per volume of glomerular filtrate during acute induced alteration of glomerular filtration rate in dogs, but unpublished observations by Frick and associates are stated (487, page 232) to show "almost perfect" glomerulo-proximal tubular balance with respect to HCO_3^- in animals (species unstated) subjected acutely during saline diuresis to acute aortic constriction above the renal arteries.

It must be added that expansion of the extracellular fluid volume of rats has recently been reported to result in a substantial reduction of glucose *Tm* (25).

tion of the absolute rate of HCO_3^- reabsorption, as plasma HCO_3^- is raised in such experiments, ensues eventually because both proximal tubular Na^+ reabsorption and acid secretion in the proximal tubules are inhibited as a result of an expansion of effective extracellular fluid volume (324, 483, 325). This expansion appears to be inevitable when $NaHCO_3$ is administered at a sufficiently rapid rate to complete the observations necessary to define *Tm*, unless countermeasures, such as bloodletting, are simultaneously undertaken; when they are, additional HCO_3^- is reabsorbed (487, 483, 325). Acid secretion is inhibited, quite inappropriately, also when extracellular volume is expanded by means of infused NaCl (473, 264, 487). The mechanisms by which expansion of the volume of extracellular fluid (131), or of vascular volumes at critical locations, inhibits Na^+ reabsorption and H^+ secretion, while volume contraction can stimulate both, are incompletely understood and continue to be intensively investigated. Saline-induced natriuresis is associated with decreased fractional reabsorption of filtered Na not wholly explicable in terms of hemodynamic factors and aldosterone secretion, and is believed to be effected by an elusive natriuretic hormone (487, 642, 656a). Sustained alteration of Na balance is said (656a), however, not to affect fractional Na reabsorption in the proximal tubule.

The *increased* absolute reabsorption of HCO_3^- occurring when urinary HCO_3^- excretion rises in response to an increasing *physiological* base load is the result of *increased* renal acid secretion. It is formally analogous to the area of "splay" in the glucose titration curve, but it has nothing to do with dishomogeneity of the nephron population; instead, it signifies that additional acid is being secreted into the nephrons (II-8D1). In spite of the increased acid secretion, renal HCO_3^- clearance varies directly with physiologically increasing base loads and increasing base excretion. From the homeostatic standpoint, the significance of the renal HCO_3^- clearance is that in the presence of a base load it is, to a near approximation, a measure of the efficiency with which the kidney protects the organism against accumulation of excess base (498). Base is eliminated largely as urinary HCO_3^-, and the higher the renal HCO_3^- clearance, the greater the increment of base which is being excreted as urinary HCO_3^- for any given increase in plasma HCO_3^-, i.e., for any given degree of alkalinization of extracellular fluid. The renal HCO_3^- clearance offers only an approximate, not an exact, evaluation of the homeostatic efficiency of the kidney in maintaining base balance for two reasons: (i) a small amount of base is excreted as "titratable alkalinity" (II-4) when the net base load is large, and (ii) urinary P_{CO_2} exceeds P_{CO_2} of glomerular filtrate when there is net renal base excretion. That portion of the HCO_3^- clearance which is attributable to the purely physicochemical effect of raised urinary P_{CO_2} on urinary [HCO_3^-] (II-4C) can accordingly not be counted toward net renal base excretion (498).

NOTES

1. Titration is an analytical process; confusion becomes likely when the term is applied to determination of the maximal rate of a cellular transport process. Can the notion that glucose is physiologically an acid, which crops up now and again in the clinical literature (Appendix G), be blamed in some part upon the expression "glucose titration"? The extension of this usage to the "HCO_3^- titration curve" compounds confusion, because what the "titration curve" displays is not at all the quantity of acid required to reduce the pH of a given volume of glomerular filtrate by some stated amount, but the rate of a physiological transport process.

It is nevertheless probably too much to hope for that the expression "titration curve," used in this now conventional sense in renal physiology, as well as the related "titration study," should be discarded, in view of a generation of entrenched usage and the eminent auspices under which it was introduced. The terms seem to have originated with Homer Smith, who wrote of the renal handling of glucose that "We may speak of . . . a *titration curve,* in the sense that we have titrated the tubules to saturation by progressive elevation of the load" (577, page 88).

2. Since neither the mechanism of glomerulotubular balance nor that of the coupling between transported acid and sodium is very clear, this inference may be thought to advance our knowledge about as much as the explanation given by Molière's honors medical student (408) for the hypnotic action of opium:

Quia est in eo
Virtus dormitiva

eight

ACIDIFICATION OF THE URINE (II)

A. TRANSTUBULAR ELECTROCHEMICAL POTENTIALS

A particle may (but does not necessarily) move passively across a membrane if there is a difference in the *activity* of the substance (I-4B) on the two sides of the membrane. Also, less energy will be required (other things being equal) for the active transport of a substance when a difference in its activity across the membrane favors passive movement in the direction of of active transport, and vice versa. In the case of uncharged particles dissolved in extracellular fluid and urine, activities are essentially identical to concentration (I-4) except in the case of gases, where activities are more exactly related to the partial pressures; movements of ions, however, are affected as well by any net electrical field existing in the area of transport.

Whether renal secretion of acid results from movement of protons into the lumen of the nephron or from movement of OH^- out of it (II-7D), the electronegativity of the luminal fluid relative to the peritubular blood must favor the passive development of luminal acidity relative to blood. Whether observed net acid secretion into the luminal fluid of the nephrons *could* be a passive process—one requiring no energy expenditure beyond that supplied by the events responsible for establishing the existing net transtubular potentials—can be decided with the aid of the Nernst equation:

$$E = 61 \times \text{pH difference} \tag{1}$$

where E is the transtubular potential difference in millivolts of luminal electronegativity relative to peritubular blood, and the transtubular pH difference is reckoned as blood less luminal fluid pH (487). This equation exhibits the thermodynamic relation between transtubular pH gradient and electrical field at equilibrium by stating the voltage necessary to maintain any observed transtubular pH difference. As one example, no work would have to be done, given 30.5 mv transtubular potential difference with lumen negative, to maintain luminal contents at a pH of 6.9 when blood is pH 7.4 (pH gradient 0.5 unit); as another, given 3.0 mv transtubular potential difference, electrochemical equilibrium is present when luminal pH is 0.05 units below blood pH. In order for H_3O^+ to diffuse passively into the lumen (or for OH^- to diffuse out), given such equilibrium conditions, either the transtubular pH gradient would have to be decreased, or luminal electronegativity relative to the peritubular blood would have to be increased; if these conditions are met, passive transport is thermodynamically possible. Unfortunately, in order to know whether transport at an observed rate is likely to be passive, much more information is needed, particularly about the resistance offered by the membrane to passage of the ions, and the mathematical treatment required becomes much more elaborate (487, 68).

The nature and location of the ion-transporting mechanisms of the renal tubular cell have been under extensive investigation, for upwards of 15 years now, by micropuncture methods, including transtubular potentiometry (194, 659, 195, 200, 198, 375). For some years it was believed that substantial electrical potential differences existed across the cells of the mammalian proximal tubule (194, 464). Frömter and Hegel, however, have provided convincing evidence that these reported potential differences were artifacts resulting mainly from inadvertent placement of electrode tips inside damaged proximal tubular cells rather than in the lumen (178). When care is taken to ensure that the electrode is passed through the cell and brush border into the lumen, the proximal transtubular potential differences are very low, about 0 to −2 mv (178, 435, 375). Such differences are inadequate to account for the rates of acid secretion against transtubular pH gradients of 0.4 and 0.8 observed when catalysis of luminal H_2CO_3 formation by carbonic anhydrase is pharmacologically inhibited (487). Acidification of the glomerular filtrate in the proximal tubules must therefore be an active process supported by expenditure of energy, presumably largely derived from oxidations within the proximal tubular cells. Disclosure of consistent gross errors attaching to transtubular voltage measurements upon which rather sweeping inferences had rested for some years has prompted some investigators experienced in micropuncture technics (331) to express disquiet at the level of critical standards which have prevailed in the field.

In the distal tubule, substantial transtubular electrical potential differences are reported (487), averaging 55 mv in rats subjected to large alkali loads; such voltages are large enough to generate and maintain the transtubular pH gradients observed under these circumstances. Rector (487) entertains the possibility that the existing transtubular voltage might account for the passive acidification of distal tubular fluid at the rates observed; Malnic, de Mello Aires, and Giebisch (373), however, consider it established that distal as well as proximal acidification is active.

B. THE ROLE OF INTRACELLULAR CO_2

Among influences markedly affecting the rate of acidification of glomerular filtrate, none is more striking than *alterations of arterial* P_{CO_2} *and administration of carbonic anhydrase or inhibitors of this enzyme*. All of these phenomena have directed attention to the importance of the CO_2 buffer system of renal tubular cell water for the normal operation of the renal cellular transport mechanism responsible for secretion of acid into the nephrons.

Pitts and Alexander (471) demonstrated in 1945 that, when dogs are excreting large acid loads in urine containing very large amounts of phosphate buffer, the rate of acidification exeeds the "quantity of acid filtered through the glomeruli," i.e., CO_2 and NaH_2PO_4. They therefore postulated that the urine is acidified by secretion of protons in exchange for Na^+ reabsorbed. These authors also extended their interpretation of their findings to intracellular events by including among their conclusions the statement that "... the cellular source of H^+ ions is carbonic acid formed by the hydration of carbon dioxide produced metabolically within the renal cells and derived by them from the postglomerular blood." This formulation has been endorsed repeatedly by subsequent writers on acidification of the urine. Rector, for example, is entirely orthodox in writing (487, page 224): "The quantity of hydrogen involved in bicarbonate reabsorption far exceeds the supply of hydrogen arising from metabolism. The only means capable of supplying this demand is the diffusion of plasma CO_2 into the cell."

Statements in this vein tend to obscure the inescapable facts (i) that, whenever acid is being excreted rapidly in bicarbonate-free urine, metabolically produced CO_2 cannot normally supply the organism as a whole with *any* protons, via H_2CO_3, for transfer to the urinary buffers; (ii) that there is no sense in which CO_2 can ever supply protons unless it is first hydrated to H_2CO_3, in which case it is the H of water which really supplies them; and (iii) that cell water can provide a virtually limitless source of protons directly (i.e., not necessarily via H_2CO_3) for a cell which secretes acid in one direction and base in another. We need consider in the present context only the situation where a sizable physiological load of nonvolatile acid is being excreted in essentially bicarbonate-free urine in the steady state; the relation of renal acid secretion to renal maintenance of base balance as well as acid balance is discussed in detail elsewhere (II-2B; II-4B; II-5). Renal excretion of nonvolatile acid, which depends entirely upon secretion of acid into the nephrons, signifies regeneration of base in the blood returned by the kidney to the systemic circulation; that is the homeostatic significance of the renal net acid excretion (II-2B; II-6), and we can make no distinction, in terms of the Brønsted-Lowry theory, between renal tubular removal of acid from the peritubular blood and generation of base in it.

The renal tubular cell is accordingly not required to produce, or take in, CO_2 or any nonvolatile acidic *compound* to serve as the ultimate source of secreted protons—as we might, if we chose, say that skeletal muscle "secretes" lactic acid into the blood. Any active transport process which effects a spatial separation of the H^+ and the OH^- of the water molecule within the cell can be made the basis of net proton transport from the peritubular blood into the urine in exchange for Na^+ transported from lumen to peritubular blood. No "source" of either the acid secreted into the lumen or the base secreted simultaneously into the peritubular blood is needed, other than the abundant potential source of both constituted by the cell water. These considerations, originally invoked by Davies (121) in relation to acid secretion by the stomach, are perfectly relevant (*mutatis mutandis*) to acid secretion into the nephrons, as Berliner (52) evidently recognizes. The large influence of variation of arterial P_{CO_2} upon the rate of renal acid secretion establishes the great importance of CO_2 for the ionic transport system supporting renal tubular acid secretion, but does not in the least oblige us to assume that H_2CO_3 is the only compound which can be made use of to supply protons for the transport system through expenditure of metabolic energy. Moreover, the effects (Section D; also Figures II-8.3 and II-8.4) of administered carbonic anhydrase upon the renal HCO_3^- clearance are the result of catalysis exerted on the luminal, not the intracellular, bicarbonate buffer system (428, 498; Section D); and it now appears (487, page 228) that the same may very plausibly be said of the effects of inhibitors of carbonic anhydrase given to the intact animal (see also below).

Since presumably no one wishes to contest that renal acid secretion has a homeostatic function, we may take it that discussions offered in the current micropuncture literature of the source of the protons transferred (in effect) across the tubular cell to the luminal fluid are inspired by the desire to define a particular molecular process (or processes) within the cell as the primary event(s) from which, by a chain of linked physicochemical reactions, the overall transport of protons and sodium ions in opposite directions across the cell is brought about. The question being raised, in other words, lies in the realm of renal molecular biology; however, the source, in this sense, of the acid secreted into the lumen appears in the context of contemporary renal physiology to be less an investigative problem than a semantic puzzle. The literature discloses, in rather more detail that we can here be concerned with, that a very large number of models can be devised, according to each of which some particular ion transfer is assumed to be the primary event, or a primary event; because of the linked character of ionic equilibria, many consequences can then be deduced. Thus we can posit that the primary event is active extrusion of protons into the lumen in exchange for sodium ions taken in; the Na^+ can then move passively across the cell and be extruded in exchange for H^+ taken into the cell. Or again the latter process, Na^+-H^+ exchange at the peritubular border, can be made the primary active process; or, if we like, it may become a second active process. Alternatively, OH^- can be actively brought into the cell from the luminal fluid together with Na^+, acid being thereby generated in luminal fluid; the incoming OH^- can be "neutralized" within the cell by protons derived ultimately from the peritubular blood in exchange for Na^+. The same overall exchange of Na^+ for H^+ across the cells results in all cases. H_2CO_3 might function as an intermediary, donating protons and then recovering them from the peritubular blood, anywhere along the transtubular proton transport chain.

The possible proposals are no doubt finite in number, but it is idle to confound model construction of this sort with scientific activity. Such speculations do not become serious hypotheses except insofar as a genuine possibility of falsification by experiment exists. Because the participation of CO_2 in uncatalyzed acid-base reactions results in a relatively very slow overall reaction, it was possible for Rector and his colleagues (Section C) to falsify the hypothesis that the principal primary event underlying urine acidification is active reabsorption of luminal HCO_3^-; but there is no indication that any of a variety of purely ionic transfers can be excluded as a primary event with the aid of any of the methods now being deployed by renal physiologists, specifically by estimations of transepithelial pH and electrochemical gradients and related micropuncture and microperfusion data. At most, these methods might permit the inference that renal acid secretion depends upon more than one primary active cellular transport process.

Further molecular biological inquiry into the source of protons secreted into the nephrons is not theoretically meaningless, since it is in theory possible that temporal primacy at one or another locus in the chain of linked acid-base reactions within and at the borders of the cell might be demonstrable as the rate of renal acid secretion is experimentally altered. These ionic reactions are not instantaneous; but the temporal sequence of secretory events—whether, for example, H^+ is actively extruded into the lumen, or rather provided on the cellular side of the secretory membrane to react with OH^- actively transported into the cell—could be disclosed only by application of physicochemical technics as sophisticated as those used in ascertaining the velocity constants of unidirectional ionization reactions (I-1C). If, and only if, such methods are to be applied can it be productive to discuss contemporary experimental data as if they were contributing to identification of a primary intracellular proton-transport system responsible for exchange of luminal Na^+ for peritubular protons.

Recently, what appears to be a pseudo-problem has been superimposed upon molecular biological discussions of the source of acid secreted by the renal tubules, a problem arising out of the contention that hydration of CO_2 cannot proceed rapidly enough, within the renal cells of animals treated with full doses of inhibitors of carbonic anhydrase, to supply protons for acidification of the urine at rates at which their urine is in fact acidified. This premise, itself questionable (see below), is now held to be at variance with the supposedly established doctrine that CO_2 is the only possible source of the secreted acid. Malnic and Giebisch (for example) have recently written (375) that "... the uncatalyzed velocity of CO_2 hydration ... is much too low to supply hydrogen ions at a rate adequate to account for the observed magnitude of bicarbonate reabsorption after carbonic anhydrase inhibition. A mechanism must then exist to account for bicarbonate reabsorption via hydrogen ion secretion in the absence of catalytic acceleration of the hydration reaction of CO_2." These authors then proceed to specify means by which the protons might be supplied from the outside to cell water: diffusion of H_2CO_3 from lumen to cell water, as suggested by Rector, or perhaps entry of protons into the cell at its antiluminal border by cationic exchange.

From what has just been said, it is clear that this is not a genuine dilemma; peritubular exchange of protons for Na^+ was always at least as likely as transport of the relatively slow-moving HCO_3^- into the blood with $Na.^+$ Furthermore, the additional assumption which gave rise to this pseudo-problem is itself of doubtful validity. This assumption is the thesis of Maren (379) that administration of acetazolamide or related drugs brings about virtually total inactivation of carbonic anhydrase activity within the renal tubular cells, and that only 10 to 20% of the observed rates of disappearance of HCO_3^- from glomerular filtrate could be accounted for on the basis of the uncatalyzed hydration of CO_2. Theoretical extrapolations from the kinetic behavior of simplified enzyme systems as observed *in vitro* to the actual functioning of enzymes within the highly nonhomogeneous intracellular milieu are inherently shaky, and Maren's thesis has not achieved general acceptance. Rector (487), for example, believes that structural orientation of the enzyme within the cell may account for some continued catalysis by carbonic anhydrase within the renal cells of animals given acetazolamide. In addition, our knowledge of the ionic milieu in which cellular carbonic anhydrase functions is unsatisfactory, a particularly relevant point since ions also catalyze the hydration of CO_2.

In summary, we do not know why urine acidification is accelerated as arterial and cellular P_{CO_2} rise. It is conceivable that steep pH gradients arising in small volumes of fluid within the cell, perhaps through active splitting of water, inhibit the transport of protons, and that transport velocity is sustained by effective buffering which is in turn dependent upon rapid transport of CO_2 by diffusion to the site of splitting. This mechanism becomes more plausible if varying speed of the fundamental processes is postulated. Alternatively, the concentration of CO_2 gas may directly influence the velocity of some step in the transport chain as cofactor or source of H_2CO_3. The importance of cellular CO_2 for renal acid secretion is evident, but we know too little of the molecular biology of the process to specify the manner in which CO_2 participates in it.

C. ROLE OF CARBONIC ANHYDRASE

Carbonic anhydrase is an intracellular enzyme which is not present in normal urine. The first physiological evidence that the luminal fluid in the distal nephron is not subject to catalysis by this enzyme came from observations of the CO_2 pressure of urine. Urinary CO_2 pressures greatly exceeding arterial values attracted attention upwards of a century ago (II-10), but the first important step toward eliciting the significance of this finding was taken by Pitts and Lotspeich (473), who recognized that it is only in higher-pH urine that raised CO_2 pressures are found; evidently the elevated pressure is related to a substantial HCO_3^- content of the luminal fluid of terminal nephron segments. These authors proposed that secretion of acid, presumably into the distal tubules, allowed progressive rise of luminal CO_2 or H_2CO_3 concentration to occur as a result of delay in the overall process of dehydration of H_2CO_3 and reabsorption of CO_2. They considered that a consequence of this rise of $CO_2 + H_2CO_3$ concentration in the luminal fluid of the distal nephron was the raised P_{CO_2} of alkaline bladder urines: acid urines were unaffected because too little filtered HCO_3^- reached the CO_2-impermeable terminal nephron segment to produce appreciable elevation of luminal $[CO_2]$.

Ryberg (534) drew attention to the relevance of the kinetic studies by Faurholt (158) of the uncatalyzed dehydration of H_2CO_3, and reckoned roughly that the half-time for dehydration of H_2CO_3 in urine at 37° C would be a matter of seconds. Ryberg considered it unlikely that a substantial CO_2 pressure gradient could develop anywhere across the nephron, and accordingly proposed that H_2CO_3 synthesized in the distal tubules by reaction of luminal HCO_3^- with secreted H^+ accumulated because the dehydration of H_2CO_3 was slower than the formation of H_2CO_3.

$$H^+ + HCO_3^- \rightarrow H_2CO_3 \xrightarrow{\overset{\text{Blockade}}{\downarrow}} CO_2 + H_2O \qquad (2)$$

Decomposition of H_2CO_3 at sites distal to its synthesis would continue to generate more CO_2 as Reaction Chain 2 proceeded. Ryberg did not attempt to specify how far distally the generation of excess H_2CO_3 and its continuing dehydration continued; it now appears that delayed decomposition of the H_2CO_3 synthesized from secretion of acid into bicarbonate-rich urine continues to generate additional CO_2 throughout the collecting ducts and beyond the papilla in the urinary tract as well (II-10A). The presence of ample nonvolatile buffer in the urine is essential to the development of high urinary [CO_2]. As the above reaction chain proceeds toward equilibrium, driven by the raised [H_2CO_3], the rate of generation of additional H_2CO_3 and CO_2 from HCO_3^-, for any given rise of pH, increases with increasing amounts of buffer (I-6C); in the absence of any urinary nonvolatile buffer, the H^+ shown at the left of Reaction Chain 3 can be supplied only by H_2O, and so hardly any H_2CO_3 (and CO_2) can be generated before rapid rise of pH halts the reaction.

The postulation that delayed dehydration of H_2CO_3 newly synthesized in the nephrons accounts for the high P_{CO_2} of alkaline urine was given strong experimental support through the demonstration by Ochwadt and Pitts (428) that intravenous infusion of carbonic anhydrase, at a rate sufficient to perfuse the nephron lumen with the enzyme, markedly lowered the raised P_{CO_2} of alkaline urines. Direct proof of accumulation of H_2CO_3 above equilibrium in the distal convolutions and collecting ducts was subsequently furnished by the micropuncture studies to be considered shortly. The entire nephron, including the collecting duct, appears to be so permeable to CO_2 (and NH_3) that the gas pressures within the lumen remain equilibrated with the gas pressures in the circumjacent tissues; delayed dehydration of H_2CO_3 synthesized in the cortex raises the P_{CO_2} of higher-pH end urine in two ways: (i) by releasing additional CO_2 in the medullary collecting ducts, where it serves to raise medullary tissue P_{CO_2}; and (ii) by liberating CO_2 beyond the papilla into the urine in the urinary tract (II-10A). At the low urine flow rates most frequently found in the mammal, water reabsorption from the collecting ducts causes still greater rise of urinary P_{CO_2} (II-10A).

Observations of the effects of administered carbonic anhydrase, or of inhibitors of the enzyme, have continued to provide a fruitful approach to the study of the mechanism of urine acidification, especially in conjunction with micropuncture technics. The most important investigation of the mechanism of urine acidification was published in 1965 by Rector, Carter, and Seldin (489), who examined the acid-base composition *in situ* of luminal fluid within the nephrons of rats caused by intravenous bicarbonate infusion to secrete alkaline urine. This investigation made clear that carbonic anhydrase, its absence from glomerular filtrate and urine notwithstanding, does catalyze the decomposition of the H_2CO_3 synthesized by acidification of filtrate HCO_3^- in the proximal tubule. The following observations were reported.

1. The pH, measured *in situ* by a glass microelectrode placed in the luminal fluid of the distal convolutions, was lower by an average of 0.7 than the pH of the same fluid aspirated and equilibrated with CO_2 gas at a pressure of 40 mm Hg. Since the P_{CO_2} is certainly close to 40 mm Hg every-

where throughout the renal cortex, the finding indicated that luminal $[HCO_3^-]$ was lowered or luminal $[H_2CO_3]$ high. When carbonic anhydrase was infused intravenously at a rate sufficient to perfuse the nephrons, the low distal disequilibrium pH was no longer demonstrable. This could only mean that generation of additional HCO_3^- in the normal luminal fluid during its withdrawal had been driven by continuing dehydration of H_2CO_3; the administered enzyme would prevent the normal disequilibrium *in situ* between CO_2 and H_2CO_3.

2. No disequilibrium pH was demonstrable by the same technic in the proximal convolutions. But when the animals were given an inhibitor of carbonic anhydrase, a disequilibrium pH averaging 0.4 was observed there, indicating that under normal circumstances it is effective catalysis by the enzyme of the dehydration reaction in the luminal fluid which prevents development of a disequilibrium pH.

These observations have been interpreted to mean that urinary acidification in proximal and distal tubules is effected at least in large part by transport of hydrogen ion from cell water to glomerular filtrate. In fact, they are not in any way incompatible with acidification of luminal contents by reabsorption of OH^- with Na^+ (II-7D). What they do show is that reabsorption of HCO_3^- with Na^+ cannot play the central role in urinary acidification in the nephron segments studied. HCO_3^- reabsorption with Na^+ reabsorption would necessitate conversion of luminal H_2CO_3 to HCO_3^-, and the H_2CO_3 disappearing would have to be replaced through hydration of luminal CO_2; kinetic blockade at this step would lead not to a lowered, but to a raised disequilibrium luminal pH, because the ratio $HCO_3^-:H_2CO_3$ would be raised above its equilibrium value:

$$CO_2 + H_2O \xrightarrow{\overset{\text{Blockade}}{\downarrow}} H_2CO_3 \rightarrow \underset{\substack{\downarrow \\ \text{Reabsorbed}}}{HCO_3^-} + H^+ \tag{3}$$

It must be remembered that a process of acidification is represented here (indicated by the H^+ at the right) in which the ratio $HCO_3^-:CO_2$ is being lowered (assuming the alkalinizing effect of water reabsorption to be overbalanced by acidification). However, pH would not be as low, while the reaction is proceeding and $[H_2CO_3]$ is below its equilibrium value, as the equilibrated pH which will be attained only after hydration of additional CO_2. This reaction, occurring after completion of HCO_3^- reabsorption, would raise the ratio $H_2CO_3:HCO_3^-$; some of this H_2CO_3 will transfer protons to be nonvolatile buffers, lowering the pH and also leaving behind more of the conjugate base HCO_3^-.

These observations by Rector and his colleagues have been confirmed qualitatively, with an antimony electrode, by Malnic and his colleagues (634, 374); and Uhlich, Baldamus, and Ullrich (618) have presented convincing evidence that a low disequilibrium pH also is present in the medullary collecting ducts of rats during bicarbonate diuresis. No data are available on Henle's loop. In respiratory and nonrespiratory acidosis (Part III) a disequilibrium pH may be absent in the distal tubules and in respiratory acidosis there may be a low disequilibrium pH in proximal tubular fluid (374).

It seems improbable that carbonic anhydrase would be present in solution in the luminal fluid of the proximal tubule, only to be removed (or inactivated) before the filtrate reaches the distal nephron. More probably, the brush border of the proximal tubule, by constituting a seive-like reticulum through which the entire volume of glomerular filtrate passes in close proximity to the cell membrane, ensures adequate contact of the filtrate H_2CO_3 with molecules of the enzyme anchored in the fimbriated cell border. Biochemical (477, 379, 380) and histochemical studies (233) alike show that carbonic anhydrase is densely concentrated in the peritubular membrane of the proximal tubule, and there is evidence (233) that enzyme in this location is capable of exerting its catalytic activity in the adjacent luminal fluid. Whether the luminal fluid within the loop of Henle is subject to catalysis by carbonic anhydrase remains uncertain at present.

It is also not established that HCO_3^- is removed from glomerular filtrate exclusively by conversion to H_2CO_3. From the velocity constant of the uncatalyzed dehydration reaction at 37° C, Walser and Mudge (643) estimated that the steady-state rate of proximal HCO_3^- reabsorption observed in rats during enzyme inhibition could be accounted for entirely in terms of prior conversion of HCO_3^- to H_2CO_3 only at markedly elevated values for luminal $[H_2CO_3]$ corresponding to a reduction of 1.3 pH unit *in situ* below equilibrium values. This degree of disequilibrium is significantly greater than the depression of 0.85 pH unit below equilibrium observed in proximal tubular fluid *in situ* by Rector *et al.* and much greater than the depression of 0.4 pH unit reported by Vieira and Malnic. These considerations have been taken to suggest that a considerable portion of the filtered HCO_3^- might be reabsorbed by some mechanism other than neutralization of HCO_3^-. The validity of such a calculation depends, however, upon the unproved assumption that no H_2CO_3 diffuses out of the lumen when $[H_2CO_3]$ is elevated. It is certainly still possible that substantial contribution to the acidification of glomerular filtrate in the proximal tubules is made by reabsorption of HCO_3^- with Na^+, but there is no adequate evidence to show that this is so.

D. REGULATION OF THE RATE OF ACID SECRETION

1. Normal Circumstances

Table II-8.1 lists factors known to be capable of influencing markedly the rate of acid secretion by the nephron as a whole. Those listed in Group II are of interest primarily in relation to pathophysiology, including states

Table II-8.1. *Factors influencing the rate of acid secretion into the nephrons*

I. Everyday variables
  a. Glomerular filtration rate
  b. Acid-base load (via plasma [HCO_3^-])
II. Abnormal circumstances
  a. Altered transtubular electrochemical gradient
  b. Altered effective extracellular fluid volume
  c. Renal disease
  d. Altered arterial P_{CO_2}
  e. Administered carbonic anhydrase
  f. Administered inhibitors of carbonic anhydrase
  g. K^+ depletion and K^+ excess
  h. Parathyroid hormone excess
  i. Mineralocorticoid excess or deficit

experimentally induced by administration of pharmacological agents. The rate of acid secretion into the nephrons, apart from glomerulotubular balance (II-7F) which adjusts it to normal circadian fluctuations of the glomerular filtration rate (651), appears to be influenced under normal circumstances principally by the acid-base load. Paradoxically, more acid is being secreted into the nephrons as a whole when alkaline urine is being elaborated than when the urine is markedly acid (51, page 69; 553, page 1234; 259; 255; 52, page 73). The paradox is explained, of course, by the fact that secretion of acid urine is normally always associated with relatively low plasma concentrations of HCO_3^- and hence of glomerular HCO_3^- filtration (II-6), so that less acid need be secreted into the nephrons to remove all the filtered HCO_3^- under circumstances when markedly acid urine is being secreted.

The reduction of filtered HCO_3^- associated with a sizable acid load is ample to cause urinary pH to decline, even though the rate of acid secretion into the nephrons is simultaneously conspicuously reduced, as shown by Table II-8.2. Small fluctuations of plasma [HCO_3^-], well within the normal range, about the mean normal value in response to changing acid-base load result in fluctuations of filtered HCO_3^- quite beyond the scale of the changes of urinary excretion of acid or base actually seen as the load varies physiologically. How does the varying acid-base load bring about this homeostatically inappropriate variation (cf. Figure II-6.3) of the rate of acid secretion into the nephrons?

A major determinant of the variation of renal acid secretion under these circumstances is clearly the influence automatically exerted (II-6C2) by the varying rate of HCO_3^- filtration upon the pH of luminal fluid throughout the nephron, and hence upon the transtubular electrochemical gradient, especially in the distal tubules. The pH of the luminal contents certainly varies considerably more, with varying acid-base load, in the distal than in

TABLE II-8.2. *Effect of fluctuation of plasma [HCO_3^-] upon glomerular filtration of HCO_3^-**

Plasma [HCO_3^-]	Filtered HCO_3^-	Δ
mEq/liter	*mEq/24 hours*	
26	3900	+300
24	3600	
22	3300	−300

* In the middle row, plasma [HCO_3^-] of 24 mEq/liter is taken as a conventional midpoint, appropriate to zero acid-base load (II-6); the glomerular filtration rate is assumed to be 150 liters/24 hours throughout. Deviations of plasma [HCO_3^-] listed above and below this midpoint are, respectively, appropriate to a modest base load sufficient to raise plasma [HCO_3^-] 2 mEq/liter (top row) and, for a moderate acid load, sufficient to depress it by 2 mEq/liter (bottom row). Should such changes in HCO_3^- filtration result in no change in the rate of acid secretion into the nephrons, they would effect the enormous increments of 300 mEq/24 hours, respectively, of renal net excretion of base or acid. In fact, the effect on renal acid-base excretion is much less, since acid secretion into the nephrons is increased when base is being excreted by the kidney.

the proximal tubules (Tables II-7.3 and II-7.4); and, though there is no doubt (Section C) that proximal acid secretion will be markedly depressed if filtrate pH is substantially lowered by administration of inhibitors of carbonic anhydrase, it is uncertain whether normal variation of the acid-base load affects proximal luminal pH sufficiently for it to exert much influence normally upon the rate of proximal acid secretion. By contrast, a substantial influence is clearly exerted by varying acid-base load upon distal acid secretion via the effect of varying HCO_3^- filtration upon the distal transtubular pH gradient.

Gross evidence of the importance of this gradient upon the overall rate of acid secretion into the nephrons is provided by the rather inflexible lower limit to urine pH, conventionally 4.4 or 4.5 under ordinary circumstances (II-6A). The physiological lower limit of urinary pH certainly reflects the development of a distal transtubular electrochemical gradient against which further net transfer of protons into the lumen is no longer possible, because measures which increase luminal electronegativity permit secretion of urine of lower pH than is possible under normal circumstances (549, 31). Moreover, acid secretion into the nephrons can be increased more or less indefinitely, after urine pH has been reduced to its physiological minimum, by provision of additional urinary buffer (539, 467); evidently, then, acidification has not been arrested by any intrinsic overall limitation upon the velocity at which acid can be secreted, but rather by the steepness of the electrochemical gradient reached toward the terminus of the distal nephron when the urine is maximally acidic (487, 489, 486).

However, it is in the higher ranges of urinary pH that the inhibitory effect of an increasing transtubular pH gradient upon acid secretion is most striking. Micropuncture observations (487, 374, 375) regularly show that the rate of acid secretion into the distal nephron segments varies directly with the HCO_3^- concentration of the luminal fluid; transtubular voltage does not change, and so the transtubular electrochemical gradient varies under these circumstances directly with the transtubular pH gradient and inversely with the luminal [HCO_3^-] (373, 374). To be sure, considerable wariness is in order before applying micropuncture data on this subject directly to the physiology of acid-base regulation by the kidney of the intact animal, for data on the pH of distal tubular fluid obtained in the same species in different laboratories are not in very good accord. For example, Rector (487) estimates that in rats in average normal circumstances 10% of the glomerular filtrate containing 1% of the filtered HCO_3^- reaches the distal tubules, corresponding to pH less than 6.3, lower than the lowest values found in the distal tubules of acidotic rats by Malnic, de Mello Aires, and Giebisch. The latter authors report pH values (6.39 to 6.52) in the distal convolutions in both respiratory and nonrespiratory acidosis which are surprisingly high in view of the small contribution of the collecting ducts to acid secretion (487, 486) and the fact that there is no net acid excretion in urine of pH 6.4 or more (II-6). Perhaps the superficial cortical nephrons which were the exclusive source of the samples collected by Malnic *et al.* differ in acidosis, in respect to luminal fluid [HCO_3^-], from the deeper nephrons.

There are, however, data available *in vivo*, in the dog and in man, which show clearly that the rate of acid secretion into the nephrons as a whole is inversely related to the transtubular pH gradient in the distal nephron, especially when high-pH urine is being elaborated. There are two ways in which the pH of the luminal fluid of the distal nephrons can be markedly and rapidly influenced in the intact animal: (i) by infusing carbonic anhydrase intravenously at a rate sufficient to perfuse the nephrons, and (ii) by lowering the rate of flow of high-pH urine. Neither enzyme infusion nor physiological variation of urine flow appreciably affects either the glomerula filtration rate or conditions in the proximal tubule, so that the observed effects of these maneuvers upon HCO_3^- excretion principally reflect alterations of the rate of the acid secretion of the distal tubules and collecting ducts.

Antidiuresis usually affects the urine pH markedly (138, 32, 668, 495). The value declines as the flow of acid urine of high ammonia content declines because of reabsorption of NH_3 from the distal nephron; but in ammonia-poor, bicarbonate-rich urine the pH rises because of distal CO_2 reabsorption down the gradient created by osmotic water reabsorption (II-10). If acid secretion into the distal nephron varies inversely with the distal transtubular pH gradient, then antidiuresis should cause the rate of acid secretion into the distal nephrons to increase whenever enough HCO_3^- is present in the luminal fluid for water reabsorption to cause the pH of the luminal fluid to rise, thus diminishing the distal transtubular pH gradient. Renal net base excretion (II-4) should accordingly decline as the flow rate of high-pH urine declines. Data obtained in healthy human volunteers show that this is the case; a representative experiment with bicarbonate-rich urine is depicted graphically in Figure II-8.1. Two diuretic peaks and two antidiuretic troughs were effected by the subject's drinking water at a variable rate as prescribed by the experimenter over the course of the experiment, which lasted approximately 4 hours. Urinary pH varied inversely with urine flow over the approximate range 7.9 to 7.2 (*squares*). Variation of the rate of renal net base excretion was estimated by two different technics (see legend); provided the base load and glomerular filtration rate are stable during the experiment, a change in renal net base excretion with sign

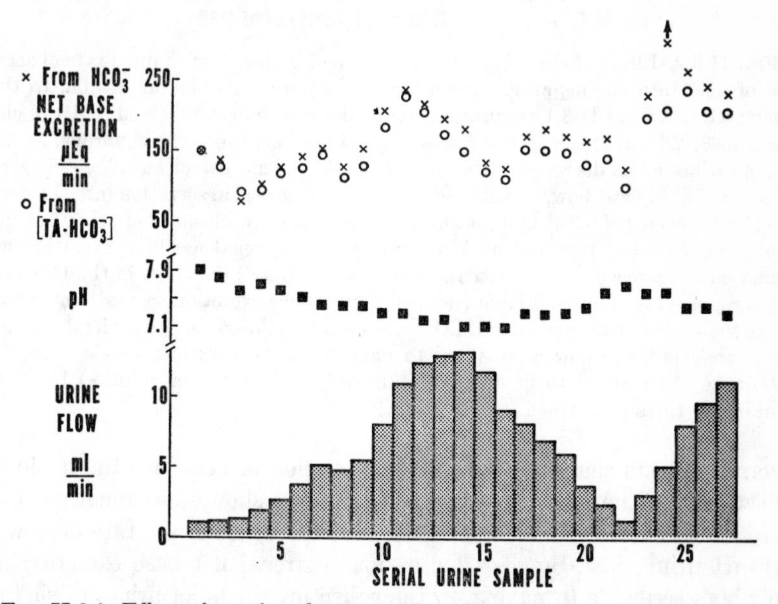

FIG. II-8.1. Effect of varying the rate of flow of bicarbonate-rich urine upon urinary pH and renal net base excretion in a healthy volunteer. Reproduced from Hills (255) with permission of the publisher. Variation of urine flow was effected by alteration of the rate of water drinking at the discretion of the experimenter. Renal net base excretion was estimated by two independent methods (II-4C) as indicated on the ordinate. *Circles* are derived from measurements of "titratable-acidity-minus-CO_2"; that is, they represent $-(TA - HCO_3^-) - NH_4^+$. *Crosses* show $HCO_3^- - TA - NH_4^+$, where TA represents titration of acidified and aerated urine between pH 7.4 and the pH at which $[HCO_3^-]$ of the sample was estimated (i.e., the pH of the collected sample). Water diuresis approximately doubled renal net base excretion over low-flow values in this experiment; agreement between the two estimates of base excretion must be considered satisfactory.

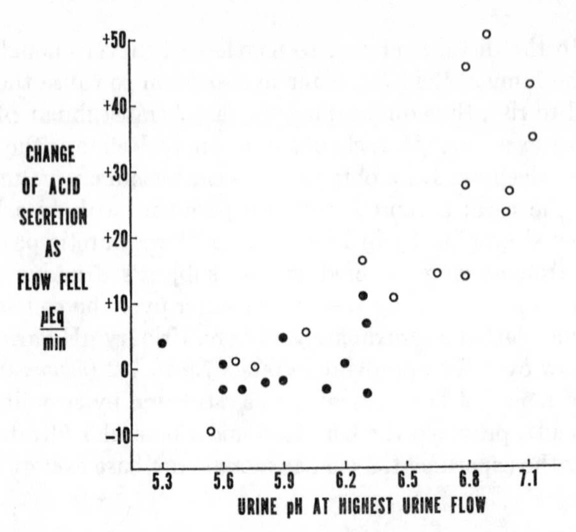

FIG. II-8.2. Effect of changing urine flow rate of high-pH urine upon rate of secretion of acid into the nephrons. Twenty-seven experiments similar in plan to that illustrated in Figure II-8.1 are presented; the data have been reported in detail elsewhere (668, 495). Increase in the rate of secretion of acid into the nephrons, as flow fell, is estimated as decrease in the rate of renal net base excretion. When urine pH exceeds 6.2 at highest flow, antidiuresis clearly increases acid secretion into the nephrons; below urine pH 6.0 at highest flow there is no discernible effect. In the 12 experiments (*solid circles*) reported by Woeber *et al.* (668), renal net base excretion was estimated as increased urinary (titratable acid − HCO_3^-) + NH_4^+; in the 15 experiments reported by Reid and Hills (495), change in base excretion was estimated from paired values for urinary pH and HCO_3^- and from calculated change in titratable acid (*open circles*). Urine flow was caused to vary from brisk water diuresis (flow >11 ml/minute) to marked to moderate antidiuresis (0.5 to 2.0 ml/minute) by water drinking at rates prescribed.

reversed reflects change in the rate of secretion of acid into the nephrons collectively as the urine flow varies. The figure shows how renal net base excretion increases (i.e., acid secretion decreases) when the rate of flow of high-pH urine rises. Smaller fluctuations of renal net base excretion are probably ascribable to mixing, in the collecting ducts, of urines of slightly different pH as urine flow changes (305), and perhaps to minor random fluctuations of the glomerular filtration rate.

The effect of the transtubular pH gradient on the rate of secretion of acid in the nephrons as a whole is particularly large in highest-pH urine. Figure II-8.2 relates the change in the rate of renal acid secretion, observed in 27 healthy human volunteers as the rate of urine flow was varied, to the acid-base composition of the urine, which is of course determined primarily by the acid-base load. The experiments represent moderate to submaximal antidiuresis; even so, lowering the urine flow from high values increased the rate of acid secretion into the nephrons by upwards of 50 μEq/minute in bicarbonate-rich urines with very high pH. The effect of antidiuresis diminishes rapidly as the HCO_3^- content of the urine declines, but it is still regularly visible when the urine pH at highest flow is 6.3, and perhaps still is visible when it is 6.1 at highest flow. When more acid urine is being secreted, antidiuresis exerts no significant effect on the rate of renal acid secretion; under such conditions the urinary pH falls instead of rising, as flow falls (II-10C), but the decline probably occurs almost entirely in the collecting ducts and so would not be expected to exert much influence on renal acid secretion.

A second method by which the transtubular electrochemical gradient can be altered experimentally in the intact organism, in a portion of the nephron, is by abolition of the low disequilibrium pH normally present in the distal tubules of animals excreting base loads. During bicarbonate diuresis there is under normal conditions a substantial accumulation of H_2CO_3 above equilibrium in the distal tubules but not in the proximal tubules; it can be abolished by infusing carbonic anhydrase intravenously at a rate sufficient to perfuse the nephrons with the enzyme. When this is done, as illustrated in Figure II-8.3, an abrupt decrease in the renal excretion and clearance of HCO_3^- is observed, which can only reflect sudden increase in the rate of acid secretion into the distal part of the nephrons, where the disequilibrium conditions have been abolished.

Figure II-8.4 shows the effect of intravenous carbonic anhydrase upon the renal HCO_3^- clearance in dogs receiving intravenous sodium bicarbonate infusions; the *open squares* represent control animals and the *solid circles* represent animals receiving the enzyme. Since renal HCO_3^- clearance is normally directly related to the arterial plasma [HCO_3^-], as is evident in the figure, it is necessary always to compare the clearances with and without enzyme at comparable values for arterial [HCO_3^-]. It is apparent on inspection that the clearances are higher, at any value for arterial plasma [HCO_3^-], in animals receiving no enzyme; as in Figure II-8.3, the effect of enzyme is clearly to increase acid secretion into the distal tubules, causing increased conversion of luminal HCO_3^- to CO_2 and decreased loss of base as HCO_3^-. The difference in HCO_3^- clearance between control and enzyme-infused dogs in highly significant (legend, Figure II-8.4).

It is possible that the effect of varying acid-base load upon the rate of secretion of acid into the nephrons is not entirely mediated by antidiuresis-induced increase of the transtubular pH gradient; a contribution of altered P_{CO_2} is difficult to demonstrate or to exclude. Arterial P_{CO_2} rises in nonrespiratory alkalosis (III-1), and if a physiological base load raises arterial P_{CO_2} at all appreciably, an increased rate of renal acid secretion would be likely. Such a chain of physiological and physicochemical causation can be represented as follows:

$$\text{base load} \rightarrow \uparrow \text{plasma } [HCO_3^-] \text{ and pH} \rightarrow \uparrow P_{CO_2} \rightarrow \uparrow \text{acid secretion} \quad (4)$$

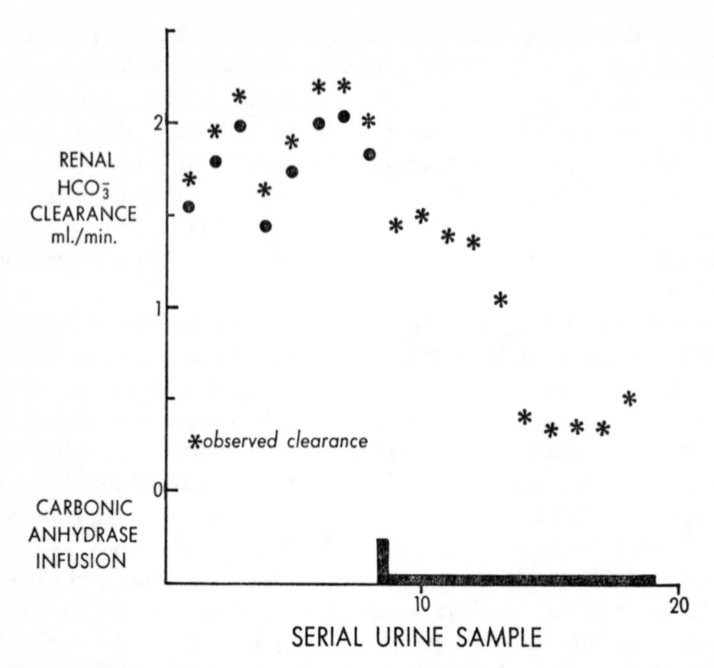

FIG. II-8.3. Effect of intravenous carbonic anhydrase upon renal bicarbonate clearance. Reproduced from Reid and Hills (498), with permission of the publisher. A dog served as its own control for this experiment; enzyme was infused at the rate of 1 mg/minute after a prime of 100 mg. Such clear-cut effects of infused enzyme are best seen when the arterial P_{CO_2} of a lightly anesthetized animal is held constant by controlled ventilation before and during the infusion. The experimental design is not well adapted to quantitative evaluation of the effect of infused enzyme on conditions within the nephron, because the steady state of acid-base balance of the animal established prior to the control period ceases to obtain as soon as the large decrease in HCO_3^- clearance is experimentally effected.

The *solid circles* represent renal HCO_3^- clearances "corrected" to urinary P_{CO_2} of 40 mm Hg. Correction is unnecessary for the clearances during enzyme infusion, when urinary P_{CO_2} was close to the P_{CO_2} of glomerular filtrate (40 mm Hg). "Correction" gives an improved estimate of the effect of infused enzyme on urinary HCO_3^- excretion and clearance, since the difference in the control period between values for *solid circles* and *asterisks* is due entirely to the physicochemical effect of the raised urinary P_{CO_2} on urinary [HCO_3^-], and is without physiological significance (II-4C).

It is doubtful, however, whether the precision of experimental measurement of arterial P_{CO_2} and especially of renal HCO_3^- clearance, the variance of which is sizable, would suffice to demonstrate, let alone quantify, increase of acid secretion into the nephrons due to a postulated physiological increase in arterial P_{CO_2} accompanying a high-normal base load (cf. II-6C).

Ordinary loads of excess alkali are eliminated effectively by the "fundamental system" that regulates acid-base balance (II-6); but acid secretion into the nephrons increases as filtered HCO_3^- increases, and to the extent

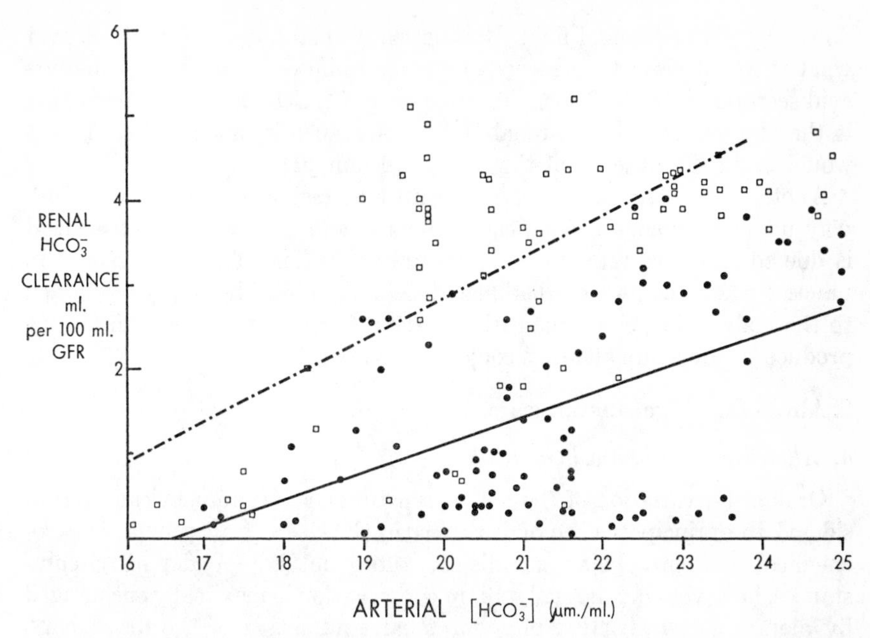

FIG. II-8.4. Effect of intravenous carbonic anhydrase on the renal bicarbonate clearance. Reproduced from Reid and Hills (498), with permission of the publisher. Carbonic anhydrase was infused intravenously into eight mongrel dogs (*solid circles*) which were meticulously matched with control animals (*open squares*) in respect to variables likely to influence the renal HCO_3^- clearance (body weight, mean urine flow rate, glomerular filtration rate (*GFR*), and arterial P_{CO_2}). Comparison of HCO_3^- clearance at any value of plasma HCO_3^- under conditions of moderate physiological base loading during moderate water-saline diuresis is afforded by the best-fitting linear regressions for control animals (*dotted line*) and enzyme-infused animals (*solid line*). The difference in elevation of the regression lines reflects the superior efficiency with which the kidneys of the control animals defend them against base excess, i.e., the lesser alkalinization of extracellular fluid required to secure base balance under conditions of alkali loading. The low disequilibrium pH normally present in the distal part of the nephron, increasing as the base load increases, serves homeostasis by inhibiting loss of acid from the body as acid secreted into the nephrons.

that it does so, the homeostatic defense against large base loads is weakened. Several accessory defenses against alkali excess play an important role in the effective elimination of unusually large alkali loads. One of these is normally present, and has an automatic character like the fundamental defense itself; it consists in the development of the low disequilibrium pH in the distal tubules and collecting ducts (Section C). Because the extent of disequilibrium increases with increasing glomerular HCO_3^- filtration—i.e., with increasing base loads —, the disequilibrium pH increasingly serves to sustain a substantial transtubular pH gradient, as the base load increases, in spite of the high HCO_3^- concentration of the luminal fluid, by holding down the pH of luminal fluid. Raising the transtubular pH gradient over what it would be in the absence of the disequilibrium pH in turn inhibits acid secretion into the lumen; regeneration of HCO_3^- in peritubular blood is thereby minimized, and renal HCO_3^- clearance is much higher than it would be in the absence of the disequilibrium pH (498).

Another accessory renal defense against excessive base load comes into play under pathological circumstances only, and only when the base load is due to excessive intake of alkaline salts (Section D2b). This means in practice when alkaline, or alkalinizing, salts of Na are being rapidly added to the body fluids, since quantities of the other alkali cations sufficient to produce severe acute alkalosis could not be tolerated.

2. Abnormal Circumstances

a. Altered Electrochemical Gradient

Ordinarily, variation of the distal transtubular pH gradient can be considered the principal cause of variation in the distal transtubular electrochemical gradient. This generalization does not hold under all circumstances, however, and, if renal sodium conservation is maximally encouraged by dietary salt restriction plus hormonal stimulation of the distal Na^+-reabsorbing mechanism, urinary pH can be lowered to 4.0 (549). This phenomenon presumably represents increase in the normal transtubular electrochemical gradient in the distal nephrons that results from active Na^+ reabsorption (II-7A; II-8A).

b. Alteration of Extracellular Fluid Volume

An accessory defense against severe alkali excess and alkalosis which is not present at all under normal circumstances results from the abnormal expansion of extracellular fluid volume inevitably attending very large loads of base, excepting that due to loss of acid from the body, as when HCl is vomited in quantity (III-2). As already noted (II-7F), the expanded extracellular fluid volume causes increased rejection of filtered Na^+ together with decreased acid secretion into the nephrons. This type of severe alkalosis, and *a fortiori* this accessory defense, are encountered more often in the experimental laboratory than in the clinic; it has been a source of considerable misinterpretation of experimental data. The belief that there is a true HCO_3^- *Tm* was originally entertained because the effect of volume expansion had not been discovered (II-7F). The accessory defense against pathological alkali excess offered by this effect of volume expansion is physiologically nonspecific; expansion of the extracellular fluid volume by NaCl is as effective as expansion by $NaHCO_3$ in depressing acid secretion, and can therefore result in some degree of nonrespiratory acidosis. Contraction of extracellular fluid volume stimulates renal acid secretion as well as reclamation of filtered Na^+.

c. Primary Renal Disorders

Whether functional, as in K deficiency, or structural, due to intrinsic renal disease, primary renal disorders can markedly affect the rate of secretion of acid into the nephrons (Part III). This function may be deranged more or less selectively by intrinsic kidney disease (renal tubular acidosis) (III-B2); it may also be affected as one of a variety of abnormalities in many types of renal disease. Marked lowering of glomerular filtration rate in renal disease generally diminishes the rate of filtration of urinary nonvolatile buffer. Less NH_3 is made, and less nonvolatile buffer reaches the urine, so that the total rate at which acid can be secreted into the nephrons at the minimal attainable urinary pH is much reduced (III-3). In renal tubular alkalosis due to K depletion and related causes, acid secretion into the nephrons is inappropriately high in relation to the plasma $[HCO_3^-]$ (III-2).

d. Altered Arterial P_{CO_2}

Raised arterial P_{CO_2} stimulates tubular acid secretion and so decreases renal HCO_3^- clearance at any value for plasma $[HCO_3^-]$ (or for HCO_3^- filtration); the HCO_3^- pseudo-*Tm* is raised (II-7F). Lowered arterial P_{CO_2} inhibits acid secretion, increases HCO_3^- clearance, and lowers the pseudo-*Tm*. From the standpoint of homeostasis, the most important aspect of these relations is that any malfunction of the lungs resulting in altered arterial P_{CO_2} automatically results in parallel change in extracellular total base ($[HCO_3^-]$); this is the mechanism of "compensation" by the kidneys (minimization of the deviation of extracellular pH) in acid-base disturbances of pulmonary origin (III-1). Arterial P_{CO_2} may participate in the modification of the rate of acid secretion into the nephrons which normally accompanies changing acid-base load (Section D-1).

e and f. Administration of Carbonic Anhydrase or Its Inhibitors

Alterations of the rate of renal acid secretion produced by administration of carbonic anhydrase (an increase if the urine contains enough HCO_3^-) or its inhibitors (a decrease) have served as tools for fundamental investigation of the mechanism of acid secretion into the nephrons (II-8B), and in the investigation of the renal tissue and urinary pressures of the buffer gas CO_2 and NH_3 (II-9; II-10). Acetazolamide and related inhibitors of the enzyme are rather frequently used as diuretic agents, especially in patients with hypercapnia.

g. K^+ Depletion

The increased renal acid secretion associated with K^+ depletion is an important cause or contributory cause of most clinical cases of nonrespiratory alkalosis (III-2). Excess K^+ and hyperkalemia suppress acid secretion.

h and i. Influence of Hormones on Renal Acid Secretion

In some but by no means all instances, primary hyperparathyroidism causes hyperchloremic acidosis with urinary bicarbonate waste, and the condition may disappear upon parathyroidectomy (413). Large doses of parathyroid hormone produce changes simulating the proximal type of renal tubular acidosis, and the hormone appears to be essential for the full expression of renal tubular acidosis (411a). Recent suggestions that the hormone is instrumental in the normal renal response to changing acid-base load appear to underestimate the fact that the response of the acid-secreting mechanism of the nephrons to changing acid-base load is counter-productive (II-8D1). Mineralocorticoid deficit, like parathyroid hormone excess, may be associated with renal tubular acidosis (411a) (III-2B), and administered mineralocorticoids increase renal acid secretion (page 276).

nine

URINARY AMMONIA EXCRETION

A. SIGNIFICANCE: RELATION TO N METABOLISM AND ACIDITY REGULATION

Very extensive investigations of the significance of the urinary ammonia prior to the present century are completely ignored by contemporary reviewers of the subject (463, 465, 658, 466, 474, 468, 469, 470, 28), and Balagura-Baruch (28) goes so far as to assert that "it was only in 1911 that the presence of ammonia in the urine was correlated with acid excretion." In fact this correlation dates back nearly to the origins of chemical physiology as that subject developed out of the chemical revolution of the late 18th century. Justus Liebig, who had previously noted the constant presence of ammonia in the urine of carnivores, wrote in his textbook of physiological chemistry: "There can be no more decisive evidence in favor of the opinion that the soda of their . . . blood is very far from sufficing to neutralize the acids which are separated, than the presence of ammonia in their urine. This urine, moreover, has an acid reaction. In contradistinction to this we find, in the urine of the herbivora, soda in predominating quantity" (350, page 155). Liebig nevertheless entertained doubts whether urinary ammonia represented a genuine product of metabolism; and in fact the metabolism of N differs from that of C and H, whose oxidation to CO_2 and H_2O provides the principal metabolic wastes, in that in NH_3 (and urea) N is in the reduced, not the oxidized, state. Liebig thought that the urinary ammonia might be derived from the spontaneous breakdown of nitrogenous compounds in the body (349, 350), and indeed the ammonia in shed blood and in putrefaction does arise in this way. He also realized that the values of urinary ammonia obtainable in his day were to some extent artifactual, since the analytical method used (platinum chloride) resulted in some liberation of ammonia from other nitrogenous compounds present in the urine.

To elucidate the full significance of the urinary ammonia has subsequently required well over a century of continuing investigation. The problem was complicated, during the early part of this period, by the undeveloped state of acid-base theory, and throughout by the inherent complexity of the subject and especially by the unfortunate happenstance that from the beginning extensive use was made of NH_4Cl in investigating the physiological consequences of ammonia administration. In logical sequence, the questions to be answered were: is the urinary ammonia an artifact or the result of spontaneous decomposition of other N compounds? If not, it is a product of the metabolism of dietary N rather than simply a preformed dietary constituent? If a product of metabolism, is its rate of excretion determined by the N content of the diet, or by acid-base economy? If the latter, is the urine pH or the pH of the body fluids the relevant variable? In practice, naturally, overlap of these problems created difficulties in interpretation of experimental data.

1. Ammonia: A Significant Urinary Constituent

Schlösing (541) introduced a reliable method for determining urinary ammonia of which all subsequent methods are essentially modifications. The principle involves conversion of ammonium salts to NH_3 by addition of strong alkali and trapping of the gas diffusing off in strong acid, where it is converted to NH_4^+. Conditions favorable for complete evolution of the gas were created by disposing the alkalinized urine aliquot in a very thin layer in an open flat vessel; the whole was placed under a bell jar with another vessel containing strong acid. After a period sufficient for complete transfer of the ammonia, the acid could be analyzed for its ammonia content by the Nessler procedure or any other colorimetric method. Shaffer (557), reviewing methods for determining urinary ammonia in 1903, concluded that the Schlösing method gave satisfactory results for clinical purposes when correctly carried out. He believed that his own contemporaries had become careless in using the method, and was impressed with the meticulous technic used by the earlier investigators. He may have underestimated the precision of Schlösing's method, since in comparative studies he assumed Folin's method to give the "ammonia actually present" and he also carried out distillation for more than 48 hours, whereas earlier users of the method had established to their satisfaction that ammonia diffusing off urine after 48 hours does not represent preformed urinary ammonia.

Neubauer (421) was the first to apply Schlösing's method to evaluation of urinary ammonia excretion. He reported average values for urinary ammonia of 835 and 614 mg/24 hours in two healthy young men, and found that when he fed them NH_4Cl by mouth he could recover most of the excess N as urinary ammonia over control values. Lohrer (356) fed himself NH_4Cl and followed his 24-hour urinary ammonia excretion. Though in an initial 3-day experiment he recovered only a third of the fed ammonia as extra urinary ammonia N above control values, urinary ammonia excretion continued to increase with additional experimental periods in which more NH_4Cl was given, one period following a day or two after the previous one. In the second period he recovered half of administered dose of ammonia, and in the third and fourth periods the excess urinary ammonia actually exceeded the quantity of ammonia ingested. With the aid of hindsight we can interpret his data as reflecting the stimulation of renal ammonia production by an acidifying salt; they quite naturally seemed to him to indicate that administered ammonia is retained in the body and only slowly excreted. That this could hardly be the case, however, was soon brought out by Lange (326), who showed in cats that injections of ammonia salts even into bilaterally nephrectomized animals did not lead to any accumulation of ammonia in the blood, and that the ammonia did not escape through the lungs.

2. Relation to N Metabolism

Neubauer and Lohrer were the last experimenters to ignore Liebig's insight that urinary ammonia excretion was related to acid-base balance (Section A-3), but they were not the last who supposed it to bear some relation to the nitrogen content of the diet or of administered NH_4Cl. It is not possible to give a meaningful figure for the normal proportion of the urinary N excretion as ammonia, because the proportion varies widely with the urinary pH as well as with the urinary total N, the latter being largely determined in the steady state by the protein content of the diet. An extensive literature summarized in the reviews of Folin (174) and of Peters and Van Slyke (454) establishes that, when the protein intake is high enough in the mammal to maintain N balance with an ample margin, excess N over that metabolized over special pathways for specialized functions is excreted as urinary urea. These special functions include the formation and excretion of creatinine (and sometimes of creatine) as end products of muscle metabolism, or uric acid as the end product of purine metabolism, and urinary ammonia, which is allocated to the urine by the responsible mechanism (Section B) at varying rates appropriate to the acid-base load to serve as a buffer for excess acid. As the dietary protein intake declines, the proportion of urinary N excreted as urea declines; when the protein intake is very low and adequate calories are provided, urinary urea excretion falls to extremely low levels and the N balance becomes markedly negative. Ammonia excretion in relation to urinary pH, like the absolute rate of excretion of creatinine and uric acid, is relatively unaffected (Table II-9.1). Balagura-Baruch's statement (28, page 292) that urinary ammonia represents 20% of the

TABLE II-9.1. *Urinary urea, ammonia, uric acid, and creatinine excretion expressed as percentages of total N of human urine in relation to substantial and minimal protein intake**

Dietary N	Total urine N	Percentage of total N as:				
		Urea	Ammonia	Uric acid	Creatinine	Undetermined
	g/day					
Unrestricted........	12.03	83	4	1	6	6
0.36 g/day..........	1.63	20	12	7	34	27

* Data from Smith (580), reproduced by Peters and Van Slyke (454, page 644).

urinary N must apply, if at all, only to persons on protein-restricted diets. The relation between urinary urea and ammonia is discussed further in relation to the ultimate sources of the urinary ammonia (Section D).

Two observations conspired, however, to keep open for many years the question whether increasing the N load resulting from protein catabolism might increase urinary ammonia excretion; these observations are that either (i) increasing the protein intake or (ii) feeding NH_4Cl does in fact increase the urinary ammonia. Protein, however, is an acid-ash food; it is metabolized to acids, largely mineral acids, and it is its acidifying effect which (unless counteracted) increases urinary ammonia excretion by acidifying the urine. Despite unguarded assertions of authorities as eminent as Henderson and Palmer (243), and an occasional more recent short-lived claim to the contrary (205), it has been established since the beginning of our century (174, 454) that neither the N content of the diet nor accelerated protein catabolism affects the rate of urinary ammonia excretion except by way of the influence exerted on the acid-base load.

It was by ill luck that NH_4Cl administration was used extensively from the beginning in investigations of the precursors of the urinary ammonia, for administration of NH_4Cl increases urinary ammonia excretion because it is an acidifying salt, not (as the early investigators understandably supposed) because of its N content. The proportion of fed NH_4Cl recoverable as urinary ammonia in any experiment is variable, depending upon the immediate as well as the continuing effect of the dose administered upon the urine pH, and also (in the case of large continued doses) upon the extent of the slower increase in the renal ammonia production produced by incipient or established acidosis (II-10). It was natural to search for analytical sources of error in attempting to reconcile what at first appeared to be discrepancies between different reports concerning the fate of administered ammonia N (316, 159, 222b, 112, 223, 224). Pioneer investigators of this problem could not be expected to realize that it is impossible to furnish a valid figure for the proportional conversion of administered

NH_4Cl to urinary urea without specifying the experimental conditions in detail.

Interpretation of experimental data was still further complicated for a time by the contention that NH_4Cl, or acid loads, might increase urinary ammonia excretion by stimulating protein catabolism (136, 86). This is probably true under very unphysiological conditions (e.g., acidosis and dehydration sufficient to produce fever), but not otherwise. Not until the meticulous studies of children and adults by Gamble and his associates (183, 182) did it become incontrovertible that the adaptive increase in urinary ammonia excretion produced by NH_4Cl in clinically conventional doses has nothing whatsoever to do with the N content of this salt, nor with any effect of acidifying substances on protein catabolism. Their studies showed that the acid-base effects of administration of equivalent sizable quantities of $CaCl_2$, NH_4Cl, $(NH_4)_2SO_4$, and $MgSO_4$ are indistinguishable; acidification of the extracellular fluids is comparable, as manifested by decreased plasma $[HCO_3^-]$ and correspondingly increased plasma $[Cl^-]$, and all produce similar increases in urinary ammonia excretion (182). In the case of $CaCl_2$, a salt which acidifies the urine and body fluids because its Ca^{++} is excreted as carbonates in the stool (II-5), the excess urinary buffer ammonia was excreted at the expense of N excreted as urea during the control period. In the case of NH_4Cl, the N content of the compound simply increased the total excreted N, so that the same increased ammonia excretion was seen as in the experiments with $CaCl_2$ and $MgSO_4$, but with less reduction of urinary urea excretion (182).

3. Relation to Acid-Base Economy

The earliest observations of the effects of administered mineral acids (II-4), in showing that these could pass through the body and blood in substantial quantity with no apparent effect on the blood's alkalinity, led to the conviction that withdrawal of some of its alkaline salts could constitute only a minor part of the mechanism by which the kidney "neutralized" the acid in excreting it in the urine. Salkowski (536, 538, 537) revived Liebig's suggestion that ammonia partially neutralizes the acid of the urine of carnivores, and was able to show that, when the urine of a dog was rendered alkaline by administration of an alkaline salt (sodium acetate) to the animal, the urinary ammonia promptly fell to levels as low as those normally characteristic of the guinea pig. Soon thereafter Walter (645), in Schmiedeberg's laboratory in Strassburg, showed that a large increase in the rate of urinary ammonia excretion could be produced by continuing administration of mineral acid, an observation which has been confirmed consistently in man and in all carnivores and (to a lesser extent) in herbivores (537, 661). The observations by Walter and Salkowski are the foundations of our realization that urinary ammonia excretion is controlled by the acid-base economy.

There remained, however, a problem which was not completely solved for three more generations: that of distinguishing the two independent ways in which the acid-base load may influence urinary ammonia excretion. Under normal circumstances, it is the urinary pH which determines urinary ammonia concentration and hence (at any given flow rate) the ammonia excretion rate (Section B); ammonia excretion varies inversely with urine pH, and the effect of a change in urine pH is immediate. But acid loads large enough to produce acidosis stimulate large increases in renal ammonia production, causing over some days large progressive increases in the rate of urinary ammonia excretion at any urine pH; and base loads large enough to produce alkalosis reduce ammonia excretion. Salkowski and Munk (538) ascribed the reduction of urinary ammonia excretion in dogs given alkali to the alkalinizing effect of the compound on the urine, but nevertheless expressed with admirable circumspection the dimensions of the problem of interpretation.[1] It was, in fact, only the pathophysiological stimulation of urinary ammonia excretion by acid excess (661, 311, 245a), and the opposite effect of administration of alkaline salts (585, 586, 229), which were generally recognized over the half-century following these studies by Walter and Salkowski (II-4A). This question was not entirely clarified until Ferguson (161) succeeded in demonstrating in a single experiment the independent effects of urine pH and of acidosis upon urinary ammonia excretion; he did so with the aid of an inhibitor of carbonic anhydrase. Such compounds initially diminish urinary ammonia excretion by raising urine pH; later, acidosis develops and raises the ammonia excretion.

B. NORMAL REGULATION OF URINARY AMMONIA EXCRETION

Urinary ammonia excretion varies in health in relation to two factors: (i) the urinary pH (Section B) and (ii) the urinary flow rate (II-10). Of these, much the more conspicuous effect is exerted by urinary pH. Reduction of urine pH increases urinary ammonia excretion, not because of any change in urinary ammonia production, but because of more effective trapping of NH_3 entering the luminal fluid of the nephrons, which favors the urine in its competition with the renal venous outflow for the renal ammonia supply. Increasing the urine flow has a similar effect; and though the magnitude of the effect of changing urine flow rate is less than that of varying urinary pH, it is of considerable theoretical importance. The effect of varying urinary flow rate on the rate of urinary ammonia excretion (668), and the related influence of urine flow on urinary pH (668, 495), are analyzed subsequently (II-10). In Figures II-9.1 to II-9.4 are shown the effects of varying urinary pH on the rate of urinary ammonia excretion at high

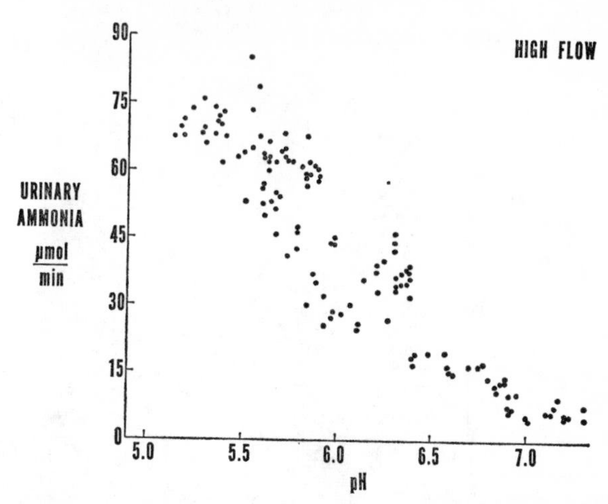

FIG. II-9.1. Rate of ammonia excretion as a function of urinary pH in normal human urine during brisk water diuresis. See legend to Figure II-9.5. Note also the restricted range of urinary pH during water diuresis in this and in the succeeding figure as compared with the larger-pH range represented in low-flow samples (Figures II-9.3 and II-9.4). See also II-10C.

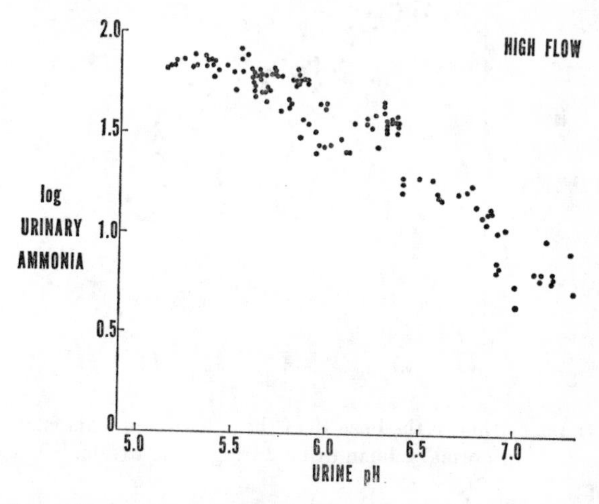

FIG. II-9.2. Logarithm to the base 10 of the rate of ammonia excretion as a function of urinary pH in normal human urine during brisk water diuresis. See legend to Figure II-9.5.

and low rates of urine flow. Figure II-9.5 is a nomogram which gives the expected rate of urinary ammonia excretion by healthy adults as a simultaneous function of the pH and flow rate of the urine.

In 1921 Nash and Benedict (419) ascertained that the concentration of ammonia in the renal venous blood always exceeded that of the vena cava,

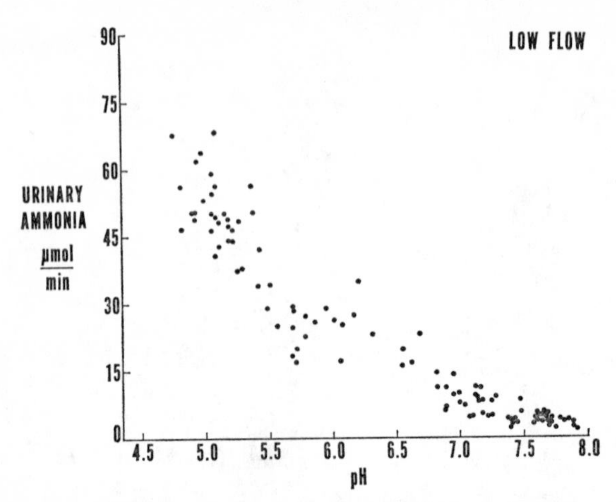

Fig. II-9.3. Rate of ammonia excretion as a function of urinary pH in normal human urine during antidiuresis. See legend to Figure II-9.5.

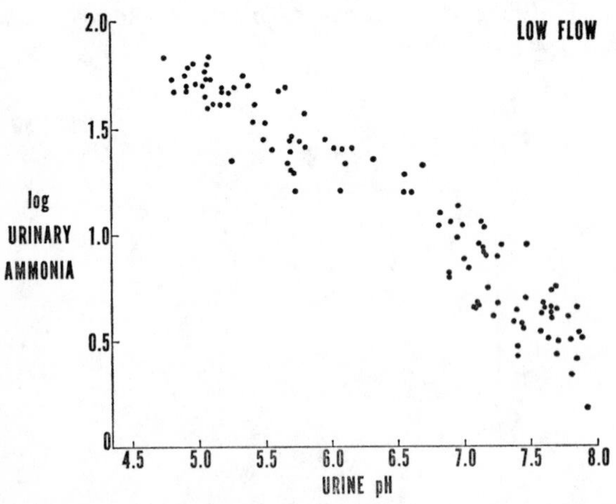

Fig. II-9.4. Logarithm to the base 10 of the rate of ammonia excretion as a function of urinary pH in normal human urine during antidiuresis. See legend to Figure II-9.5.

i.e., that ammonia was always flowing out of the kidney via the renal veins. This observation has been abundantly confirmed in all mammalian species studied. Some ammonia flows into the kidney via the renal artery, but the quantity is insufficient to account for the renal venous outflow, to say nothing of the urinary ammonia; it is therefore evident that the kidney always produces ammonia. Van Slyke and his associates (629) demonstrated in normal, acidotic, dogs and alkalotic dogs that the precursor of most of this ammonia is L-glutamine, and this observation, too, has been

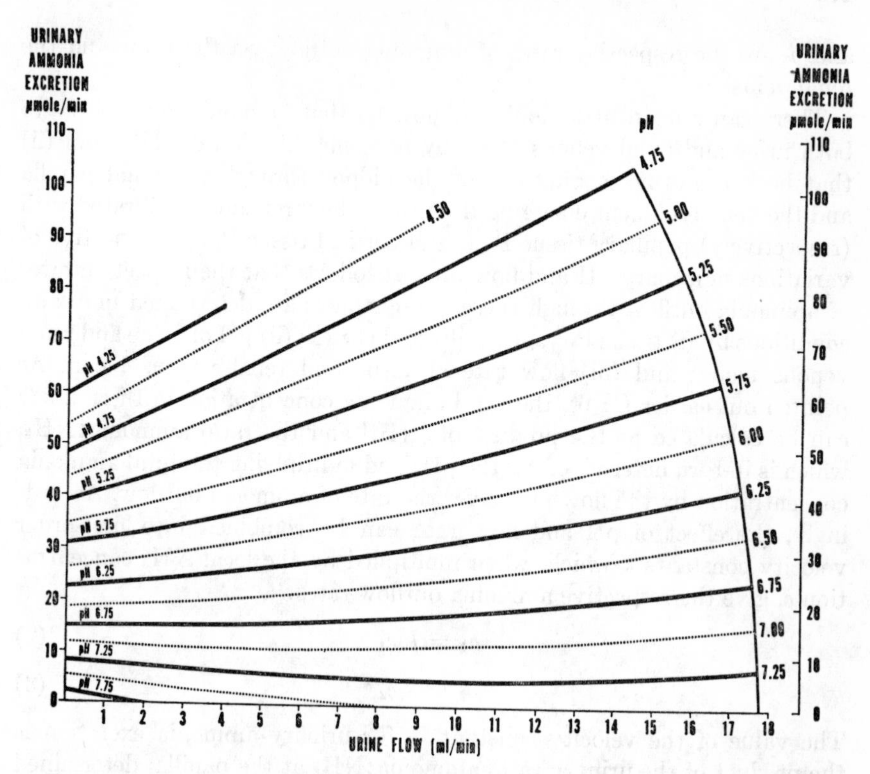

FIG. II-9.5. Urinary ammonia excretion of healthy men as a simultaneous function of urinary pH and flow. The figure is based on 507 samples derived from 27 previously reported experiments investigating the effect of water diuresis in healthy adult volunteers (668, 495) plus 4 additional experiments of the same type. The figure is a nomogram representing in two dimensions the curved surface best fitting the data; the equation expressed is:

$$\text{Ammonia excretion} = -116 - 8.23\ \text{flow} + 0.0289\ \text{flow}^2 + 1132/\text{pH} - 1611/\text{pH}^2 + 55.5\ (\text{flow}/\text{pH})$$

Figures II-9.1 and II-9.2 are based upon 129 of these urines of which the flow rate was 11.0 ml/minute or greater, and Figures II-9.3 and II-9.4 are based upon 107 samples of which the flow rate ranged from 0.5 to 3.0 ml/minute.

abundantly confirmed in all mammalian species studied (Section C). It follows that for each kidney at any steady state of renal ammonia balance:

$$s = i + p = o_1 + o_2 \tag{1}$$

where s is the renal ammonia supply, defined as the sum of the rates of the arterial inflow i and the renal ammonia production p, and where o_1

and o_2 are the respective rates of ammonia outflow via the urine and the renal veins.

There can remain little doubt by now (i) that ammonia gains entry to both urine and renal venous blood by nonionic diffusion of NH_3, and (ii) that both the urine passing out of the kidney through the renal papilla and the venous effluent emerging from the kidney remain equilibrated with (respectively) papillary tissue P_{NH_3} and cortical tissue P_{NH_3} irrespective of variations in urinary pH and flow rate. It follows that the respective rates of ammonia outflow through these two channels are determined under all conditions by (i) tissue P_{NH_3} of papilla and cortex; (ii) pH of urine and renal venous blood; and (iii) flow rate of urine and renal venous blood. As pointed out earlier (II-6), the total ammonia concentration in these fluids can be calculated as the product of $[NH_3]$ and the ratio ammonia:NH_3, which is in turn determined by the pH; and multiplying the total ammonia concentration by the flow rate gives the rate of ammonia outflow. Accordingly, the effect of pH and flow rate can be combined into first-order velocity constants k which, when multiplied by the local NH_3 concentration c, give the respective ammonia outflow rates:

$$o_1 = k_1c_1 \tag{2}$$

$$o_2 = k_2c_2 \tag{3}$$

The value of the velocity constant k_1 for urinary ammonia excretion is the product of the urinary ratio ammonia:NH_3 at the papilla, determined essentially by the pH of papillary urine, and the urine flow rate. Similarly, the value of k_2 is the product of the renal venous blood flow and the blood ratio ammonia:NH_3, again determined essentially by the pH. Combining Equations 1, 2, and 3 gives an entirely general equation relating tissue ammonia pressures and ammonia outflow rates to the pH and flow rate of urine and renal venous blood:

$$s = i + p = k_1c_1 + k_2c_2 \tag{4}$$

The rate of ammonia production p by the kidney of man (258; Appendix J) and by dog kidney (261) does not appreciably vary as the urine pH and flow are varied over their ranges of normal. Also, since neither the pH nor the flow rate of the renal venous blood is affected by normal variation of urinary pH and flow, k_2 does not fluctuate as a function of changing normal acid-base loads; and changes of i are small and have little effect on s (258). Accordingly, we can rewrite Equation 1 with little error in the following form, where capital letters indicate invariant quantities:

$$S = k_1c_1 + K_2c_2 \tag{5}$$

The values of c_1 and c_2 are obtained as the products of P_{NH_3} and the molar solubility factor for ammonia in urine and blood, respectively. It is evident that the only one of the variables whose value is primarily affected by the fluctuating conditions occurring under physiological circumstances is k_1, which is increased by falling urinary pH and (to a lesser extent) by rising urine flow. The result of such increases must be secondary declines of c_1 and c_2.

Some authors have represented urinary ammonia excretion as a linear function of the urinary pH (461, 28), while others have depicted the logarithm of urinary ammonia concentration or excretion as a linear function of pH (139, 671, 591, 105, 673). Neither relation has fundamental significance; each is an empirical approximation suitable for particular conditions of observation. Figure II-9.6 shows the approximately linear relation observed by Wrong and Davies (673) between the logarithm of the urinary ammonia excretion and the urinary pH; observations of urine samples collected in the clinic and from healthy volunteers are almost invariably low-flow samples (<2.0 ml/minute), and under these circumstances the logarithm of the ammonia excretion is nearly linearly related to the urinary pH (Figure II-9.4), whereas the ammonia excretion itself is not (Figure II-9.3). The fortuitous character of the linear relation between the logarithm of the ammonia excretion and the pH is brought out, however, when comparison is made with Figures II-9.1 and II-9.2; these data were obtained from the same subjects and the same experiments as those in Figures II-9.3 and II-9.4, but the data selected for presentation

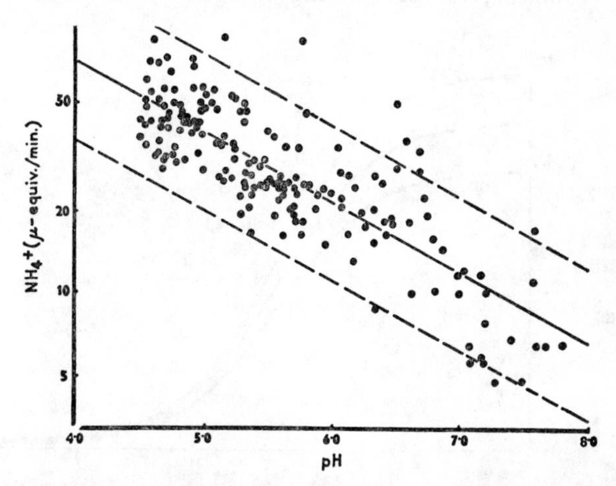

FIG. II-9.6. The urinary ammonia excretion of healthy adults plotted as a logarithmic function of urinary pH. Reproduced from Wrong and Davies (673). The data were obtained from normal volunteers 15 to 64 years old, two of whom took small amounts of sodium and potassium bicarbonate so that alkaline urines could be represented. The *lines* shown are the best-fitting regression and upper and lower 95% confidence limits.

are confined to brisk water diuresis (11 to 17 ml/minute). At these high flows, a plot of the logarithm of the urinary ammonia excretion against the urinary pH gives a curvilinear function to which no straight line could be fitted without gross imprecision (Figure II-9.2). The relation between the urinary ammonia excretion itself and the urinary pH at high flow is also curvilinear in the upper-pH range (Figure II-9.1), but below pH 6.5 approximates the linear relation reported in experimental acidosis in the dog by Pitts and his associates (461, 28), no doubt because quite high rates of flow are likely to be associated with acidosis and administration of intravenous fluids. At the low flows characteristic of most human urines, linearity of the relation between urinary ammonia excretion and urine pH is very approximate indeed (Figure II-9.3).

If the simplifying assumption is made that the tissue ammonia pressure is always normally uniform throughout the kidney, then $c_1 = c_2$ in Equation 5, and the equation reduces to:

$$S = c(k_1 + K_2) \tag{6}$$

Figure II-9.7 shows that Equation 6 yields a symmetrical mathematical function when ck_1 is plotted as a function of k_1; the physiological range over which pH can be varied, occupying, however, only a limited part of the curve. The values for k_1 and K_2, the first-order outflow velocity constants for ammonia in urine and renal venous blood, respectively, have been evaluated, in constructing the figure, for high-flow human

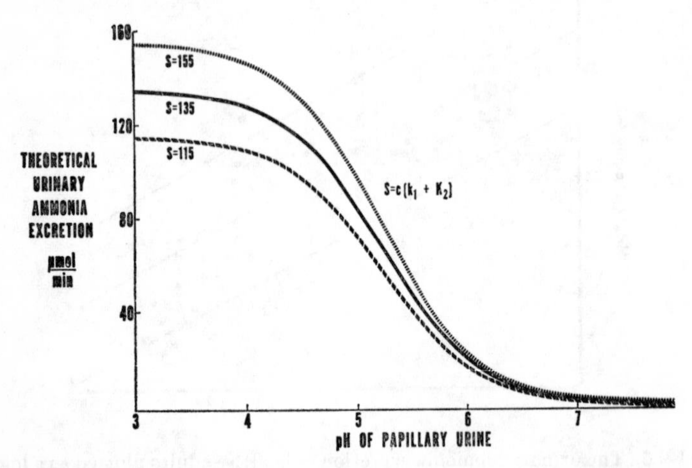

Fig. II-9.7. Theoretical relation between urinary ammonia excretion and the urinary pH, reckoned according to Equation 6. Three curves are shown in order to display the effect of varying the value of S. Note that the effect of changing the urinary pH is very small below pH 4.0 and above pH 6.5, and that the quasi-linear portion of the curve for $S = 135$ µmoles/minute extends only from about pH 4.5 to 6.0.

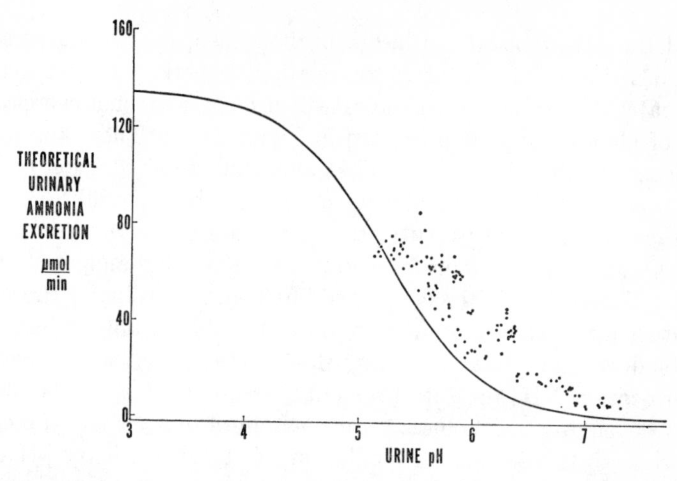

FIG. II-9.8. Theoretical relation between urinary ammonia excretion and urine pH, reckoned according to Equation 6, taking S to be equal to 135 μmoles/minute (Appendix J). The function consistently underestimates the experimentally determined relation.

urine (flow rate 13 ml/minute); the derivation of these values, to which little error can be supposed to attach provided urinary pH equals the effective pH at the papilla (II-10), has been presented elsewhere (258). The influence of the value assigned to S upon the function is visible in the three curves presented (S = 155, 135, or 115 μmoles/minute, respectively); the best currently available estimate for S as an average in healthy male adults 25 to 50 years of age is 135 μmoles/minute (Appendix J).

Figures II-9.7 and II-9.8 exhibit the symmetrical character of the function described by Equation 6 if the effective urinary pH is extended well beyond the limits actually attainable physiologically. Because the effect of declining urinary flow upon renal net acid and base excretion is to a considerable extent opposed by the widening of the normal range of urinary pH during antidiuresis (II-10), the location of the curves on the pH scale in Figure II-9.7 is not greatly altered by antidiuresis. It will be observed that, whereas the urinary pH range 6.0 to 4.5 is located in that part of the curve where reduction of urine pH is maximally efficient in increasing urinary ammonia excretion, depression of the pH much below the physiological limits would be an increasingly inefficient means of increasing the renal excretion of acid buffered as NH_4^+. A much more effective means of increasing acid excretion in very acid urine is the increase of S resulting from the increased renal ammonia production which develops in acidosis (II-10).

Figure II-9.8 shows that the very simple relation based on Equation 6, though it approximates the form of the relation between ammonia excretion and the pH of bladder urine in healthy men, does not give a satisfactory description of the absolute relation, experimentally determined during water diuresis, between the rate of urinary ammonia excretion and the pH of bladder urine as presented in Figure II-9.1. There appear to be two reasons for the discrepancy. First, continuing dehydration of H_2CO_3 present in urine at the papilla at a concentration above equilibrium (II-8D; II-10A) results in rise of pH after the urine passes into the urinary tract, where CO_2 and NH_3 pressure gradients between urine and tissue can no longer be dissipated effectively by gas diffusion. Accordingly, the velocity constant k_1 for ammonia outflow in the urine should not be based upon pH of bladder urine, but upon the pH of urine at the papilla. Second, a papilla-to-cortex NH_3 pressure gradient was observed by Hills and Reid (261) to develop in greyhounds during mannitol diuresis as a function of rising urinary pH. Assuming a similar effect of rising urinary pH to be a general phenomenon in diuretic mammals, including man, Equation 6 becomes inexact; for accurate results Equation 5 must be substituted, and the ratio c_1:c_2 must be known as a function of the urinary pH. In the dog infused with carbonic anhydrase, this ratio increases as an exponential function of increasing urinary pH; in order to obtain the theoretical function presented in Figure II-9.9, it has been assumed that in diuretic man c_1:c_2 increases exponentially from unity at lowest urine pH to 2.6 at highest urine pH, and k_1 has been reckoned on the basis of estimated pH of papillary urine rather than the measured values for bladder urine (Appendices

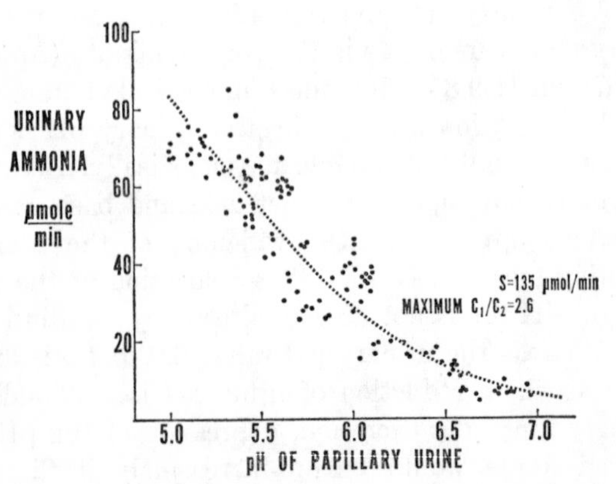

Fig. II-9.9. Urinary ammonia excretion represented as a function of the pH of papillary urine. The experimental values (*solid circles*) are based on the urine samples represented in figure II-9.1, but have been replotted to show the ammonia excretion as a function of the pH of papillary urine. Derivation of papillary urine pH from bladder urine pH is presented in Appendix K. Note that the physiological range of pH of papillary urine is somewhat less (5.1 to 7.0) than that of bladder urine (5.1 to 7.3). The theoretical *line* has been calculated according to Equation 5 by assuming that c_1/c_2 increases from unity at papillary urine pH 5.1 to 2.6 at pH 7.0.

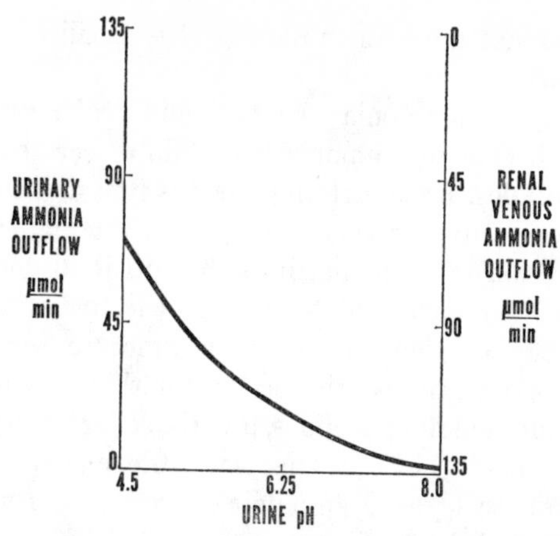

FIG. II-9.10. Distribution of the urinary ammonia supply *S* in healthy adults under physiological conditions as a function of the pH of bladder urine. Schematic. The smooth curve is derived from Figure II-9.3, and *S* is taken as invariant at 135 µmoles/minute. Approximately half the renal ammonia supply is excreted in the urine at top physiological acid loads; when the urine is alkaline, almost the entire renal ammonia supply is returned to the systemic circulation.

J and K). The excellent agreement of this function with experiment suggests that during water diuresis the papilla-to-cortex NH_3 pressure gradient is 2.6 in man when highest-pH urine is being secreted. The construction of the theoretical function shown in Figure II-9.9 is discussed in greater detail in Appendix J.

It is not profitable at present to extend the analysis to antidiuretic conditions because of the complications introduced by terminal dehydration of the urine in the medullary collecting duct and the experimental difficulties associated with very high urinary CO_2 pressures (261). Figure II-9.10 shows the distribution of the renal ammonia supply between urine and the renal venous outflow in man as a function of bladder urine at the ordinary low rates of urine flow.

C. RENAL SOURCES OF URINARY AMMONIA

1. Renal Cellular Sources

Which of the renal cells produce ammonia is not known with certainty. The proximal convolutions are the site of reabsorption of filtered amino acids (533); but the most important of the precursors of ammonia produced in the kidney, L-glutamine, is extracted by the renal cells at a rate exceeding its filtration (558, 458), and it has not been demonstrated which cells take it up from the peritubular blood. The distribution of the deaminating enzymes in kidney tissue may suggest that ammonia is normally produced throughout the kidney in a number of mammalian species, but that adaptive increase in renal ammonia production is localized in kidney cortex (507, 650, 295, 270).

The topography of ammonia secretion and reabsorption is not much better understood, though ammonia is certainly secreted into both proximal and distal tubules. Renal arterial blood is first distributed to the renal cortex and there equilibrates very rapidly with the higher P_{NH_3} of the cortical tissue. Secretion of ammonia into the proximal convolutions, which cannot always be demonstrated because of the inadequate sensitivity of analytical methods applied to the micropuncture samples (203a, 234, 374, 98, 60, 430, 430a), necessarily occurs for this reason and also insofar as the pH of the luminal fluid declines, a variable factor depending normally on the acid-base load (II-7E). The percentile contribution of proximal ammonia secretion to urinary ammonia excretion, derived by comparing excretion to proximal ammonia trapping (203a, 234), is substantial; but the figure probably varies inversely with the urinary pH, in view of the much smaller fluctuation of the pH of proximal tubular fluid with changing acid-base load. The percentile contribution of the distal tubule is less (374).

2. Precursors

With the aid of compounds suitably labeled with ^{15}N, Pitts and his associates (463, 465, 658, 466, 474, 467, 469, 470, 28) have gathered extensive quantitative data on the renal precursors of the urinary ammonia in the acidotic dog. Their studies have consistently confirmed the original finding of Van Slyke and his associates (629), who reported that L-glutamine is the principal renal precursor of glutamine in normal, acidotic, and alkalotic dog. For some years Pitts and co-workers assumed in their calculations that the ammonia flowing out of the kidney in their experiments was all derived from glutamine plus four other precursor amino acids present in the arterial inflow; but Stone, Balagura, and Pitts (595) abandoned this assumption after Reid and Hills pointed out (496, 258; see also 466) that the arterial inflow of performed ammonia makes an appreciable contribution to the renal ammonia supply, a fact also originally overlooked by Hills (251). The effect of the recalculations by Stone, Balagura, and Pitts is to raise the estimated proportion of renal ammonia production which is derived from arterial glutamine over the value previously reported; almost all the urinary ammonia is now reckoned to be derived from the two N's of glutamine (595, 28). In the acidotic dog 33 to 50% of the renal ammonia supply is derived from the amide N of arterial glutamine, and 17 to 25% each from the amino N of glutamine and from the arterial ammonia inflow; the latter, however, is derived from circulating ammonia and hence—at least in the resting animal (Section D)—also largely from glutamine (cf. Figure II-9.12). Less than 9% of the urinary ammonia is derived from renal deamination of any other precursor, and this comes almost entirely from alanine. Additional details of these experiments are considered by Balagura-Baruch (28).

Relatively few data are available in the dog or in other species in states other than nonrespiratory acidosis. However, a careful study of human subjects by Owen and Robinson (442), who measured urinary ammonia excretion together with renal venous-arterial amino acid differences, indicates that glutamine is the principal precursor of the ammonia of human urine, both under normal conditions of acid-base balance and in nonrespiratory acidosis produced by administration of NH_4Cl. Only glutamine was always extracted by the kidney under normal circumstances, whereas serine was always added to renal venous blood; in acidosis, additional glutamine plus some glutamate was regularly extracted. Ample amide N and α-amino N were extracted to account for the urinary ammonia, assuming that some of the glutamate derived from glutamine had been deaminated.

Aside, therefore, from the contribution, always small, of the renal amino acid oxidases which split ammonia from other amino acids, it is the reactions by which the amide N of glutamine and the amino N of glutamic acid are removed which are responsible in man as in the dog for renal ammonia production. The reactions yielding the renal ammonia supply (466, 28) are represented in Figure II-9.11; the enzyme systems capable of

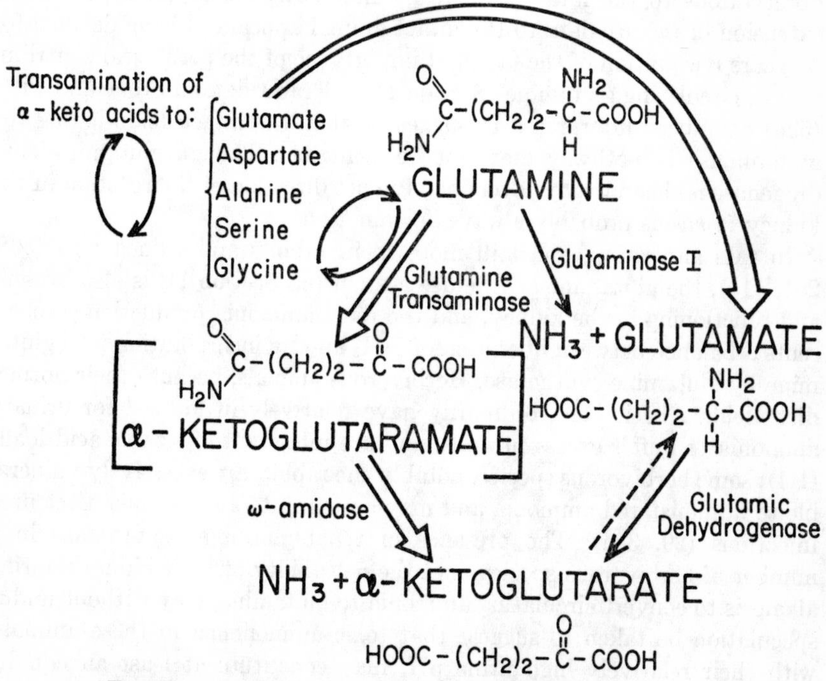

FIG. II-9.11. Sources of ammonia released within the kidney.

catalyzing these transformations are glutaminase I, glutaminase II, and glutamic dehydrogenase.

1. Glutaminase I: An enzyme that catalyzes deamidation of glutamine to yield glutamic acid and ammonia was identified by Krebs (318) in 1953; two isoenzymes have been identified, one "phosphate-dependent" and the other "phosphate independent." The remaining amino N can be removed by the glutamic dehydrogenase reaction.

2. Glutaminase II (206): Glutaminase II is known as glutamine-keto acid transaminase-ω-oxidase, and consists of two linked reactions as shown at the *left-hand side* of Figure II-9.11. The transamination reaction is rate-limiting; the transferred amino group provides glutamate, by direct or indirect transamination of α-ketoglutarate, as substrate for glutamic dehydrogenase. In addition, the ketoglutaramate resulting when the amine N is removed from glutamine is immediately deamidated by ubiquitous tissue ω-oxidases, yielding additional ammonia.

3. Glutamic Dehydrogenase: The reaction is linked with reduction of the flavin co-factor nicotinamide adenine dinucleotide or nicotinamide adenine dinucleotide phosphate (319).

The glutaminase II reaction is generally assumed, on the basis of various *in vitro* estimates of activity, to contribute only a very small proportion of the renal ammonia production p (28). Such extrapolations of *in vitro* observations to the intact animal are inherently shaky (II-1); a similar extension of *in vitro* data to the intact animal appears to have delayed for 25 years recognition of the fact and importance of the peripheral contribution to circulating glutamine (Section D-2). Tracer data of Stone and Pitts (596) have been interpreted to suggest that in the intact acidotic dog the glutaminase II pathway may not be negligible. Though glutamic dehydrogenase is theoretically reversible, the net direction of the reaction in the kidney *in vivo* is probably always deamination.

In rats and sheep, and still more so in rabbits and guinea pigs (368, 291, 115), the glutamine synthetase system (cf. Section D) is also present and functioning in the kidney, and the renal ammonia production p represents a balance between liberation of NH_3 and its incorporation into glutamine by glutamine synthetase. Herbivorous animals, because their normal diet yields a basic load, ordinarily have relatively little need for urinary ammonia to buffer excess protons, and are intolerant of sizable acid loads (I-4); some herbivorous species, notably the rabbit, are excessively vulnerable to administered ammonia and may succumb to ammonemia after urea injections (29, 33a). The presence of renal glutamine synthetase in a number of herbivorous species, and the net ability of their kidneys during alkalosis to convert circulating ammonia to glutamine, may without undue speculation be taken to suggest that toxic ammonemia in these animals, with their relatively high urine pH, may constitute at least an equally common threat to their survival as is offered by their relatively low capacity to produce extra ammonia to serve as urinary buffer for excess acid. The kidney is after all the principal contributor of ammonia to the systemic blood; and even in man and carnivores, whose predominantly acid urine diverts much of the renal ammonia supply into the urine, normal blood ammonia levels are close to half the values at which ammonia toxicity begins. Herbivores whose urine is unremittingly alkaline would be gravely menaced should renal ammonia production rise without prior acidification of the urine.

It is not known whether the human kidney can synthesize glutamine, and the literature on adaptive change of p has so far concerned itself principally with increases due to accelerated deaminations. Adaptive responses of the renal ammonia production to acidosis and alkalosis are discussed subsequently (II-10).

D. ULTIMATE CELLULAR SOURCES OF THE URINARY AMMONIA

1. The Circulating Ammonia Pool

Figure II-9.12 shows schematically the sources and distribution of the renal ammonia supply. The immediate sources of the urine ammonia are

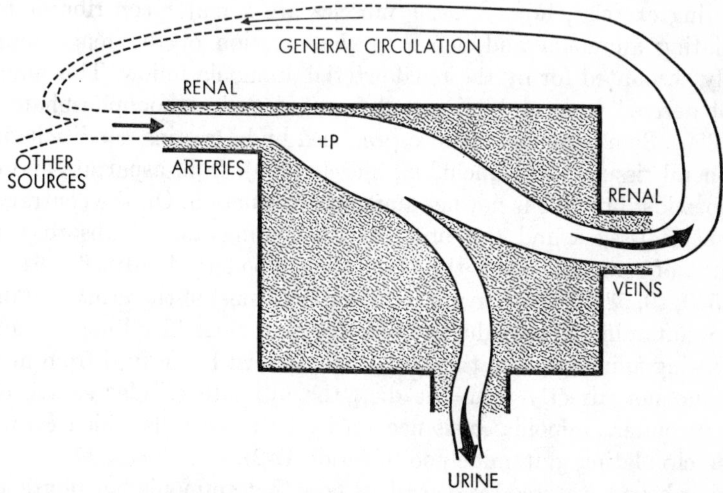

FIG. II-9.12. The circulating ammonia pool: immediate sources of urinary ammonia (schematic). In the resting subject, the kidney is principal source of circulating ammonia; during physical exercise, skeletal muscle is also an important source. The central importance of the normal liver in keeping the blood ammonia level at its very low normal value becomes evident in clinical hepatic failure; but the brain, and at rest the peripheral tissues, make an important contribution to removal of ammonia from the blood, especially when the blood ammonia rises. At rest, most of the ammonia entering the renal artery has been contributed to the circulation by the kidney. *Stippled area* represents kidney; p is net renal ammonia production.

the ammonia production *p* and the arterial inflow of preformed ammonia. Approximately 55 μmoles/minute, or 40% of the ammonia supplied to the kidney, is derived from the arterial ammonia inflow in man (257). Preformed ammonia entering the kidney is accordingly by no means a negligible contributor to the urinary ammonia excretion (cf. Section C-2), but the entire arterial inflow would not suffice to supply urinary ammonia at rates appropriate to large physiological acid loads (cf. Figure II-6.2 and Figures II-9.1 to II-9.5), let alone pathological ones.

Moreover, the mammalian organism is still more dependent, for an adequate supply of urinary buffer ammonia, upon renal ammonia production than appears at first from the fact that only 60% of the renal ammonia supply is directly derived from simultaneous release within the kidney of ammonia from precursor material, principally L-glutamine. As indicated in Figure II-9.12, the kidney is itself the principal source of the circulating ammonia in the resting mammal (481), and most of the ammonia entering the kidneys in the renal artery has previously been released within the kidney by deamidation and deamination of glutamine. That is why these processes, though directly accounting for only about 55% of the ammonia being supplied at any moment to the kidney, account indirectly and directly for about 90% of the ammonia under resting conditions.

During exercise, however, the muscles are a major contributor to the circulating ammonia and hence to that portion of the renal ammonia supply accounted for by the renal arterial ammonia inflow. This ammonia is not derived from glutamine, but from adenosine monophosphate (398, 192, 359). Some ammonia is also produced under resting conditions in the peripheral tissues (416), including muscle (359), from aspartate and other amino acids; but this is not net ammonia production. On the contrary, the peripheral tissues and the brain of the resting animal absorb a small amount of ammonia from the circulating blood (544, 610, 62, 648, 101, 617, 520, 49, 588, 635). The ultimate source of most of the urinary ammonia is the glutamine N brought to the kidneys in their blood supply, and, in the fasting animal, where this glutamine N must be derived from metabolism and not directly from the diet, the ultimate cellular source of the urinary buffer ammonia must necessarily be those cells which contribute to the circulating glutamine pool (Section D-2).

It has been suggested by several authors that ammonia has physiological functions other than its centrally important one of serving as volatile urinary buffer for excess nonvolatile acid. There is no doubt that NH_3 is enzymatically fixed into organic compounds, but the utility of this fixation appears to reside in its serving as (i) a mechanism protecting the organism against toxic ammonemia (Section D-2), and (ii) a means of sustaining somatic N conservation in the face of dietary protein shortage. The blood ammonia concentration necessarily fluctuates passively in relation to systemic acid-base regulation, rising with rising urinary pH (because then the kidney returns more of the ammonia supply to the systemic circulation) and with rising renal ammonia production; and it is very difficult to see how such fluctuations could be harnessed so as to regulate any more specific cellular process in the interest of the animal economy. It seems, for example, more likely that depression of insulin secretion by ammonia *in vitro* (160) is a toxic effect.

Lund, Brosnan, and Eggleston (365) reckoned from tissue ammonia analyses that there is probably more than a 31-fold gradient of the total ammonia concentration between liver cell water and hepatic vein in the rat, which is reduced to 3.3 by anaerobiasis; the data were taken to suggest active transport of NH_4^+ into the liver cells. They calculated that cell pH would have to be improbably low (5.89) to account for the ammonia ratio normally observed; but they did not allow, in this estimate, for the much higher pK′ to be expected in cell water because of its high ionic strength (I-4B; Appendix C), which must raise the equilibrium cell-to-blood ammonia ratio. Two serious obstacles to accepting a normal high cell-to-blood ammonia ratio as evidence of active ammonia transport into cells lie (i) in the great difficulty of reckoning the tissue P_{NH_3}, and hence the tissue-to-blood NH_3 pressure gradient, from total tissue ammonia measurements even if these are accorded full credence (261); and (ii) in the lack of economy implied by an active ion transport process which would either be unnecessary or else rendered inefficient by back diffusion of NH_3.[1] The effects of simultaneous anoxia, hypercapnia, and cyanide intoxication are unphysiological and difficult to interpret. Until more decisive evidence to the contrary is adduced, it seems justifiable to regard the presence of ammonia in the blood stream as primarily representing transport of this waste product of N metabolism to the kidney for excretion. This is of course not to deny the well established fact that both ammonia and urea, including administered material, partially provide NH_3 for synthetic reactions. As in the case of CO_2, NH_3 *molecules* are fixed, but protein catabolism inexorably results in net production of urea plus NH_3 (544, 454, 365).

2. Circulatory Transport of Ammonia as Glutamine

In all probability (cf. Section C-2) the human kidney continuously extracts glutamine from the circulation and continuously produces ammonia under all conditions, though the production rate may be quite low in chronic alkalosis (II-9; II-10B) and when the kidney is diseased (III-B3). The splanchnic circulation also takes up glutamine in the fasted animal (see below); evidently these continuous withdrawals from the circulating glutamine pool must be balanced by continuous contributions from other tissues. Addae and Lotspeich (1, 2, 358) reported, on very scanty and suboptimal evidence (263), that the liver does, and the peripheral tissues do not, contribute glutamine to the circulation. On no sturdier direct evidence than this, plus the supposed lack of glutamine synthetase in mammalian skeletal muscle (see below), Cahill and Owen (85) asserted in 1970 that "the liver is undoubtedly the principal producer of glutamine," a view voiced also by other recent reviewers (416, 417, 435). The relation of the liver to the circulating glutamine pool is in fact at present writing in need of clarification.

There seems, however, to be no doubt that the peripheral tissues are a major source, probably the principal source, of the glutamine that reaches the circulating pool. In 1968 Kerr, Reid, and Hills (307) reported that the hind legs of fasting mongrel dogs, greyhounds, and spider monkeys continuously put out glutamine; Figure II-9.13 shows that this release of glutamine N always exceeds the constant but small uptake of ammonia from the circulation. These quantitative observations of peripheral glutamine output have now been confirmed qualitatively (i.e., by measurements of arteriovenous glutamine differences without flow estimations) by Marliss *et al.* in man (381) and by Ruderman and Lund (532) in the rat.

In the fasted animal the glutamine N continuously released from the peripheral tissues can be derived only from the catabolism of whole body

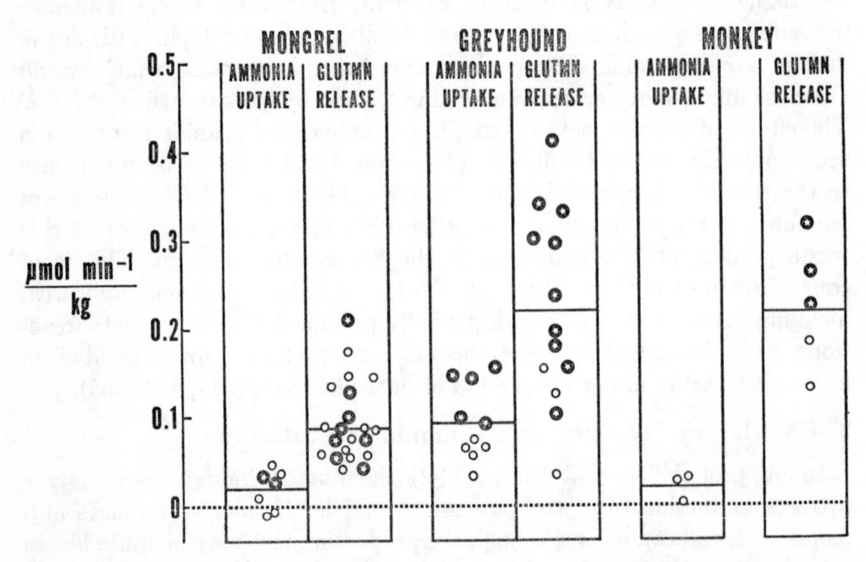

FIG. II-9.13. Uptake of ammonia and release of L-glutamine (*Glutmn*) by the peripheral tissues (hind leg) of fasted mammals. Reproduced from Hills, Reid, and Kerr (263). *Circles* represent mean values from individual experiments; *starred circles* indicate that the mean was significant. *Horizontal lines* indicate the average values of the individual means; all were significant. Ammonia is always taken up, but the quantity absorbed is insufficient to provide one N, let alone both N's, of the glutamine that is simultaneously released. Values are per kilogram of body weight.

protein. Only a small portion of this catabolic N could be made available in the form of the ammonia simultaneously being taken up from the blood (Figure II-9.13); accordingly, in 1968 we propounded the thesis that local protein catabolism in the peripheral tissues supplies the released glutamine N (307, 257, 263). Not only is the grossly negative N balance of the skeletal muscles obvious, to gross inspection, as wasting after a relatively short period of fasting; there is also known to be no net peripheral uptake of α-amino N during fasting, rather release. Also no N-containing compounds brought from elsewhere in the body which could furnish the peripheral cells with the glutamine N which they release are known to be taken up during a fast (627b, 332, 391a, 416, 417, 85). Our proposal has recently attracted the support of Marliss *et al.* (381).

It was further suggested (307, 254, 263) that release of local catabolic N as glutamine by the peripheral and other glutamine-releasing tissues represents NH_3 formed as the end product of protein catabolism and immediately incorporated, with great efficiency, into glutamine. This argument rests in the first instance upon relevant aspects of the fitness of chemical compounds for particular biological purposes, and upon the evolution of nitrogen excretion (254). NH_3 is the natural and immemorial end product of protein catabolism for the same thermodynamic and molecular reasons which render CO_2 the natural and immemorial end product of carbon metabolism, and H_2O that of hydrogen metabolism: no further bond energy can be extracted from these small, stable molecules (254). Of the three, only NH_3 has ceased to serve as bulk end product of N metabolism in the mammal, and the reason for the change, as well as its location in geological time, is well understood. It was the colonization of dry land, first by the adult amphibians and then definitively by the immediate ancestors of the protomammals, which necessitated replacement of gills by lungs; this adaptation in turn enforced the change from ammonia to some other form of bulk waste N excretion (420a, 28b). Gills receive most of the cardiac output, and they allow the ammonia concentration of the blood passing through them to be reduced practically to the normally very low level of the ambient water; by contrast, lungs can excrete almost no ammonia. Pulmonary elimination of ammonia by diffusion is ineffective because the solubility of NH_3 in water is extremely high (254). Lung-bearing, terrestrial animals must rely entirely on the kidneys to excrete any ammonia which must be eliminated as such, and the kidneys cannot compare with gills in their ability to excrete ammonia rapidly at low concentrations in blood.

The threat of toxic ammonemia was warded off, during the evolution of mammalian pulmonary ventilation, principally by the development of ureotelism. Low levels of circulating ammonia are maintained in the mammal because of highly efficient hepatic trapping of ammonia added to the portal venous blood by urea and glutamine lysis by intestinal bacteria with or without protein digestion in the intestines (543a), and of ammonia reaching the systemic blood from the kidney and other sources like contracting muscle (359). The waste ammonia taken up by the liver is largely converted to urea, which is secreted into the hepatic veins and then conveyed as urea to the kidneys for excretion. Activation of the hepatic arginine-ornithine enzyme system, which brought about evolution of mammalian ureotelism, can be observed during the metamorphosis of contemporary amphibians (415, 47); George Wald has called it the most striking instance of biochemical recapitulation known to us. Development of ureotelism did not, however, provide for a continuing supply of adequate urinary buffer to accept the excess protons claiming transfer from the buffers of extracellular fluid to the environment.

It is clear (Section D-1) that the toxicity of ammonia is sufficiently great that the mammalian organism could not derive sufficient urinary buffer ammonia under all circumstances by simple diffusion of arterial ammonia into the urine at tolerable blood ammonia concentrations. Therefore, if nonionic diffusion was to be retained as the mechanism of ammonia excretion, as gills gave way to lungs, it was necessary that a means be evolved of raising the blood ammonia pressure within the kidney above that of the systemic blood. We accordingly suggested (307, 257, 263) that the development of glutamine as a chemical vehicle for circulatory carriage of the organism's waste ammonia N to the kidney must have been an essential factor that permitted retention of ammonia excretion by nonionic diffusion. (An ancillary factor, as indicated in Section B, is the increase in papillary over cortical P_{NH_3} made possible under most conditions by the countercurrent structures of the mammalian renal medulla.) The fact that the mechanism of elimination of ammonia as such from the body, nonionic diffusion, has remained unchanged in the line of descent leading from unicellular organisms all the way down to man is presumably a testimonial to the economy of this process and to the difficulties of abandoning passive transport of so diffusible a substance. Catabolism of nitrogen-containing compounds all the way to NH_3 is also among the most venerable of the fundamental metabolic characteristics of animal cells; the adaptation by which ureotelism arose in the Devonian era among the amphibian ancestors of the protomammals is by contrast an extremely recent event.

Evolutionary considerations would then seem to favor the view that net glutamine synthesis and release by the peripheral tissues represents, like hepatic urea synthesis, an evolutionary adaptation providing a means of transporting ammonia N. In this manner the end product of peripheral protein catabolism is carried in a harmless chemical vehicle to the kidneys. However, it had been widely supposed for a generation that the principal peripheral tissues, the skeletal muscles, were incapable of net glutamine synthesis, though it was realized that isotopic ammonia N is rapidly incorporated into muscle glutamine. Net glutamine synthesis by skeletal muscle was discounted on enzymological evidence (391a), and the question was not directly examined by measuring peripheral arteriovenous differences until the mid-1960's. The glutamine synthetase system (318)—perhaps systems would be more accurate (282)—is widely supposed (391a) to be the only means by which NH_3 is directly incorporated into glutamine; and activity of this enzyme had been found in skeletal muscle neither by Krebs (318), who discovered the enzyme, nor in an extensive later survey of many species (674). Recently, however, Lund and Goldstein (366) showed that accumulation of inhibitory adenosine diphosphate in the conventional *in vitro* system interfered with demonstration of glutamine synthetase activity in muscle. They demonstrated activity in the skeletal muscle of lower vertebrates, and Iqbal and Ottaway (282) have now shown activity in mammalian muscle.

There is therefore no remaining enzymological objection to assuming that net incorporation of NH_3 into glutamine occurs via the glutamine synthetase pathway. Only a small portion of the glutamine released by the peripheral tissues of the fasting animal can be derived from ammonia simultaneously taken up (Figure II-9.13), and the rest of the released glutamine may all be derived from ammonia catabolically released into cell water. Color was lent to this supposition by the demonstration (307, 257, 263) that raising the P_{NH_3} of the arterial blood which supplies the peripheral tissues, by infusing ammonia salts into the femoral artery, regularly increases the rate of ipsilateral glutamine output (Figure II-9.14). The continuous net uptake of ammonia by the peripheral tissues of the fasted animal, and the increased uptake when blood P_{NH_3} is raised, indicate that NH_3 diffuses into the peripheral cells at a rate proportional to the NH_3 pressure gradient existing between the capillary blood and a pool of ammonia in cell water. Increased glutamine release into the circulation during femoral arterial ammonia infusion shows that the rate of glutamine release is responsive to the rate at which ammonia is being added to the cellular ammonia pool, and that any NH_3 added to this pool by local protein catabolism must contribute to the continuous release of glutamine N into the circulation. Indeed, most of the ammonia added to the cell water appears to be incorporated with great efficiency into glutamine, for Hills, Reid, and Kerr (263) have estimated that most of the additional net peripheral uptake of ammonia N observed during femoral arterial ammonia infusion can be accounted for as additional release of glutamine N during the infusion, plus a smaller quantity of N released as glutamine and as ammonia after cessation of the infusion.

Though it is a plausible speculation that development of the circulatory carriage of waste N as glutamine and as urea were contemporaneous

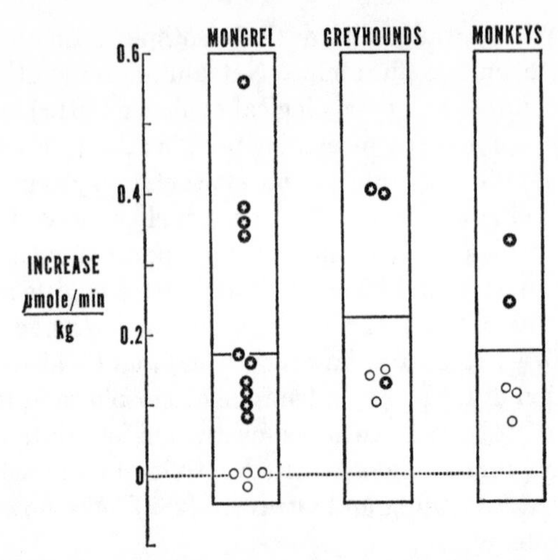

FIG. II-9.14. Increase (over control values) of glutamine release by mammalian hind leg during infusion of ammonium acetate or ammonium carbonate into the ipsilateral femoral artery. Reproduced from Hills, Reid, and Kerr (263). Symbolization is the same as in Figure II-9.13. Averages of the mean increases (*horizontal lines*) were all significant.

events in the evolution of mammals, evidence from comparative physiology which could bolster a specific historical reconstruction is completely lacking in the case of glutamine. Aside from output of glutamine by the liver of the carp (451, 452), there seem to be no reports of arteriovenous glutamine differences in any lower vertebrates. Perhaps, like urea synthesis, carriage of ammonia as glutamine from the tissues to the excretory organ will be found to have evolved more than once as an adaptation to environmental aridity. The subject deserves systematic investigation.

Present knowledge of glutamine uptake and output in the several regions of the mammalian circulation has been reviewed by Hills, Reid, and Kerr (263). It is noteworthy that the only two divisions of the mammalian vascular system in which glutamine uptake has been demonstrated are the renal (Section C-2) and hepatic-splanchnic circulations (1, 2, 93, 381, 263; Figure II-9.15). Both these circulatory beds are metabolically specialized in the very relevant respect that there is effective net removal of ammonia from them—respectively, by renal excretion and by highly effective hepatic trapping of ammonia and its conversion to urea, which is then carried to the kidneys in the circulation and excreted in the urine. Harmless circulatory conveyance of NH_3 as urea to the urine is strikingly analogous to conveyance, in particular, of that portion of the peripherally released glutamine N which represents peripheral uptake of circulating ammonia. The increasing uptake of circulating ammonia by the glutamine-releasing tissues as blood ammonia rises must be considered an ancillary defense against systemic ammonemia.

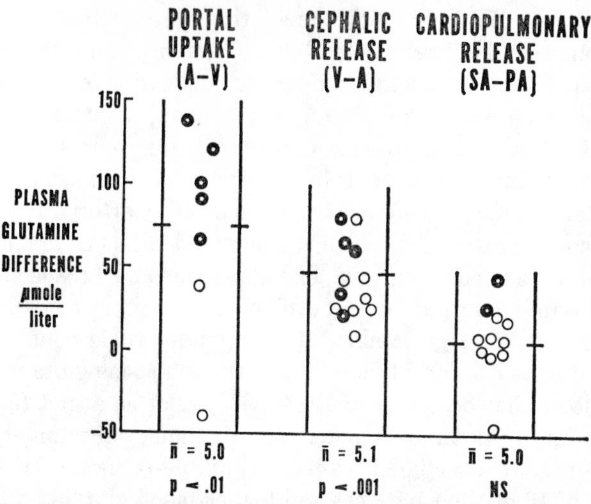

FIG. II-9.15. Portal uptake and cephalic and cardiopulmonary release of glutamine in the fasted greyhound. Reproduced from Hills, Reid, and Kerr (263). Symbolization is similar to the two preceding figures. Mean cardiopulmonary glutamine release was significant in two experiments; the mean of the mean cephalic releases was significant, as was the mean of the mean glutamine uptake by the tissues (intestines, spleen) drained by the portal vein. *A-V:* artery less portal vein; *V-A:* external jugular vein less systemic artery; *SA-PA,* systemic artery less pulmonary artery; $\bar{n}$*:* mean number of estimates per experiment.

Leaving aside the hepatic-splanchnic and renal circulations, it appears that release of glutamine into the circulation, as a conveyor of local waste N, may be a general property of mammalian tissues which are not specialized in ammonia disposition. In addition to the extremities, both the head and the cardiopulmonary circulation release glutamine (Figure II-9.15). The glutamine N released from the head is not derived from circulating ammonia, since little or none is taken up except when the arterial ammonia is raised (263). Much of this glutamine is evidently provided by the brain, for Ruderman and Lund (532) have recently demonstrated directly that the glutamine concentration of blood taken from the cranial sinuses of the fasted rat exceeds the arterial concentration. Glutamine is synthesized by cerebral glutamine synthetase (318, 391a); the N of administered ^{15}N-ammonia is incorporated into cerebral glutamine (544, 49, 135a, 635), and when in dogs the blood ammonia level rises after hepatectomy (171), or ammonium salts are infused into the carotid artery (103), the brain glutamine content increases. It may be that cerebral glutamine synthesis helps protect the brain tissue against ammonia toxicity. Conversely, it has been suggested that the mechanism of impaired cerebral function during ammonia intoxication might reside in the diversion of Krebs-cycle intermediates from provision of energy to synthesis of glutamate and glutamine (538a).

The relation of the liver to the circulating glutamine pool is at present unclear. Carlsten, Hallgren, Jagenburg, Svanborg, and Werkö (93) observed hepatic venous-arterial differences indicative of net splanchnic-hepatic glutamine uptake in postabsorptive man, a finding recently confirmed (381); but in the absence of simultaneous portal venous data it is impossible to judge from such findings whether or not the liver is producing glutamine, i.e., detracting from the positive arterial-hepatic venous glutamine concentration difference which would result from spleno-gastro-intestinal glutamine consumption. Though Addae and Lotspeich (1, 2, 358) reported glutamine output by the canine liver, indirect evidence points to net hepatic uptake of glutamine, at least under some conditions. Hepatectomy in dogs is regularly followed by a rise in plasma glutamine (17, 18); and liver disease has been reported (644, 675) as often as not (551, 118a) to cause elevation of blood glutamine. Indeed, though Seegmiller, Schwartz, and Davidson (551) concluded that blood glutamine does not rise in hepatic cirrhosis, 3 of 13 of their patients had higher blood glutamine than any of their normal control subjects. It is of course possible that portal hypertension interfered with splanchnic glutamine consumption. On the other hand, perfusion of the isolated rat liver with various substrates (364) has suggested that glutamine synthetase plays a minor role in the disposition of ammonia taken up by this organ.

Whereas waste CO_2 is always available in quantities far beyond any possible need for urinary buffer for excess base, this is not invariably true in the case of the principal urinary buffer for excess acid, NH_3. When the diet is adequate in protein, excess waste N is available, and acidification of the urine, with its associated obligatory increase in ammonia excretion, results merely in excretion of correspondingly less nitrogenous waste as urea (182). But when the diet is marginally adequate in protein intake in man (521, 598) and experimental animals (460, 8, 370a), negative N balance is respectively precipitated or increased by urine acidification: effective defense of extracellular fluid pH takes precedence over N conservation. The defense is automatic, resulting from the control exerted by urinary pH over the passive partitioning of the renal ammonia supply between renal veins and urine (Section B). Evidently, hepatic urea production cannot be further throttled sufficiently to compensate for the increased loss of ammonia N when, under these circumstances, the urine is acidified.

NOTES

1. Net ammonia fixation in the liver (or elsewhere) would certainly be expected to cause unidirectional net ammonia transport by nonionic diffusion down the resulting NH_3 pressure gradient. What is difficult to accept without conclusive supporting evidence is that consequential NH_3 pressure gradients would be maintained across cell membranes with or without net ammonia transport.

ten

P_{CO_2}, P_{NH_3}, AND THE MEDULLARY COUNTERCURRENT SYSTEMS

A. URINARY AND INTRARENAL PRESSURES OF CO_2 AND NH_3

Diffusion is an extremely effective means of transporting highly diffusible gases like CO_2 and NH_3 over microscopic distances; the high surface-to-volume ratio of fluid contained in capillary-sized structures is the physical factor which permits massive gas transport between blood and either tissue or alveolus by diffusion down gas pressure gradients, which may be so small as to be unmeasurable, across the capillary wall. It has always been thought unlikely that the nephrons would offer any greater barrier than the capillaries to equilibration of the pressures of CO_2 and NH_3 between glomerular filtrate and the adjacent renal tissue (535, 621, 489); and Oelert, Uhlich, and Hills (430) have shown that filtrate P_{NH_3} is in fact equilibrated with cortical tissue P_{NH_3} in both proximal and distal convolutions of the rat nephron. Whether gas pressure gradients develop in any part of Henle's loop has not been directly studied; there is evidence (see below) that the P_{CO_2} and P_{NH_3} of luminal fluid within the collecting ducts remain equilibrated with those of the surrounding medullary tissue.

Raised urinary pressures of CO_2, however, have attracted attention for a century (597, 261); in low-flow alkaline urines, values over twice those encountered in venous blood are measurable (II-8C), and in such urines P_{NH_3} can rise to at least 150 times the venous blood values (261). Ochwadt and Pitts (428) showed that intravenous infusion of carbonic anhydrase at a rate sufficient to perfuse the nephrons with enzyme, as shown by its recovery in the urine, regularly reduces the high CO_2 pressures of alkaline urine; evidently H_2CO_3 synthesized by acidification of HCO_3^--rich glomerular filtrate normally continues to be dehydrated beyond some point at which the resulting CO_2 cannot escape by diffusion down a gas pressure gradient into the tissues. Nevertheless, in animals in which delayed H_2CO_3 dehydration is prevented by intravenous infusion of carbonic anhydrase (II-8C), the pressures of both CO_2 and NH_3 in high-pH urine are still higher than the venous pressures (Figures II-10.1 to II-10.3).

Under these circumstances, the raised urinary CO_2 pressures can only be ascribed to countercurrent exchange of gas between the ascending and descending limbs of Henle's loop and of the vasa recta, resulting in rise of medullary above cortical P_{CO_2}. Countercurrent systems are in fact the only structures in the mammalian organism that we know of which can provide for the development of a substantial gradient of concentration or pressure of substances which rapidly diffuse across cellular and other biological membranes; it was this consideration which led Werner Kuhn to postulate that

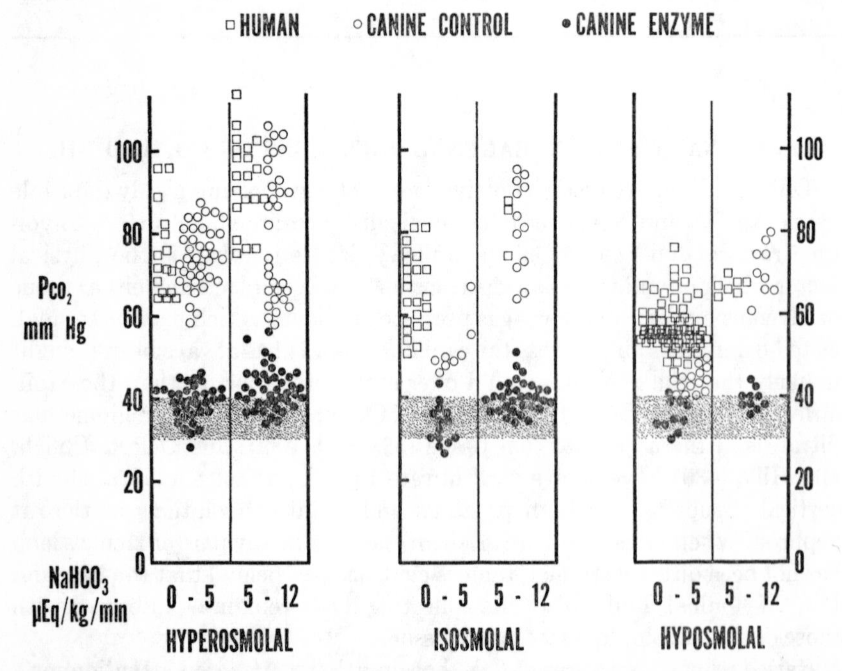

Fig. II-10.1. The P_{CO_2} of mammalian urine related to its water content and its bicarbonate content. Reproduced from Hills and Reid (261). *Solid symbols*, carbonic anhydrase infused intravenously. *Open symbols*, no enzyme. Values are similar for dogs (*circles*) and man (*squares*). *Shaded zone* shows range of arterial P_{CO_2} in the dogs, whose ventilation was assisted during light anesthesia.

Cortical P_{CO_2} is always close to arterial P_{CO_2}. Papilla-to-cortex CO_2 pressure gradients are always proportionately small compared with NH_3 pressure gradients, but do develop in the enzyme-infused dog (as reflected here in the urinary P_{CO_2} values) provided alkali is infused at a rate rapid enough to cause mild acute nonrespiratory acidosis.

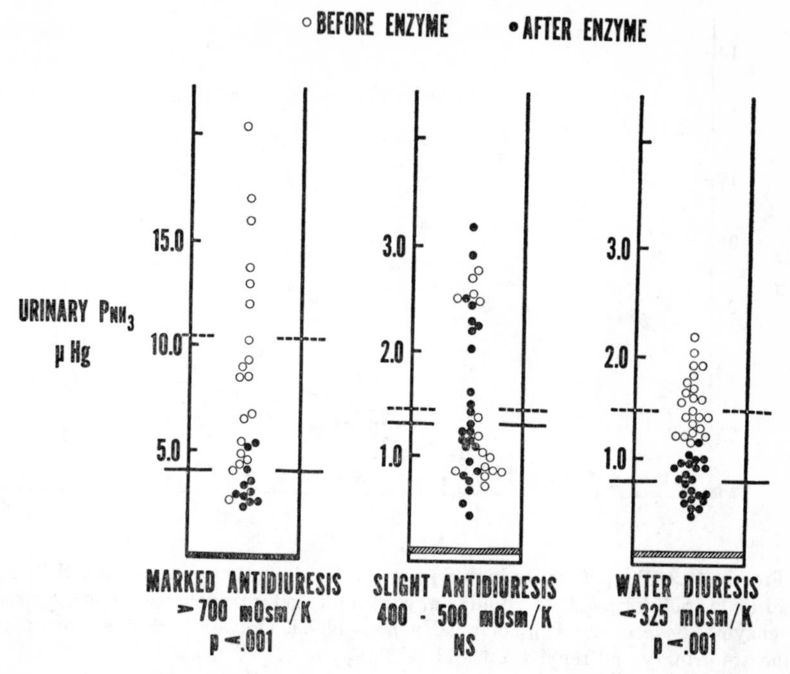

FIG. II-10.2. P_{NH_3} of alkaline canine urine. Reproduced from Hills and Reid (261). *Solid circles*, carbonic anhydrase infused intravenously. *Open circles*, no enzyme. Note larger scale of *left-hand bar*. Because of the difficulty in obtaining maximal water diuresis in the anesthetized animal, the low-osmolality samples (*water diuresis*) were procured by adding mannitol to the infusion solutions.

medullary osmotic stratification, effected by countercurrent exchange and multiplication, was the means by which urine that is hypertonic to plasma is elaborated (323, 666, 322, 213, 621).

The effectiveness of countercurrent exchange in creating an osmotic or concentration gradient between cortex and medulla is in general directly related to the diffusivity of the substance being exchanged (621); considerably larger cortex-to-papilla pressure gradients (in proportion to the absolute pressures) are seen in the case of the more diffusible NH_3 as compared with CO_2 (Figures II-10.1 to II-10.3). Axially directed medullary gas pressure gradients in which the pressure is higher at the papilla than in the cortex—these are referred to throughout this book as papilla-to-cortex gradients—represent steady states resulting from countercurrent trapping of gas added along some or all of the looped structures themselves; the magnitude of the pressure gradient which develops is naturally a direct function of the rate of entry of the gas into the countercurrent systems as well as a function of the diffusivity of the gas. However, the details of the process are not at present known for either CO_2 or NH_3. CO_2 is clearly reabsorbed from the

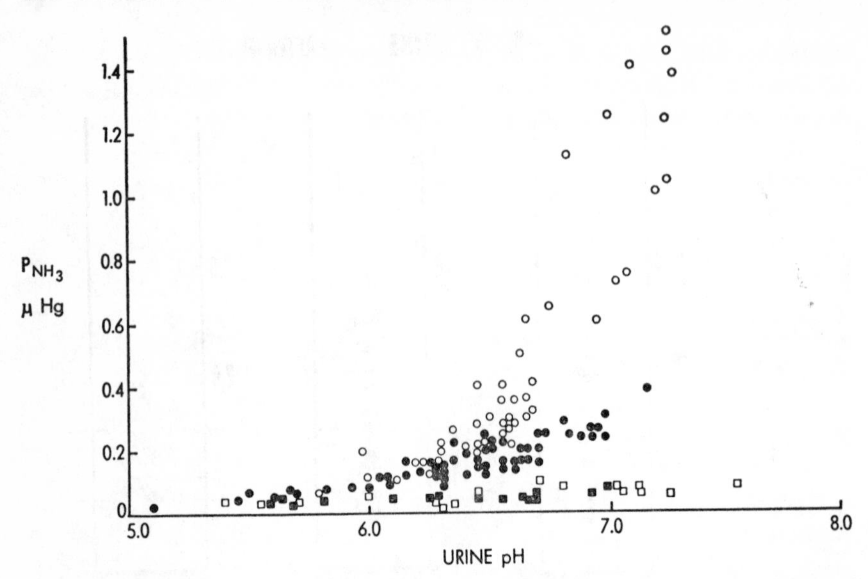

FIG. II-10.3. P_{NH_3} of canine high-flow urines (*circles*). Reproduced from Hills and Reid (261). *Solid symbols*, carbonic anhydrase infused intravenously. *Open symbols*, no enzyme. *Squares* are P_{NH_3} of renal venous blood. Only in high-flow, lowest-pH urine are urinary and renal venous blood P_{NH_3} equal.

collecting ducts into medullary tissue as a result of osmotic water reabsorption (Section C) from the HCO_3^--containing fluid within the collecting ducts during antidiuresis (495, 259). Though inflow into the medulla of the NH_3 contained in the glomerular filtrate that descends from the proximal convolution and pars recta would of itself create a cortex-to-papilla rather than a papilla-to-cortex NH_3 pressure gradient, it is very possible, as suggested by Gottschalk, Lassiter, and Mylle (214), that NH_3 is generated within the descending limb as a result of rising pH of luminal fluid; the effect would be to create a NH_3 pressure gradient from papilla to cortex. But whether in fact there is a progressive rise in the pH of luminal fluid as it passes down the descending limb of Henle's loop remains uncertain at this time.

Such a rise was considered probable by Gottschalk *et al.*, especially during antidiuresis, because progressively increasing concentration of the luminal fluid, generally reaching a maximum at the hairpin bends, would of itself markedly raise [HCO_3^-] and hence pH. The pH of the fluid would then again decline in the ascending limb if any of the Na^+ extruded is exchanged for protons; and in fact Gottschalk *et al.* found the pH of luminal fluid in the early distal convolution of the rat to be similar to that of samples from the last accessible portion of the proximal convolutions. However, we have at present writing no direct knowledge of the pH of luminal fluid in any portion of Henle's loop, and it remains possible, not to say likely, that a low disequilibrium pH develops within the descending limb of the loop, as in the distal tubule, because of delayed dehydration of H_2CO_3 synthesized by reaction of luminal HCO_3^- with secreted acid. NH_3 is not generated in fluid proceeding down the descending limb of the vas rectum, at least under diuretic conditions; the ratio $HCO_3^-:CO_2$ declines in both saline and HCO_3^- diuresis (618). Active transport of CO_2 and NH_3, as of H_2O, is improbable in the nephron or elsewhere in the body because of the ease with which these small uncharged particles pass through membranes; we may assume movement of CO_2 and NH_3 between the various medullary structures to be passive, but available information about the direction and magnitude of such movements is incomplete. We do know, however, a good deal about the axially directed gradients of P_{CO_2} and P_{NH_3} between canine papilla and cortex as a function of varying acid-base loads and urine flows, and from available information it appears that the relations are very similar in man.

Much higher pressures of both the buffer gases are observed in high-pH urine under normal conditions than when carbonic anhydrase is being induced intravenously (Figures II-10.1 to II-10.3). In the case of CO_2 the difference might be due to the fact that, in the normal absence of carbonic anhydrase catalysis in the distal convolutions, much of the H_2CO_3 synthesized there reaches the medullary collecting ducts before it is dehydrated, increasing the total rate of delivery of CO_2 to the countercurrent structures. This does not seem to be a very effective way of raising medullary P_{CO_2}, however, for Uhlich, Baldamus, and Ullrich (618, 28a) found CO_2 pressures in the vasa recta of the acutely alkalotic rat which had been treated with carbonic anhydrase to be essentially the same (50 mm Hg) as in the absence of enzyme. A higher equilibrium P_{CO_2} was observed in the medullary collecting duct as compared with that in vasa recta blood, indicating delayed dehydration of H_2CO_3 present in the luminal fluid at a concentration above equilibrium. These findings indicate that, in the diuretic acutely alkalotic rat, postpapillary dehydration of H_2CO_3 is the principal cause of the raised values for urinary P_{CO_2} present under normal circumstances over values observed during infusion of carbonic anhydrase.

Hills and Reid (261) have also presented evidence that the much higher pressure of NH_3 in canine alkaline urine observed during antidiuresis and at high urinary flow rates under normal conditions, as compared with values in animals infused with carbonic anhydrase (Figure II-10.3), results from rise in urinary pH beyond papilla. Rise in pH must cause rise in the ratio $NH_3:NH_4^+$ and, hence, if ammonia is not correspondingly reabsorbed, rise in P_{NH_3}. By comparing papillary and bladder urine, they showed that little or no ammonia is reabsorbed from the urinary tract of the dog, nor from the bladder of man. Because of the much higher absolute urine-to-tissue pressure gradient, however, CO_2 is reabsorbed at a significant rate from low-flow, high-pH urine; and some of the postpapillary rise in P_{NH_3} results in such urines because of the rise in pH produced by reabsorption of CO_2 from the urinary tract.

Too little CO_2 is reabsorbed at higher urine flows to account, to any appreciable extent, for the higher values for urinary P_{NH_3} seen in the normal dog as compared with the carbonic anhydrase-infused animal. It is therefore clear that during water diuresis the difference between urinary P_{NH_3} with and without carbonic anhydrase infusion is the result of continuing dehydration of H_2CO_3, under normal circumstances, beyond some point at which NH_3 cannot diffuse into the tissues down a NH_3 pressure gradient. Physical and anatomical considerations speak for the identification of this point as the urinary papilla, where the urine abruptly moves from a conduit of microscopic dimensions, favoring equilibration, into a hollow viscus of macroscopic dimensions, in which the surface-to-volume ratio of the urine is abruptly and greatly reduced, and in which diffusion is known to be ineffective in dissipating urine-to-tissue osmotic and gas pressure gradients (258, 261). On the basis of these considerations, and the experimental confirmation presented in the rat by Uhlich and his colleagues (28a, 618), it appears established that rise in urinary P_{CO_2} and P_{NH_3} beyond the papilla is the principal cause of the higher pressures of the buffer gases in isosmotic and higher-flow urines in the normal animal as compared with the enzyme-infused animal.

Impermeability of the terminal nephron segment, including the medullary collecting duct, to NH_3 was originally postulated by Orloff and Berliner (434) to explain their finding that water diuresis does not accelerate the rate of ammonia excretion in canine lower-pH urine. This argument would not be valid, however, if the observation were correct (II-9C; Appendix J); and it is now established (Appendix H) that water diuresis regularly increases the rate of urinary ammonia excretion in the mammal, irrespective of urinary pH.

Finally, on the assumption that the P_{NH_3} of urine and tissue is equilibrated at the papilla in man, it becomes possible to calculate theoretically the distribution of the renal ammonia supply between urine and renal venous blood, in any steady state, from the respective pH and flow rates of renal venous blood and urine and the cortical and papillary P_{NH_3} (II-9C). The agreement of observed and theoretical values for urinary ammonia excretion as a function of urinary pH during brisk water diuresis (Figure II-9.8) constitutes synoptic evidence for the soundness of the assumptions that the rate of ammonia outflow in the urine and renal venous blood is determined by the relative values for tissue P_{NH_3} at these locations and by the pH and flow rates of renal and venous blood, respectively.

Passive distribution of the NH_3 formed in the kidney between renal venous blood and urine ensures with admirable effectiveness that more buffer for excess acid shall be delivered to the urine as the need for it increases, with rising acid load and falling urine pH (II-6). If urinary ammonia production is little altered by physiological fluctuations of urinary pH, it is inevitable (II-9, Equation 5) that the effective urinary [NH_3] shall fall, unless the ratio of papillary-to-cortical [NH_3], c_1:c_2, can be made to rise. In fact, however, c_1:c_2 also falls as the urinary pH declines; inevitably so, if ammonia transport is wholly passive, since increasing the urinary ammonia outflow "bleeds off" NH_3 from the countercurrent systems in the region of the hairpin bend, thus diminishing the effectiveness of countercurrent exchange. Cortical P_{NH_3} is higher during antidiuresis (261), and the ratio of papillary-to-cortical P_{NH_3} probably increases very considerably under the influence of vasopressin (II-9C); the resulting increase in papillary P_{NH_3} in the antidiuretic person or animal is valuable to the organism, inasmuch as it helps to counteract the inevitable diminution, during hydropenia, of the kidneys' capacity to eliminate excess acid, buffered as NH_4^+, at a rapid rate (261).

The rate of urinary ammonia excretion would increase as an exponential function of declining urinary pH if papillary P_{NH_3} were held entirely constant (Figure II-6.2). The increase is in practice less, as pH falls, because of the lowering of the P_{NH_3} of more acid papillary urines which, as has just been noted, results from (i) a declining papilla-to-cortex NH_3 pressure gradient and (ii) the characteristics of the system that governs the passive distribution of the renal ammonia supply (II-9, Equation 5). In order for this system to stabilize papillary P_{NH_3} as well as it does, it is necessary that much of the ammonia made in the kidney always be returned via the renal venous blood to the general circulation: about half of it when the urine pH is maximally acid, almost all when the urine is alkaline (Figure II-9.10). The return of so much of the renal ammonia production to the systemic circulation is of course wasteful from the standpoint of energetics, since the original synthesis of the precursor glutamine in the tissues of the extremities and head must be recapitulated, or else urea must be synthesized, by incorporation of the ammonia contributed to the systemic circulation. An additional disadvantage is that the kidney becomes the principal contributor of ammonia to the circulating blood, where this substance is always present at fairly close to toxic concentrations. However, this element of inefficiency and potential hazard is a feature indispensable to the effectiveness of the system that provides the urinary ammonia, whose operation depends upon stabilization of the effective P_{NH_3} at the papilla in the face of increasing rates of ammonia outflow as urine pH declines. The urinary ammonia outflow never normally accounts for much more than half the renal ammonia supply, and this permits papillary P_{NH_3} to be sustained, as urinary ammonia outflow increases, so as to support the large rate of urinary ammonia excretion in acid urine by purely passive means.

The CO_2 tension of very acid urine is commonly reduced well below arterial P_{CO_2}. An early explanation given by Ullrich and Eigler (619a) has remained current, but may not be correct, since it depends upon the postulate (II-8B) that the HCO_3^- left behind in cell water when a proton derived from H_2CO_3 is secreted into the lumen is then physically transported across the acid-secreting cell and into the peritubular blood. Since CO_2 cannot be generated by secretion of acid into HCO_3^--free luminal fluid, so the argument runs, cellular CO_2 depletion results from the hydration of CO_2 to supply secreted protons, and CO_2 must diffuse into the cell from the medullary tissues in the region of the hairpin bends of Henle's loop; the result is to create a cortex-to-papilla CO_2 pressure gradient which is reflected in lowered urinary and tissue P_{CO_2} at the papilla.

Unless the secreted H^+ made available by hydration of cellular CO_2 were replaced by H^+ at the antiluminal surface, replacement of cellular H_2CO_3 by $NaHCO_3$ would progressively alkalinize cell water anywhere in the nephron (cf. II-8B). But we cannot at present distinguish between antiluminal transport of $NaHCO_3$ out of the cell and antiluminal exchange of Na^+ for protons taken into the cell from the blood (II-8B); and only the former and probably less likely process would deplete cellular CO_2. Probably, therefore, the lowered P_{CO_2} of very acid urine reflects metabolic fixation of CO_2 by the medullary tissues (621). Water diuresis raises the P_{CO_2} of very acid low-flow urine, principally by rapidly conveying some H_2CO_3 synthesized in the distal convolutions past the papilla before the excess above equilibrium has disappeared (261); a small amount of HCO_3^- still present because of the raised H_2CO_3 then reacts with protons in the urinary tract and generates enough CO_2 to raise P_{CO_2}.

B. INCREASED RENAL AMMONIA PRODUCTION IN ACIDOSIS

Figure II-10.4 shows the effect of administration of a large acid load to a dog upon the rate of urinary ammonia excretion; when the load is sufficient to produce moderate subacute or chronic acidosis, an increase in ammonia excretion to about five times the normal rate of excretion in acid urine is regularly observed. Much lesser increases are observed in respiratory acidosis (28, pages 297–298). The markedly increased ammonia excretion during nonrespiratory acidosis reflects increased renal production of ammonia, the ammonia supply being distributed between urine and renal venous blood according to the kinetic principles already presented (II-9C). This response to the threat of acidosis is present to some extent in most vertebrates, but it is not as well developed in certain herbivorous mammals as in man and carnivorous mammals. In man (and many other mammalian species) the response takes about 5 days to develop fully and then stabilizes at a plateau corresponding to about a five-fold increase in renal ammonia production in spite of continuing acidosis or excess acidic load.

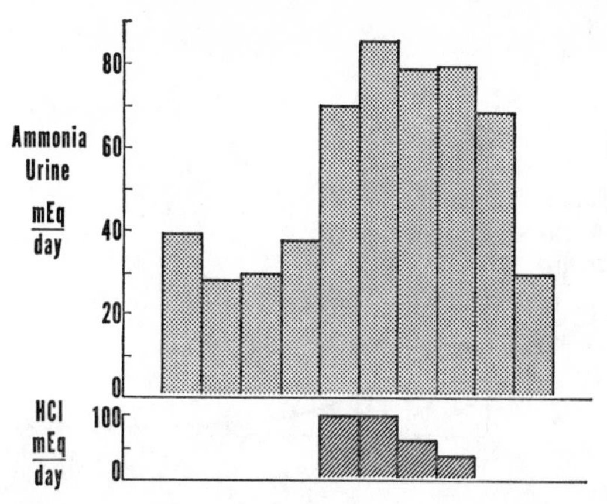

FIG. II-10.4. Effect of administered mineral acid on urinary ammonia excretion. The first such experiment reported (645). The experimental animal was a dog that weighed 11 kg and received 500 g of fresh meat daily. Each *bar* represents a 24-hour observation period. With continued acid administration, increased ammonia excretion in relation to urine pH is sustained indefinitely.

When the stimulus is removed, about the same time is required after the acidosis is abolished for urinary ammonia to return to normal levels reactive to urinary pH. The adaptive increase in renal ammonia production in acidosis entails increased renal extraction of circulating glutamine, but this is not consistently attended by rise in arterial glutamine. Practically all the glutamine filtered at the glomerulus is reabsorbed under all circumstances irrespective of the blood glutamine level and of the acid-base state; but the latter determines the fate of the glutamine reabsorbed into the tubular cells (458). In chronic alkalosis it is almost all returned unchanged to the peritubular blood; in acidosis it is largely deaminated and deamidated to yield ammonia, and there is additional glutamine extraction from peritubular blood. It appears that adaptive increase in urinary ammonia production in acidosis results from some effect of the abnormal acid-base state of the organism on the renal cellular milieu.

A truly massive investigative onslaught upon the nature of this intracellular alteration, especially in the past decade, has failed to disclose its nature; and once again the wholesale slaughter of beautiful hypotheses by ugly facts, of which W. H. Huxley regretfully spoke, has been the order of the day. The hypotheses have largely fallen into four classes: the increase might be due to (i) increase (for a variety of reasons) in the intrinsic activity of the deaminating enzymes; (ii) alteration of the concentration of cofactors, in particular an increased ratio of oxidized-to-reduced pyridine nucleotides; (iii) decrease in the concentration of the products of the deaminating reactions that yield ammonia, owing to a primary influence of acidosis upon some other reaction linked directly or indirectly to deamination; and (iv) alteration of the transport of substrate into the mitochondria.

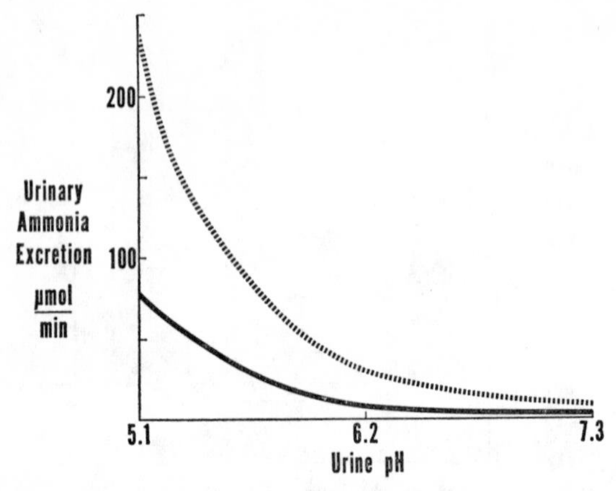

FIG. II-10.5. Urinary ammonia excretion in relation to the pH of high-flow urine under normal circumstances (*solid line*) and the effect of a tripling of the urinary ammonia supply during acidosis (*dotted line*). Schematic. Ordinarily the urine pH is low during acidosis, even in the presence of advanced renal disease, so that usually only the *far left* portion of the *upper curve* is applicable. The urinary ammonia excretion can rise so high during acidosis because the adaptive increase in renal ammonia production is superimposed on the already markedly raised excretion rate effected simply by lowered urine pH; this in turn depends upon the exploitation by the kidney of excretory buffering by a volatile buffer system (I-6; II-6C), as reflected in the nearly exponential rise in urinary ammonia excretion as urinary pH declines.

It will not be profitable to enter further here into this literature, which has recently been reviewed by several authors (435, 28, 470a). It has exemplified a frequent difficulty associated with deployment of *in vitro* technics for the purpose of elucidating the cause of phenomena observed in the intact animal: particular explanations can be supported quite extensively and rather seductively by experimental evidence, owing to the latitude enjoyed by the investigator in choosing his conditions (cf. II-1A); but the degree of relevance of the findings to actual events observable in the intact animal is difficult to ascertain with certainty.

Increased ammonia production serves to sustain the effective NH_3 pressure at the papilla, i.e., to compensate for the fall in papillary P_{NH_3} inevitably resulting from urine acidification and the consequent increased urinary ammonia outflow at constant urinary ammonia production (II-9B). Increased ammonia production is an extraordinarily effective means of increasing the rate of urinary acid excretion in the face of large acid loads, because it potentiates an increase already produced with great efficiency by the lowered urinary pH associated with large acid loads (Figure II-6.2). Figures II-10.5 and II-10.6 show schematically the great increase in the rate of urinary acid secretion effected when a three-fold increase in renal ammonia production is superimposed upon the exponential increase in

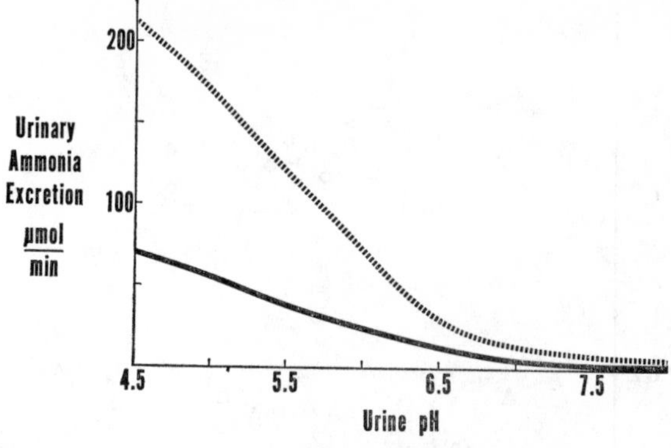

FIG. II-10.6. Urinary ammonia excretion in relation to the pH of low-flow urine under normal circumstances (*solid line*) and the effect of a tripling of the urinary ammonia supply during acidosis (*dotted line*). Schematic.

urinary ammonia excretion effected by reduction of urinary pH to minimal values.

Analogous increases in urinary HCO_3^- excretion cannot be brought about in the same way because very little CO_2 is produced (in relation to HCO_3^- excretion in high-pH urines) within the kidney. However, the increase in urinary $[H_2CO_3]$ in the terminal portion of the nephron, which occurs when high-pH urine is being secreted, is analogous to increased renal ammonia production in acidosis in that it increases the rate of base excretion as urinary HCO_3^- (II-8D).

C. EFFECT OF URINARY FLOW RATE ON URINE pH AND RENAL ACID AND BASE EXCRETION

The first report I have located indicating a connection between the urinary flow rate and renal acid-base regulation was the observation in 1854 by Becquerel and Rodier (43) that concentrated human urines are generally more acidic than dilute ones,[1] an observation later recorded by a number of authors. Dr. Grace Eggleton first noted in 1947, however, that higher-pH urines, instead of becoming more acidic as the urine flow rate declines, exhibit a rise in pH (138); her observations were confirmed the following year (32). It was not possible to explain them at the time, however, because the mechanism by which urine is concentrated was not then understood. The effect of flow upon urine pH is substantial and is mainly effected in the lower range of flow (Figure II-10.7).

Antidiuresis affects the urinary pH because the buffer gases CO_2 and NH_3 diffuse freely out of the luminal fluid in the distal tubules and collecting ducts where facultative water reabsorption occurs down the (imperceptible) gas pressure gradient created by dehydration of the fluid (668, 495, 259). The effect, necessarily, is to diminish the rate of excretion of the urinary

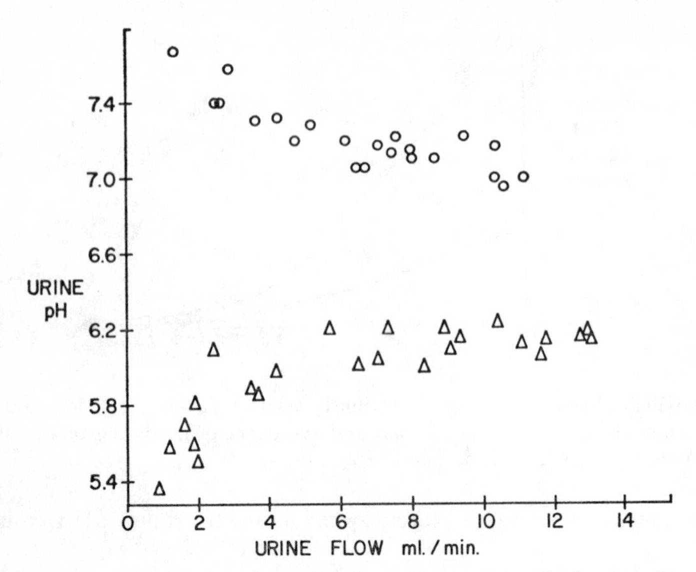

FIG. II-10.7. Effect of changing urine flow rate on the pH of urine. Reproduced from Hills and Reid (259). The experiments are typical of their respective pH ranges: markedly acid urine (pH < 6.3 at 37° C) becomes more acidic as flow declines, while the pH of higher-pH urine (pH > 6.5 at 37° C) rises with falling flow.

buffer ions HCO_3^- and NH_4^+, and so to decrease the rate of renal net acid or base excretion at any value for plasma pH. The principal effect of dehydration of higher-pH urines upon the acid-base composition of the fluid in the terminal nephron segment is the conversion of HCO_3^- to CO_2:

$$H^+ + HCO_3^- \rightarrow H_2CO_3 \xrightarrow{-H_2O} \underset{\text{diffuses off}}{\overset{CO_2}{\downarrow}} \qquad (1)$$

The loss of CO_2 down the CO_2 pressure gradient created by water reabsorption causes protons to be donated by all the nonvolatile urinary buffers (as indicated by H^+ at the left) to HCO_3^-, and urinary pH rises. In the more acid urines, which contain little HCO_3^- and plenty of NH_4^+, the principal effect of water loss upon the acid-base composition of the urine is transfer of protons from NH_4^+ to the nonvolatile urinary buffers:

$$NH_4^+ \rightarrow H^+ + \underset{\text{diffuses off}}{\overset{NH_3}{\downarrow}} \qquad (2)$$

Some of the NH_3 lost from the urine by diffusion is regenerated, but the nonvolatile buffers are all acidified and the pH declines. It has been suggested (259) that the maximal transtubular pH gradient attainable by acid secretion may be represented by the difference between blood pH and the minimal pH normally observed in high-flow urine, the remainder of the decline in urine pH observed in antidiuresis (from 5.1 to 4.5) being the result of generation of acid in the medullary collecting ducts by back diffusion of NH_3 which had previously diffused into the acidified fluid within the distal tubules and cortical collecting ducts. Terminal generation of urinary acid, if it occurs, is an ingenious method of sustaining to some degree the rate of total renal acid excretion as the urinary flow rate declines. Decline of the urinary pH to lower values than can be attained by active acid secretion could in this manner compensate to a large extent for the inevitable decrease caused by antidiuresis in the rate of excretion of acid buffered as NH_4^+.

NOTES

1. ". . . as a general rule, urines of high water content are pale, . . . of low density and low acidity, and voluminous; while those of low water content are dark-colored, very dense, very acid, . . . and always diminished in volume" (43, page 274, author's translation).

III

PATHOPHYSIOLOGY

one

ACIDOSIS AND ALKALOSIS

Naunyn originally coined the word "acidosis" (see 18) in response to the need for a term for clinical conditions resulting from overproduction of acidic compounds, specifically diabetic acidosis; it was soon extended to include accumulation of acid in the body due to base loss (cf. II-5) and to accumulation of CO_2. The term was at first made to do double duty (18, 165), but in contemporary usage (418) *deviations in the pH of the blood* beyond its normal limits of 7.35 to 7.45 are called (respectively) *acidemia* and *alkalemia*. Acidemia may be produced in either (or both) of two ways: (i) by rise in arterial P_{CO_2} (hypercapnia) or (ii) by accumulation or generation of an excess of nonvolatile acid in extracellular fluid.[1] If primary, these conditions are called, respectively, *respiratory acidosis* and *nonrespiratory acidosis*,[2] but these terms are not always applied to exactly the same processes when they occur, as is normal, as a compensatory phenomenon. Rise in arterial P_{CO_2} above 40 mm Hg is (for example) a normal response to accumulation or generation of excess nonvolatile base (nonrespiratory *alkalosis*); this compensatory hypercapnia reduces the degree of alkalemia which would otherwise result from the failure of the kidney to maintain acid-base balance. Similarly, the kidney responds physiologically to respiratory alkalosis by generating excess acid in extracellular fluid, thus moderating the alkalemia resulting from the hypocapnia. Pulmonary and renal compensatory responses also normally reduce the degree of acidemia which would otherwise accompany acidosis of (respectively) nonrespiratory or respiratory origin (Table III-1.1). The mechanisms of these compensatory responses by lung and kidney are discussed elsewhere (II-3; II-8B); the extent of clinical compensation is variable but seldom complete. Since primary disorders of pulmonary and renal acid-base regulation may coexist, blood analysis alone cannot reliably inform the physician whether a given

TABLE III-1.1. *Pulmonary and renal responses to disturbances in acid-base regulation**

Primary disturbance	Consequences	Compensatory response
1. Hypercapnia	Respiratory acidosis with acidemia	Kidney retains or generates excess base in extracellular fluid
2. Hypocapnia	Respiratory alkalosis with alkalemia	Kidney retains or generates excess acid in extracellular fluid
3. Positive balance of nonvolatile acid	Excess nonvolatile acid in extracellular fluid with acidemia	Hyperventilation and hypocapnia
4. Positive balance of nonvolatile base	Excess nonvolatile base in extracellular fluid with alkalemia	Hypoventilation and hypercapnia

* It is useful for the clinician to know approximately what degree of compensation by kidney or lung is to be expected for a given degree of functional failure on the part of the other organ (385). Woodbury (672) has reviewed the time course of events after large abrupt changes in arterial P_{CO_2}; up to 1 hour may be required for a new steady state of tissue P_{CO_2} to be reached. Schwartz, Brackett, and Cohen (547) found that in experimental hypercapnia plasma $[HCO_3^-]$ averaged about 36 mmoles/liter after arterial P_{CO_2} has been maintained for 5 days at 70 mm Hg; similar values had been reported by Refsum (491, 492, 493) and others in chronically emphysematous patients with a similar degree of chronic hypercapnia, though in these patients some potassium depletion, which is commonly attended by increased renal base conservation, is generally also present. Plasma $[HCO_3^-]$ does not usually rise above 40 mEq/liter even when arterial P_{CO_2} goes higher (493). Lowered serum K^+, and values for arterial pH and plasma $[HCO_3^-]$ exceeding those expected from the degree and duration of hypercapnia, suggest complicating nonrespiratory alkalosis and call for close monitoring of the acid-base composition of the blood, as arterial P_{CO_2} is lowered therapeutically, lest acute alkalemia develop.

Albert, Dell and Winters (4) have quantified the respiratory response to acute nonrespiratory acidosis observed clinically; the arterial P_{CO_2} declines linearly as the arterial pH falls, reaching an average minimum value of 15 mm Hg when the pH has declined to 7.10. The respiratory response to nonrespiratory alkalosis is much more variable (204a, 431a, 616a, 385). When the condition is due to K depletion, little or no compensatory rise in arterial P_{CO_2} is seen as a rule, perhaps because the chemoreceptors are affected by the associated increase in intracellular acidity (III-2B). In alkalosis due primarily to Cl depletion, arterial P_{CO_2} as high as 50 to 55 mm Hg may be compensatory to chronic nonrespiratory alkalosis, but a search for an independent cause of respiratory acidosis is recommended whenever the P_{CO_2} exceeds 50 mm Hg.

Respiratory alkalosis results from hyperpnea which may represent either hyperventilation (psychogenic, iatrogenic, or, rarely, central) or "pseudohyperventilation" secondary to hypoxemia or hypoxia (165, 385, 475a). Pseudohyperventilation occurs at high altitude and in many serious acute illnesses, especially in the elderly; in brain disorders it is a grave prognostic sign (475a). Chronic pseudohyperventilation is prominent when diminished pulmonary compliance aggravates hypoxemia, as in interstitial pulmonary diseases and pulmonary embolization, and in ventilation-perfusion disturbances (165, 385, 493, 475a). According to Masoro and Siegel (385), plasma pH may be expected to be 7.45, corresponding to plasma $[HCO_3^-]$ of 17 mEq/liter, after arterial P_{CO_2} has been reduced to 25 mm Hg for 5 days at high altitude. Eventually normal pH is restored at high altitude in spite of the continuing hypocapnia, in contrast to chronic hypercapnia due to pulmonary disease, in which the pH usually remains subnormal (493, 165).

net accumulation of nonvolatile acid or base, or abnormal P_{CO_2}, is a primary error of acid-base regulation or represents a physiological process compensatory to some existing *or formerly existing* abnormality of acid-base regulation.[3]

Ionic exchanges between the erythrocytes and the blood plasma that occur every time the blood passes through the lungs and through the systemic circulation (118b, 165) are important for the carriage of the blood gases rather than for acid-base balance. In analyzing the collaboration of lung and kidney in normal acid-base regulation, we are also little concerned with ionic transfers between the extracellular fluid and fluid inside the nucleated cells (II-2D). But under pathological circumstances the buffer substances of the fluid within the nucleated cells make a contribution which is quantitatively more important to accommodation of rapidly accumulating acid or base than that of the extracellular fluid plus the erythrocytes; in addition, the solid phase of bone may contribute toward neutralization of accumulated acid (Section A). Buffering outside the blood and interstitial fluid must therefore be taken into account in conjunction with efforts to quantify the acid-base analysis of blood in acidosis and alkalosis (Section B).

A. WHOLE-BODY BUFFERING

The acute nonrespiratory acidosis produced by rapid intravenous infusion of sublethal quantities of mineral acid has been studied many times (645, 627, 462, 548a, 144, 601, 610a, 549a, 676, 187), beginning with the classical paper of Walter nearly a century ago in which the severe hyperpnea of this type of acute acidosis and the obstinate resistance of the blood to acidification were abundantly documented (645). Van Slyke and Cullen (627) inaugurated quantitative studies of the buffering of an acute load of administered mineral acid; they infused 75 ml of normal H_2SO_4 into a collie bitch over a period of 90 minutes, and calculated from the observed depression of the total CO_2 of the blood that at the end of the infusion only one-sixth of the infused acid had served to convert blood HCO_3^- to CO_2. They concluded that there was sufficiently rapid equilibration of the buffer systems of cell and extracellular water, under the circumstances of their experiments, for blood $[HCO_3^-]$ to serve as an index of the alkaline reserve of the whole body.

Most of the administered mineral acid not reacting with HCO_3^- is accommodated at first by the nonvolatile buffers of the body fluid, and from subsequent studies it appears that the buffers of extracellular fluid plus the erythrocytes account for roughly half of the immediate neutralization of massive quantities of mineral acid (3 to 15 mEq/kg) infused over several hours. However, within 24 hours (676), and doubtless much earlier (549a), around 75% of the administered acid has reacted with the buffers of extravascular cell water. The effect of prior nephrectomy has been studied in dogs (601) and cats (610a); it does not greatly affect the manner in which the acid load is buffered, because even after 24 hours the kidneys will have eliminated only about a quarter of the administered acid. Yoshimura and his colleagues (676) furnished the only sequential study, carried out for almost a week, of the complete defense against a sublethal load of mineral acid administered over a period of several hours; they reckon that at the end of the infusion 60% has been accommodated by the buffers of extracellular fluid and 40% by the buffers of extravascular cell water, and that after 24 hours plasma $[HCO_3^-]$ has returned to low-normal values and the buffers of cell water have accommodated all but the 25% of administered acid excreted in the urine. Over an average of 6 days ensuing, the remainder of the administered acid is excreted.

Clearly major participation of the fluid within the nucleated cells in the neutralization of administered mineral acid takes place promptly, within a matter of hours. With the aid of simultaneous estimates of extracellular fluid volume by various technics—a serviceable if approximate approach—all students of the subject have concluded that protons promptly enter the water of the nucleated cells of various tissues in exchange for alkali cations moving out; and some Cl^- appears to enter cell water with the protons. Skeletal muscle is a major source of the alkali cation, mainly Na^+ and K^+, exchanged for protons (118, 417a) entering the cells, and hyperkalemia is characteristic of acute nonrespiratory (and respiratory) acidosis (549a, 187). Connective tissue and the solid phase of bone are believed to supply Ca^{++}, and Na^+, in acute (47a, 342a, 82) as well as chronic nonrespiratory acidosis (589, 334). Not much is known about the mechanism of these shifts of ions between compartments in acidosis. Gamble and his colleagues (187) showed that, in contrast to the erythrocytes, the nucleated cells contribute as usual to the buffering of administered HCl by releasing Na^+ and K^+ in exchange for protons even though the pH of extracellular fluid is held at the normal value by simultaneous reduction of arterial P_{CO_2}; they suggest that these ion transfers are incited by the altered pH gradient developing under these circumstances between nucleated cells and interstitial fluid. "Nucleated cells" should be taken to include bone and connective tissue.

The disposition of very large loads of alkali administered over short periods was first examined by Stadelmann and his associates (586), and effects of rapid infusion of sodium bicarbonate have been studied in the nephrectomized dog by Swan et al (467) and in man by Singer, Clark, Barker, Crosley, and Elkinton (574). These studies make clear that buffering reactions outside extracellular fluid account for about a quarter of the alkali administered by the end of an hour. Singer *et al.* concluded from their own data and a thorough review of the literature that after alkali administration, "Changes in extracellular buffer base that are not accounted for by renal excretion can be explained on the basis of transfer of sodium into cells in exchange for cellular hydrogen ions. . . . A similar ionic exchange pattern is found in respiratory alkalosis, and a reversed pattern in respiratory and metabolic acidosis. Changes in intracellular pH . . . must depend on respiration (CO_2 pressure) and intracellular production of organic acids as well as on any hydrogen ion transfer."

A distinction should always be observed between buffering in cell water—shifts of equilibrium of existing physicochemical systems in the various fluid compartments within the cell which are incited by entry of acid or base—and alterations of the concentration of acidic and other compounds in these compartments, and in extracellular fluid as well, that result from adaptive changes in intermediary metabolism and tend to stabilize the acidic intensity of these fluids. Effects on cellular metabolic processes resulting from alterations of the acidic intensity of extracellular fluid have been recognized for a century, beginning with Walter's recognition of the stimulation of urinary ammonia excretion by acidosis (II-4A; II-10B). The influence of changing extracellular fluid pH on cellular metabolism includes altered net production rates and steady-state concentrations of organic acids; rise in the concentrations of such compounds in cell and extracellular water in the presence of alkalemia, or their decrease in the presence of acidemia, contributes to acid-base homeostasis, but is not buffering. Increase in the steady-state concentration of blood lactic acid in respiratory alkalosis (9a, 370b, 77, 165) is a particularly striking and well known example of such an adaptive metabolic response. Changes in the steady-state concentrations of other organic acids also contribute to acid-base homeostasis, and are under active current investigation. Comprehensive description of the effects of acute hypercapnia and hypocapnia upon acid-base equilibria in the extracellular fluid of the brain appears, for example, to necessitate sequential analysis of alterations of local steady-state concentrations of a number of organic acids as well as altered ionic fluxes between compartments (567a, 567b, 395a, 396).

Metabolic contributions of the cells to acid-base homeostasis in cell and extracellular fluid are more readily defined than are the changes in position of the acid-base equilibria of cell water in response to disease or experimental manipulations. A serious obstacle to obtaining direct knowledge of the acid-base composition of cell water is its nonhomogeneity; we do not know how many fluid compartments must be reckoned with within different types of cells. Discrete compartments of cell water, including the aqueous phase of the cytoplasm enclosed only by the cell membrane ("cell sap"), the fluid contained within the mitochondria, and almost certainly the fluid within other organelles, appear to exhibit differences in pH of at least 0.5 units among themselves. These differences are believed (94a) to be maintained by active ion transport processes between different compartments of cell water. However, current orthodoxies concerning the nature of active transport across the cellular membrane (to say nothing of intracellular membranes), upon which such formulations are based, are regarded in some knowledgeable quarters (234a) as dubious inferences unjustified by the extent of our knowledge of the physical state of water and electrolytes within the living cell *in situ*.

Many reports of estimations of the mean pH of intracellular water, especially in muscle cells, have been published in recent years. Two technics have been deployed: electrometric measurement of voltage across the the cell membrane, and observation of the distribution by nonionic diffusion of certain buffer pairs in the cell and extracellular fluid—originally CO_2-HCO_3^-, more recently especially dyes, specifically the acidic compound DMO (5,5-dimethyl-2,4-oxazolidinedione) and the organic base nicotine. Detailed review of this literature lies beyond the compass of this book, and the interested reader is referred to recent reviews and articles (85a, 636, 3, 94a). Voltage measurements and dye methods alike suggest considerable non-uniformity of pH among the compartments of cell water. The cell sap is believed to be quite acid (pH near 6.1 in the muscle of the giant barnacle) and to represent only about 3% of cell water (94a); mean cell water pH, as estimated from dye distribution in rat muscle, is reported near 7.0, with some discrepancy between values based on DMO and on nicotine distributions.

It is by no means clear that reported values for the pH of the various compartments of cell water can at present be applied in any meaningful way to the deductive analysis of the equilibria between buffer pairs of interest existing in compartments of cell water. Estimates of pH remain phenomenological entities until it is possible to relate them theoretically to the ratios of all the buffer pairs in solution in terms of mass law relations, and this cannot be done until the respective pK′ values can be calculated with acceptable accuracy from pK and the ionic strength of the solution. The great advantages of extracellular fluid from this standpoint—its uniformity, its sufficiently low ionic strength, and the repeated experimental validation of expected theoretical relations (I-2B)—all seem far from realization in the study of acid-base equilibria in cell water, and may be inherently unrealizable for the cell as a whole. It is unknown to what extent cell water is structured (234a); and, in addition, the mean ionic strength of cell water is much higher than that of extracellular fluid—high enough to raise the question with what degree of accuracy the Henderson-Hasselbalch equation could relate "mean pH" of cell water to ratios of buffer pairs even on the assumption of uniform ionic strength among the several discrete compartments of cell water. Such an assumption is in any case not warranted; and even were the ionic strengths of all compartments of cell water known, together with compartment volumes, it could not be proper to derive from them a mean pK′ for use in relating mean cell water pH to mean buffer ratios in cell water. Experimental evidence that theoretical relations derivable from the mass law are approximated between pH and buffer ratios in a particular compartment of cell water when alterations of pH are experimentally induced would be a welcome development in this difficult area of investigation.

B. BLOOD ANALYSIS

Often enough the physician can base a reliable diagnosis of an acid-base disorder, together with a fairly accurate evaluation of its severity, on purely clinical evidence; and laboratory data not directly concerned with the acid-base analysis of blood or blood plasma may also be of the greatest assistance. Given, for example, a history of diabetes mellitus in a young person, and on physical examination Kussmaul breathing, severe dehydration, semistupor, and a fruity breath-odor, with sugar and ketone bodies found to be profusely present in urine and blood, the clinician will hardly be able to doubt that he is dealing with severe nonrespiratory acidosis due to diabetic ketosis. He can predict that plasma [HCO_3^-] and pH will be markedly subnormal and that the arterial plasma P_{CO_2} will also be lowered markedly, since hyperventilation is moderating the acidemia, which is nevertheless very severe. Laboratory analysis of the acid-base status will simply be confirmatory of these inferences at the outset, though valuable in assessing the subsequent response to therapy.

Sometimes, however, a severe acid-base disturbance goes unrecognized until disclosed by the laboratory; or clinical clues prove to be misleading; or the apparent acid-base disturbance is a complex one which cannot be accurately evaluated by the most skillful clinician without all possible help from the laboratory. In addition, in the emergency room, on medical, surgical, and pediatric in-patient services, and especially in intensive-care units and during anesthesia and surgical procedures, hyperacute alterations (especially of pulmonary function) are encountered, giving rise to acid-base derangements of the utmost urgency which cannot be managed rationally and aggressively without frequent recourse to a complete acid-base analysis of arterial blood. Whatever the clinical setting, any complete acid-base analysis aims at answering three principal questions: (i) what is the pH of arterial blood? (ii) what role is the lung playing in the acid-base disturbance at the moment the blood is drawn? and (iii) how much *cumulative* net excess of nonvolatile base or acid is present per liter of blood plasma? What the analysis cannot disclose is whether or to what extent deviation of either the pulmonary or the renal component is primary or compensatory, or, if primary, what its pathogenesis was or is. The correct answer to these questions depends upon the synthesis of historical information by a skilled clinician, the physical findings, and other pertinent laboratory data, as well as acid-base analysis of the blood.

Thanks to technical contributions by many workers over more than a half-century, the physician can now depend upon acid-base analyses that incorporate information obtained by the use of electrodes which give rapid and precise estimations of the pH and P_{CO_2} of small samples of either arterial or capillary blood; with their aid he obtains a direct and unequivocal answer to the first two of the three principal questions which he asks of the laboratory. Knowledge of *the arterial pH* tells him how gravely the acid-base environment of the cells is disturbed; and the functional contribution of the lung to the acid-base composition of extracellular fluid is given unequivocally by the arterial P_{CO_2}, for this variable is directly controlled by the lungs on a moment-to-moment basis. Aside from a laboratory error, there is no way in which arterial P_{CO_2} can mislead except insofar as one might fail to realize that the report based on a single arterial sample might not be representative of the patient's condition before or after the blood was drawn; for the arterial P_{CO_2} can change markedly in a minute or so. Acute hyperventilation due to anxiety or pain experienced by the patient at the time of sampling can cause temporary lowering of the P_{CO_2}; and major acute disorders markedly affecting pulmonary function, such as embolism or pulmonary edema, can of course supervene so abruptly that the report of an acid-base analysis of the blood made a half-hour earlier may sometimes already be seriously out of date. Careful clinical observations, and technical skill, care, and alertness at the time of sampling can largely obviate misinterpretations stemming from such causes.

But *there is no way in which the nonrespiratory constituent in any abnormality of the acid-base composition of extracellular fluid*—the accumulation of an excess of nonvolatile acid or base in extracellular fluid—*can be evaluated directly and quantitatively, as the pulmonary constituent and the arterial pH can be.* Because of the predominating position of the CO_2-HCO_3^- system among the buffers of the extracellular fluid (I-6; II-5; II-6), the plasma [HCO_3^-] offers a practically useful index of the extent of accumulation of excess nonvolatile acid or base in extracellular fluid; but for several reasons (see below) it does not offer a quantitative measure. Estimation of plasma [HCO_3^-] from measurement of the bicarbonate plus dissolved CO_2 evolved from acidified plasma,[4] termed either "plasma CO_2 content" or "total CO_2" of the plasma (see below), has nevertheless well served two generations of physicians in evaluating the nonrespiratory constituent of clinical acid-base disorders, even though the CO_2 content does not exactly measure plasma [HCO_3^-]. On the other hand, the plasma CO_2 content gives no information about extracellular fluid pH and is useless in evaluating pulmonary function. Partial evaluation of the electrolyte composition of the plasma, with plasma CO_2 content serving as a means of estimating [HCO_3^-], was the principal reliance of general physicians seeking laboratory information about acid-base regulation in their patients until about two decades ago. J. L. Gamble gave classical expression in his little syllabus (182) to the chemical and physiological principles on which this type of acid-base analysis of blood is based; it is considered in Section B-1. More recently, comprehensive acid-base analyses of arterial or capillary blood, yielding knowledge of its pH, P_{CO_2}, and [HCO_3^-], began to be accessible to the general physician as well as specialists in the field.

A major goal of much meticulous and productive investigation in acid-base blood analysis, both theoretical and developmental, over decades, has been *to combine with evaluation of arterial blood pH and P_{CO_2} an accurate estimate of the nonrespiratory component*, and in this way to offer the physician *a comprehensive acid-base analysis derived from a single small sample of arterial or capillary blood* (Section B-2). The roll call of contributors to this field is an illustrious one, comprising among others the names of Henderson and Van Slyke and their colleagues, Hastings and Singer, and Astrup and Siggaard-Andersen and their associates. The reader is referred elsewhere (626, 165, 18, 569) for an account of the historical development of this field; the only purpose of what follows is to orient the novice in acid-base blood analysis to the utility, as well as the limitations, of the principal methods of acid-base blood analysis widely available. Every physician nowadays absolutely requires sufficient understanding of the principles underlying these methods to make intelligent use of them in his daily practice. Today physicians and medical students (and their patients) enjoy an advantage over their predecessors in the accessibility of clinically very convenient and useful acid-base blood analysis; medical students may take additional pleasure in the circumstance that these methods of analysis give every sign of having approached a technical and theoretical optimum so nearly as to ensure their continuing application, without substantial change, to the practice of medicine.

1. Analysis of the Electrolyte Structure of Plasma[5]

By this expression is to be understood a method of evaluating the nonrespiratory component of an abnormal acid-base state, principally by knowledge of plasma "total CO_2," Na^+, and Cl^-. The type of graphic representation of the electrolyte structure of plasma which was developed by Gamble (182) is shown in Figures III-1.1 and III-1.2; it is applicable also to urine, for the purpose of quantifying renal excretion (184), by a method alternative to that presented in Chapter II-4, one which is not discussed in this book. The *twin bars* of each illustrative example given in the figures display the electrical equivalence of the total positive and negative charges as necessitated by the requirement for electroneutrality in solution, and the method of presentation is well designed to exhibit the independent regulation of the concentration of ions in solution subject only to the requirement of electroneutrality (I-3B). $[H_3O^+]$, because its concentration is less than 0.0001 Eq/liter in plasma, is invisible among the cations, nor can $[OH^-]$ be seen among the anions. Approximate inferences about $[H_3O^+]$ could be drawn from any of the bar pairs only if the assumption were made that the concentration of dissolved CO_2 gas is close to normal, and this is unwarrantable in many if not most clinical situations. On the other hand, a major advantage of electrolyte analysis of plasma is that it can be carried out on venous blood, and so is useful for screening purposes and is very generally available, whereas acid-base analysis of blood involving estimation of pH and P_{CO_2} (Section B-1) should in general be made on arterial or arterialized blood.

The clinical laboratory can determine the total protein—the principal plasma nonvolatile buffer—in grams per 100 ml, and other buffer sub-

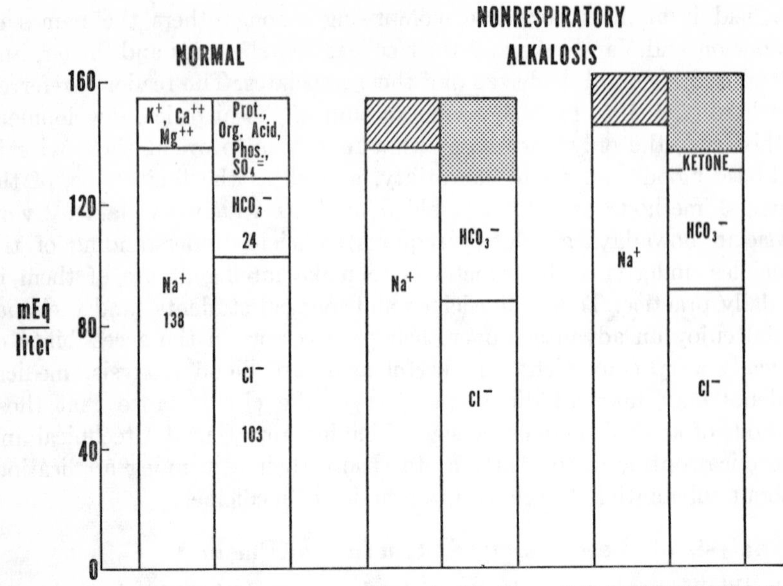

FIG. III-1.1. The "electrolyte structure of the plasma" in health (*left-hand bar pair*) and in nonrespiratory alkalosis (*center* and *right hand bar pairs*). After Gamble (182). The pathognomonic feature of nonrespiratory alkalosis is increased plasma $[HCO_3^-]$, irrespective of whether the condition is uncomplicated (*center pairs*) or not. The *right-hand bar pairs* exhibit the complications of ketonemia and increased plasma osmolality (and hypernatremia), usually due to dehydration.

stances, too, like phosphate, as total compound; but we should have to know the plasma pH in order to determine how much accumulated acid or base had been buffered by these substances.[6] However, the proportional contribution of all the plasma nonvolatile buffers to plasma buffering of nonvolatile acid or base is very small compared with the change in plasma $[HCO_3^-]$; even the most extreme changes in plasma pH compatible with survival (about $\pm$ 0.4 pH) do not permit the plasma nonvolatile buffers to contribute substantially. It is only the volatile character of the CO_2 buffer system and the high $HCO_3^-:CO_2$ ratio at pH 7.4 (I-6) which allows this system to buffer so effectively. In short, accumulation of nonvolatile base in plasma is reflected mostly in increased $[HCO_3^-]$, and accumulation of nonvolatile acid is reflected mostly in decreased $[HCO_3^-]$. For the differential diagnosis of nonrespiratory acidosis, knowledge of Na^+ and Cl^- as well is most helpful (Figure III-1.2) because it discloses whether the acidosis is chloride acidosis or acidosis associated with accumulation of other anions (cf. Table III-1.1B). $[K^+]$ and $[Ca^{++}]$ may be useful aids in identifying the etiology of one or another of the four principle primary derangements of acid-base homeostasis listed in Table III-1.1. Reduction of $[K^+]$, especially when marked, suggests that K depletion may play a prominent etiological

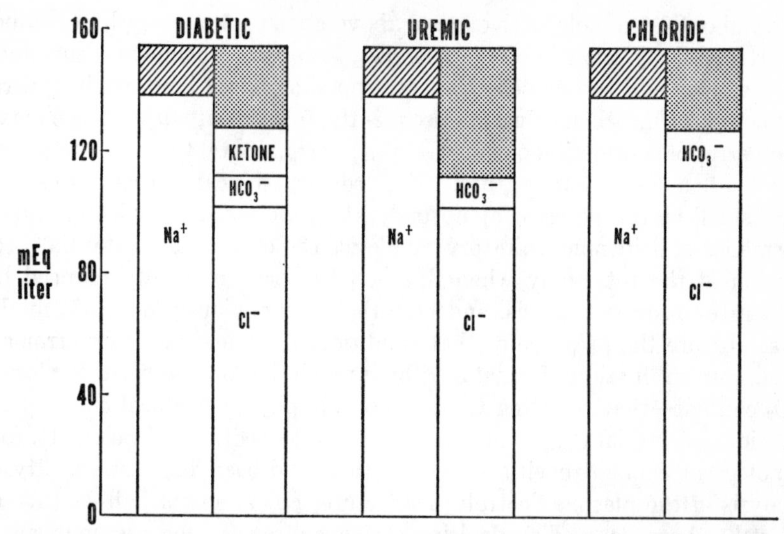

FIG. III-1-2. Three types of nonrespiratory acidosis. After Gamble (182). In diabetic acidosis, accumulation of strong organic acids (diacetic and hydroxybutyric) results from their overproduction; their concentration in extracellular fluid, normally low, rises under these circumstances until the renal threshold is exceeded, and continues to rise as long as urinary excretion does not keep pace with net production. These acids react as they accumulate—as do the organic acids accumulating in the plasma in uremia—with $NaHCO_3$ to yield Na^+ plus the anion (A^-) of the reacting acid (HA) plus H_2CO_3.

$$HA + NaHCO_3 \rightarrow Na^+ + A^-$$

The H_2CO_3 is dehydrated and the CO_2 exhaled. (A few of the protons of HA are transferred to the nonvolatile buffers since the pH of extracellular fluid is generally lowered in acidosis.)

Chloride acidosis, in which plasma HCO_3^- has been to some extent replaced by Cl^-, can result from giving HCl (676) (or an equivalent like NH_4Cl) more rapidly than the acid can be eliminated by the kidney; but clinically this type of acidosis occurs spontaneously either (i) because the diseased kidney does not regenerate enough HCO_3^- to maintain normal plasma [HCO_3^-], or (ii) because regeneration of HCO_3^- in peritubular blood by an essentially normal kidney cannot keep pace with the net acid load produced by extrarenal base losses.

role; while hypocalcemia (without hypoalbuminemia), and especially a lowered value for the product of serum calcium and phosphate, suggests chemical osteomalacia, a frequent complication of renal tubular acidosis (7). Of the additional chemical information which may be supplied by the clinical laboratory, the most generally useful is measurement either of plasma nonprotein N (NPN), or urea N (serum urea N, or SUN), or creatinine.

[HCO_3^-] is known to a near approximation from the commonly provided "total CO_2", because the concentration of dissolved CO_2 gas rarely falls below about 0.5 mmole/liter or rises above about 4.5 mmoles/liter. Values of arterial P_{CO_2} below the corresponding lower limit cannot be sustained by the lungs and those above the corresponding upper limit are in general not compatible with survival. Consequently, for clinical purposes the maximum error of considering plasma [HCO_3^-] to equal total CO_2 less 2 mmoles/liter is about ±1.5 mmoles/liter.[7] For ordinary clinical purposes this is a very small percentile error at normal values for plasma [HCO_3^-], and it is smaller still in nonrespiratory alkalosis. It becomes a substantial proportion of the total only when [HCO_3^-] has been markedly reduced by moderate-to-severe acidosis, and here the error obviously does not in the least obscure the presence of that condition. A somewhat larger error attaching to evaluation of renal acid-base regulation on the basis of plasma CO_2 content—that resulting from the purely physicochemical effect of alterations of arterial P_{CO_2} upon plasma [HCO_3^-]—is discussed below. It, too, is not great enough to vitiate the practical acid-base diagnostic utility of analysis of the plasma electrolyte structure; but almost a half-century of effort has been devoted to devising a means of estimating the nonrespiratory component of abnormal blood acid-base composition more precisely than can be done by means of the blood [HCO_3^-] (Section B2).

If an alkaline compound like NaOH were simply added to the plasma *in vivo*, both [HCO_3^-] and [Na^+] would rise; but as long as renal volume and osmolality regulation are accurate, the excess Na^+ (or any other alkali cation similarly assimilated as an alkaline salt) will be excreted in part together with some of the Cl^- already present in the extracellular fluid. *The end result of net accumulation of any nonmetabolizable (mineral) base in the plasma is therefore normally an increase in [HCO_3^-] at the expense of Cl^-* (Fig. III-1.1, *middle bar pair*). Of course HCO_3^- will also be excreted "matched" against some of the Na^+ (I-3B), but, if alkalosis is developing, not at a rate sufficient to prevent rise in extracellular [HCO_3^-]. It is Cl^-, not some other anion, which must accompany the rest of the excess Na^+ into the urine, largely because Cl^- predominates so greatly among the plasma anions, but also specifically because the concentrations of the other anions of the plasma are for particular reasons adjustable by the kidney to only a limited extent, and hardly at all in response to acid-base challenges. The plasma proteins, for example, are tenaciously conserved by the kidney under all normal circumstances; plasma phosphate is adjustable in response to the dietary phosphate load and hormonal influence, but then only to the extent of several milliequivalents per liter, and not much at all in relation to the acid-base economy.

Cl^- seems nearly—not wholly (81a)—automatically reabsorbed in parallel with the increasing or decreasing Na^+ fractional reabsorption of filtered Na^+ dictated by endocrine regulation of extracellular fluid volume (II-7D);

the link with Na^+ reabsorption is very close empirically, and has generally been represented (659), as far as the proximal tubule is concerned, in terms of the diffusion of Cl^- down the electrochemical gradient created by active Na^+ reabsorption. The molecular character of the linkage is perhaps somewhat less clear now that the proximal transepithelial voltage has been reduced to close to zero (II-8A). Since the rate of filtration of HCO_3^-, which normally largely determines the proportional reabsorption of filtered HCO_3^- from the nephrons, is predominantly controlled by the acid-base load (II-2; II-6), and since the Na^+-K^+ exchange is relatively a very small portion of total Na^+ reabsorption, it is necessarily the proportional reabsorption of filtered Cl^- which is most conspicuously affected when Na^+ reabsorption is actively increased or decreased by the kidney's homeostatic regulation of extracellular fluid volume.

In practice, therefore, nonrespiratory alkalosis results in increased plasma $[HCO_3^-]$ at the expense of decreased $[Cl^-]$. If renal regulation of extracellular fluid volume and osmolality is entirely normal, plasma $[Na^+]$ will remain normal in nonrespiratory alkalosis while $[Cl^-]$ and $[HCO_3^-]$ vary reciprocally. But alkalosis may of course be accompanied by plasma hyperosmolality where increases in $[Na^+]$ are in addition "matched" by increase in $[Cl^-]$, or by plasma hyposmolality where there are parallel reductions in $[Na^+]$ and $[Cl^-]$. Accordingly, *the essential feature of nonrespiratory alkalosis is not the lowered plasma* $[Cl^-]$ *but the raised* $[HCO_3^-]$. The magnitude of the increase in $[HCO_3^-]$ is an index of the alkalosis, while the extent of osmotic abnormality of the plasma is indicated by the degree of abnormality in $[Na^+]$ concentration.

The plasma cations other than Na^+ are present in such relatively small quantity as to contribute little to total cationic equivalence and osmolality (Figure III-1.1, *left-hand bars*). If, for example, an alkaline salt of K is taken in, there will be a tendency for $[HCO_3^-]$ to rise and $[Cl^-]$ to decline, the Cl^- being eliminated in the urine along with the excess K, not by direct suppression of the reabsorption of both, but rather by increased secretion of K^+ in exchange for Na^+ reabsorbed, together with decreased reclamation of Na^+ with Cl^-.

When the cause of acidosis is failure of the renal tubules to secrete acid into the nephrons, in exchange for Na^+ reabsorbed, at a rate appropriate to the glomerular HCO_3^- filtration (II-8D) (as, for example, in renal tubular acidosis), a parallel decline in plasma $[Na^+]$ and $[HCO_3^-]$ might be anticipated. This would occur but for the facts that (i) net Na^+ loss from extracellular fluid promptly activates the hormonal system charged with preservation of the sodium content of extracellular fluid, and (ii) renal osmolality regulation adjusts extracellular fluid volume to total extracellular Na^+ so as to stabilize $[Na^+]$ (182, 65, 467, 348, 642). Consequently, reabsorption of Na^+ from glomerular filtrate is stimulated, and plasma $[Na^+]$ usually remains within normal limits. Cl^- reabsorption from the nephrons is increased automatically in consequence of the increased Na^+ reabsorption effected by renal volume regulation, and the overall effect on the plasma electrolyte structure is that the loss of $[HCO_3^-]$ has been compensated for by an equivalent increase in $[Cl^-]$. This type of nonrespiratory acidosis is termed "chloride acidosis" or "hyperchloremic acidosis" (Figure III-1.2, *right-hand bar pair*); deformation of the normal plasma structure of this sort is the hallmark of accumulation of nonvolatile acid in the extracellular fluid secondary to net alkali loss as base bicarbonate (predominantly Na^+, but often also some K^+, with HCO_3^-).

Only in a minority of clinical instances is this "alkali-loss" type of nonrespiratory acidosis due in pure form to renal failure to conserve filtered HCO_3^- appropriately (III-3B3); the base loss is much more commonly associated with loss of the HCO_3^--containing intestinal secretions as in chronic or acute diarrhea, ileostomy drainage, suction tubes placed in the intestines, and the like. The same effect on the plasma electrolyte structure can, however, be produced by ingestion of acidifying salts of Cl^- like NH_4Cl and $CaCl_2$ (II-9A2). Addition of concentrated HCl to the plasma at constant P_{CO_2} would simulate this kind of deviation from normal plasma ionic composition (cf. I-3B).

Nonrespiratory acidosis can also result from accumulation in extracellular fluid of a variety of strongly acidic compounds other than HCl. In disorders affecting the intermediary metabolism, like diabetic acidosis and lactic acidosis, strong acid is produced from neutral fat or carbohydrate precursors and added to the body fluids at so rapid a rate that the rate of acid elimination, even by entirely normal kidneys, cannot keep pace, and excess nonvolatile acid accumulates in extracellular fluid (III-3). Naturally, net addition of this acid to the plasma, like addition of any strongly acidic compound, results in transfer of protons mainly to the principal plasma buffer, HCO_3^-, which is thereby converted to CO_2 and is exhaled; the result is that HCO_3^- is displaced, in the electrolyte structure of the plasma, by the anions of the strongly acidic compounds which have been added. Hydroxybutyrate and diacetate accumulate in diabetic acidosis, and lactate in lactic acidosis, because the corresponding acids are being added to extracellular fluid at rates too fast for excretion by the normal kidney to keep pace. In uremia (III-3B3), various ions accumulate because the failing kidney does not efficiently excrete the acidic compounds.

It was noted above that in chloride acidosis the concentrations of $[HCO_3^-]$ and $[Cl^-]$ are reciprocally altered so that the sum of $[HCO_3^-]$ and $[Cl^-]$ remains essentially unchanged by the acid-base disturbance. Even if $[Cl^-]$ is independently altered, along with $[Na^+]$, by some accompanying abnormality of plasma osmolality, $[Na^+] - [HCO_3^- + Cl^-]$ remains unaltered, since the change in plasma osmolality is effected by equal, parallel changes in $[Na^+]$ and $[Cl^-]$. The plasma value for $[Na^+] - [HCO_3^- + Cl^-]$ normally averages 11 mEq/liter with a normal variation of about ±3; this difference is variously known as Δ, undetermined anion, or the anion gap. It is very useful as an indication of the type of nonrespiratory acidosis, which can be classified into two primary diagnostic categories: (i) chloride acidosis, with Δ within the normal range, and (ii) acidosis due to accumulation in plasma of the anions of one or more acidic compounds not normally present or present only in much lower concentration. In severe acidosis of the latter type, Δ reaches values of 20 mEq/liter or greater.

As illustrated in the *right-hand bar pair* of Figure III-1.1, anions other than Cl^- may sometimes accumulate in the extracellular fluid, even though the principal abnormality is nonrespiratory alkalosis. Alkalosis resulting from vomiting of stomach and upper-intestinal secretions, as in intestinal obstruction, is very frequently accompanied by inability of the patient to take food; when this happens, and carbohydrate is not supplied parenterally, fat is increasingly oxidized, so that the concentrations of the ketone bodies, including the anions of the strong acids, diacetic and β-hydroxybutyric, in extracellular fluid rise. It is therefore not rare for nonrespiratory alkalosis to coexist with accumulation of nonvolatile acid. Should the Cl^- lost from the plasma be exactly replaced by the accumulating anions of the ketone bodies, there might be entirely normal plasma pH, though both ketosis and chloride deficit would be present.

Additional differential diagnostic clues may be obtained, in nonrespiratory alkalosis as well as nonrespiratory acidosis, from the concentrations of other constituents of the plasma likely to be estimated routinely with the Na^+, HCO_3^-, and Cl^- of venous plasma. Lowered plasma [K^+] may point to K deficiency as a cause of alkalosis (III-2); marked elevation of nonprotein N or plasma urea means functional glomerular insufficiency and is therefore a sure sign that excess anions of some organic acids are present in the plasma (III-3B3); a lowered product of the concentrations of calcium and phosphate may suggest the chemical osteomalacia often complicating renal tubular acidosis (III-3).

2. Comprehensive Acid-Base Analysis of Blood

This term means that the measurements include direct estimations of arterial pH and P_{CO_2}, whose significance was noted at the beginning of the section, plus an evaluation of the nonrespiratory component. Plasma [HCO_3^-] is only an index, not a measure, of the accumulation of nonvolatile acid or base in extracellular fluid, because it is also influenced by changes in the P_{CO_2} of the blood:

$$\begin{array}{lllll} (\text{Kidney}) \rightleftharpoons & H^+ + HCO_3^- \rightleftharpoons H_2CO_3 \rightleftharpoons & CO_2 + H_2O & & \\ & + & & & \\ & Buf^- & \downarrow\uparrow & & \quad (1) \\ & \downarrow\uparrow & (\text{Lung}) & & \\ & HBuf & & & \end{array}$$

Clinically, the largest effects on plasma [HCO_3^-] result from gain or loss of protons (H^+, upper left) resulting from any imbalance between proton transfers effected by the kidney into or out of extracellular fluid, and transfers in the opposite direction resulting from the acid-base load. But alteration of arterial P_{CO_2}, normally equilibrated with alveolar P_{CO_2}, inevitably also influences plasma [HCO_3^-]. When P_{CO_2} rises, [H_2CO_3] increases, and some protons must dissociate from this compound and be accepted by the nonvolatile buffers (Buf^-) as pH declines. This dissociation of some of the newly formed H_2CO_3, however, contributes additional HCO_3^- to the solution. Changes in [HCO_3^-] produced by change in fluid P_{CO_2} are inversely related to pH change, in contrast to nonrespiratory changes, where [HCO_3^-] declines with pH as acid is added, and rises with pH as base is added.

The more strongly buffered a neutral or slightly alkaline solution is, the less its pH declines for a given rise in its P_{CO_2}; the buffers accommodate the H^+ dissociating from H_2CO_3, so that less H_3O^+ is formed. Also, more HCO_3^- is formed for a given rise in P_{CO_2} as the buffer strength increases. In pure water practically no HCO_3^- is formed from H_2CO_3 as P_{CO_2} is raised, because the slightest release of protons lowers the pH of water to the range where hardly any HCO_3^- can be present. The more non-CO_2 buffer is present in an aqueous solution, the less the pH declines for a given rise in P_{CO_2}, and at any P_{CO_2} the ratio $HCO_3^-:CO_2$ is therefore higher in relation to increasing buffer strength. In the case of the body fluids, the only consequential buffers other than the HCO_3^--CO_2 system are various nonvolatile buffers. Interstitial fluid is relatively poorly buffered because it contains little protein; plasma is better buffered, the plasma proteins contributing much more to the buffer strength of plasma than the phosphate, urate, and other crystalloids, collectively, of the extracellular fluid. Whole blood is in turn far better buffered than plasma, because of the large amount of hemoglobin (Hbg) present and the effectiveness of this protein per gram as a buffer in the range pH 7 to 8. In fact, so dominant is Hbg among the nonvolatile buffers of whole blood that for all practical purposes buffering of CO_2 by whole blood may be treated as if the only buffer systems present were CO_2 and Hbg.

Figure III-1.3 shows the effects of altering the P_{CO_2} of whole blood by tonometry; P^- represents the anionic equivalence of protein, made up in whole blood almost entirely of Hbg. Practically all protons donated as [H_2CO_3] rises in blood are transferred to Hbg, and the protons accepted by HCO_3^- as P_{CO_2} declines are practically all furnished by Hbg. Examination of the effect of alteration of the P_{CO_2} of whole blood (or of any other fluid) upon its pH and its HCO_3^- content is termed *CO_2 titration*. When the blood [HCO_3^-] is plotted as a function of the pH, as the P_{CO_2} is varied, a nearly linear function is obtained whose slope is inversely related to the blood Hbg. Blood from an anemic patient, which contains less of the

CO_2 pressure, P_{CO_2} = 41 mm. pH_s = 7.41. Hematocrit, Vc = 0.45.

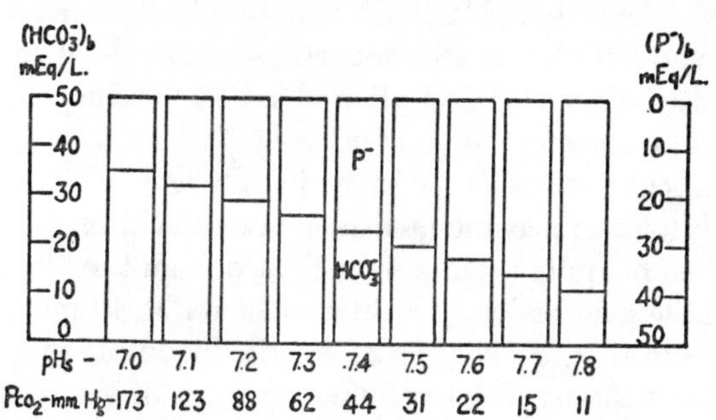

EFFECT ON BUFFER ANIONS OF VARYING CO_2 PRESSURE

Oxygenated whole blood. Hematocrit, Vc = 0.45. Whole blood buffer base $(BB^+)_b$ = 50 mEq./L.

FIG. III-1.3. The effect of varying the CO_2 tension with which the blood is equilibrated on the buffer anions of the plasma. Reproduced, with permission, from Singer and Hastings (575). Protons dissociated from additional CO_2 are transferred almost entirely to the blood protons, primarily hemoglobin. The "*whole blood buffer base*" is no longer in clinical use, and the reader is referred elsewhere for full descriptions (575, 569, 165). A "base excess" of zero for the whole blood is essentially a normal mean value for "whole blood buffer base."

principal blood buffer, will show a greater-than-normal fall in pH and a less-than-normal generation of HCO_3^- as the P_{CO_2} is raised. *It is the physicochemical effect of raising or lowering the* P_{CO_2} *of the body fluids upon their* [HCO_3^-] *which disqualifies the blood* [HCO_3^-] *from serving as a measure* (rather than an index) of the *nonrespiratory alteration of the acid-base composition of extracellular fluid.*

All efforts to improve upon the blood or plasma [HCO_3^-] as a laboratory guide to the degree of accumulation of excess nonvolatile acid or base in extracellular fluid have been based upon some form of CO_2 titration of whole blood or plasma, the idea being that, by equilibrating at 40 mm Hg CO_2 tension a sample of blood drawn from the patient at some other value for P_{CO_2}, the sample's [HCO_3^-] would be "corrected" to the value it would have had if there had been no abnormality of arterial P_{CO_2} *in vivo.* One of the first of these derived variables to achieve widespread use was the serum "CO_2 combining power" or "CO_2 capacity" of Van Slyke and Cullen (627), which is the "total CO_2" (HCO_3 + CO_2) concentration of venous serum equilibrated at 40 mm Hg CO_2 pressure; other variables based on CO_2 titration of whole blood have included the "buffer base" of Singer and Hastings, the "standard bicarbonate" of Jorgensen and Astrup, and the "base excess," or *BE*, of Astrup and Siggaard-Andersen. The objective of all these derived variables is to correct for errors arising out of the necessity of evaluating the extent of accumulation of nonvolatile acid or base in the whole body from a *sample* derived from the vascular compartment only.

If the nonvolatile buffer strengths of all compartments of body water were identical, then altered arterial P_{CO_2} would have an identical effect upon the pH and [HCO_3^-] in all compartments, and no ionic disequilibrium would be created (in respect to pH and HCO_3^-). The effect of returning the P_{CO_2} of a blood sample to 40 mm Hg *in vitro* upon its [HCO_3^-] and pH would be identical to the effect of a similar change in P_{CO_2} in all compartments of body water; accumulated nonvolatile acid or base in the whole body could then be quite accurately reckoned from the blood sample. The extra nonvolatile acid or base per liter of blood at 40 mm Hg P_{CO_2} could be reckoned from (i) the deviation of arterial blood pH at 40 mm Hg P_{CO_2} from 7.4 and (ii) the nonvolatile buffer strength of the blood; multiplication of this quantity by the total body water in liters (about 65% of body weight) would give total accumulation of nonvolatile acid or base in the body. In fact, however, extensive ionic exchange occurs between the compartments of extracellular water, and between cell and extracellular water, when arterial P_{CO_2} is changed, because of the large differences in nonvolatile buffer strength between blood plasma, erythrocyte water, interstitial fluid, and the cell water of the various tissues. Delays in ionic equilibration after change in arterial P_{CO_2} must also be taken into account in acute situations. After an abrupt change in arterial CO_2, there is believed to be very rapid equilibration of the P_{CO_2} of all compartments of body water; pH, however, changes less in heavily buffered compartments (erythrocytes, nucleated cells) than in less well buffered compartments (plasma, interstitial fluid). Equilibration between plasma and erythrocytes involves movement of HCO_3^- down the established concentration gradient and an exchange of HCO_3^- for Cl^- (the "chloride shift"); these exchanges are very rapid, as are passive movements of ions across the capillary membrane. On the other hand, ionic exchanges between the water of the nucleated cell and extracellular fluid in response to administered nonvolatile acid or base (Section A), and to altered arterial P_{CO_2}, are slower.

Ionic exchanges between compartments of body water in acidosis and alkalosis, and their relation to acid-base blood analysis, are complex, and the reader is referred to the literature for detailed consideration of the associated theoretical problems and practical applications (627, 625a, 631, 241, 575, 118b, 569, 165). Filley (165) has provided a clear and very useful discussion of "the base excess concept in historical perspective"; a lively play of controversy is given in the proceedings of a symposium sponsored by the New York Academy of Sciences in 1966 (418). A short description of the *BE* of Astrup and Siggaard-Andersen must suffice to close our discussion of acid-base blood analysis.

The instrumentation used to determine *BE* is ingenious and elegant. A microsample of capillary blood is obtained from earlobe or finger, its pH is determined, and then the pH is twice measured when the sample is equilibrated with two different CO_2-O_2 mixtures which saturate the Hbg and of which the CO_2 tensions are accurately known. Because the solubility and pK′ of CO_2 in blood are known, the $[HCO_3^-]$ at each known P_{CO_2} is calculable. A suitable nomogram (Figure III-1.4) permits the pH and $[HCO_3^-]$ of the sample to be represented as continuous functions of P_{CO_2} as this is changed in the course of CO_2 titration; the CO_2 titration is represented by drawing a straight line, projected if need be, between the points representing the two CO_2 tensions at which the three acid-base parameters were evaluated.

The buffer strength of the blood determines the slope of the linear function that results when log P_{CO_2} is plotted against pH in the CO_2 titration (Figure III-1.4); pH changes more for a given change in P_{CO_2} for plasma (*line A*) than for whole blood (*line B*), and so the slope of the function gives a rough estimate of the Hbg content of the blood. Since the buffer strength of the blood is known from the CO_2 titration, it is possible to calculate how much mineral acid or alkali would be required to bring the

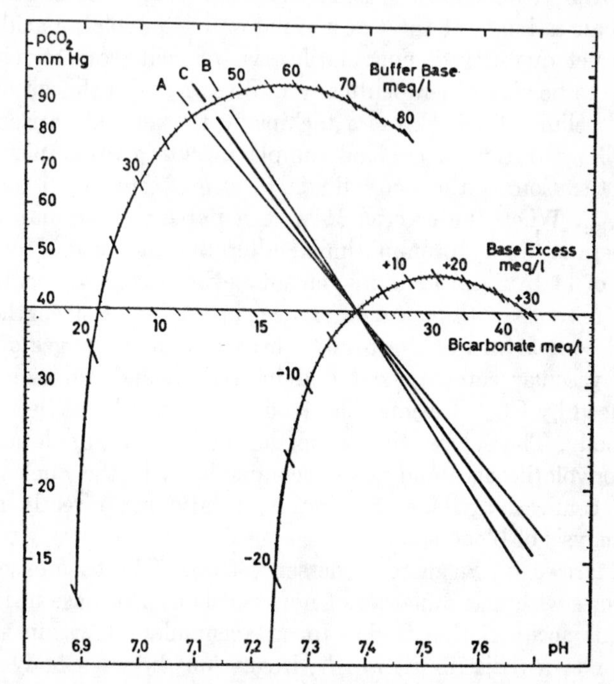

Fig. III-1.4. The Siggaard-Andersen curve nomogram. Reproduced from Siggaard-Andersen (569a). By plotting any two points with known coordinates on the pH and P_{CO_2} scales, a linear function is obtained which represents the CO_2 titration curve of the blood sample over any range of arterial P_{CO_2} encountered in the clinic. The negative value of the slope of the titration line decreases with increasing nonvolatile buffer strength of the blood, in practice with its hemoglobin content.

[HCO_3^-] of the sample at P_{CO_2} = 40 mm Hg to 23 mmoles/liter; *this is the BE* (+ or −, respectively). The P_{CO_2} and HCO_3^- of the sample when it was withdrawn can be read off on the titration line at the point where pH equals the originally measured pH of the withdrawn sample. The value for HCO_3^- when the blood sample is equilibrated at 40 mm Hg is the "standard bicarbonate."

A correction can be made for any unsaturation of the blood which may have been disclosed by the independent measurement. Since reduced Hbg is a weaker acid than is oxyhemoglobin, its presence means a greater "actual" *BE* than was estimated on the oxygenated sample; the relation is given by the equation:

$$BE_{act} \text{ (mEq/liter)} = BE_{ox} + 0.3 \times \text{unsaturated Hbg (g/100 ml)} \quad (2)$$

Positive values for *BE* give an estimate of the total accumulation of nonvolatile base in milliequivalents per liter of blood; negative values evaluate the accumulation of nonvolatile acid in the same units. As a rule of thumb, multiplication by 0.3 × body weight in kilograms has been recommended as an estimate of total accumulation of nonvolatile acid or base, i.e., of the net quantity of nonvolatile base or acid (respectively) which would have to be given therapeutically to do away with this accumulation in the extracellular fluid. Errors attaching to the estimate arise from the fact that CO_2 titration of a blood sample *in vitro* gives results different from CO_2 titration of the body fluids *in vivo* effected by manipulating alveolar P_{CO_2}. When the arterial P_{CO_2} of a patient or normal subject is altered, there is a very prompt similar alteration of the P_{CO_2} of all compartments of body fluid; ionic movement between compartments results, principally because of the large differences in the buffer strength between the several compartments. Naturally, these *in vivo* exchanges of ions between the vascular compartment and interstitial and cell water cannot be reproduced by CO_2 titration of a blood sample which has been removed from the body. The substantial proportion of large accumulations in the body of nonvolatile acid and base accommodated in the nucleated cells, bone, and connective tissue (Section A) is also not directly accessible through analysis of blood samples.

As summarized by Siggaard-Andersen (569a), "The base excess of the blood changes with accumulation of nonvolatile acid or base in the blood, whether this accumulation is due to an accumulation within the whole organism or to a redistribution of hydrogen ions between body compartments. The changes due to a redistribution as seen in hypercapnia or hypocapnia (or potassium depletion) are small and usually without clinical significance." Small corrections may be applied for severe hypercapnia. As arterial P_{CO_2} rises in the patient, protons pass from the less well buffered extravascular fluids into his blood. These protons cannot leave the blood sample when its P_{CO_2} is reduced *in vitro*, and so the measured *BE* is factitiously reduced. According to Siggaard-Andersen (418, page 111), about 3 mEq/liter may be added to the measured *BE* in patients whose arterial P_{CO_2} is twice normal or more, to obtain the corrected *BE*; little correction is needed for hypocapnia.

In many clinical laboratories, CO_2 titration of the blood sample is not carried out; instead the P_{CO_2} and pH of the arterial or capillary blood is measured, and reliance is placed, for evaluation of the nonrespiratory component, upon the $[HCO_3^-]$ calculated for the sample. Some clinicians greatly prefer this approach as compared with use of a derived variable (418); they make allowance on empirical grounds for the influence of altered arterial P_{CO_2} upon sample HCO_3^-. Schwartz, Brackett, and Cohen (546) have ascertained that the physicochemical effect on plasma $[HCO_3^-]$, at the end of 10 minutes, of raised arterial P_{CO_2} of volunteers inhaling 7 to 10% CO_2, is small, about + 3 mmoles/liter when P_{CO_2} is 80 mm Hg.

C. EFFECTS OF ALTERED pH OF EXTRACELLULAR FLUID

Study of the effects of acidemia and alkalemia on physiological functions is rendered laborious and sometimes controversial by the inherent complexity of the subject. It is necessary always to distinguish between the direct effect of altered pH of extracellular fluid and the effects of change in concentration of some particular acid or base, whether CO_2 or a variety of nonvolatile compounds, which must be manipulated primarily in order to effect pH changes. Possible effects of the changed concentration of one or both members of a number of buffer pairs secondarily effected as a result of the changed pH may need also to be kept in mind. In addition, mutually antagonistic consequences of altered pH are sometimes observed, and the time course of events must be followed carefully since some prominent effects are relatively delayed.

These complexities are illustrated in studies of the effects of respiratory and nonrespiratory acidosis upon cardiac and pulmonary function and upon O_2 transport. Both acidemia and alkalemia appear to decrease peripheral vascular resistance, an effect partially counteracted by concomitant increase in catechol hormone activity; nonrespiratory acidosis causes in addition marked systemic venoconstriction with redistribution of some of the venous blood into the systemic arterial and pulmonary vessels. Alkalemia, especially when due to hypocapnia, causes cerebral vasoconstriction and reduces cerebral blood flow. Acute changes in blood pH markedly affect the curve relating blood O_2 saturation (on the ordinate) to P_{O_2} (on the abscissa)—the Bohr effect (626); the curve is shifted to the right by acidemia and to the left by alkalemia. These immediate effects of altered blood pH on the curve are, however, increasingly reversed, after 6 to 8 hours, by delayed effects on the erythrocyte content of 2,3-diphosphoglycerate, which attaches itself to reduced Hbg, shifting the O_2 dissociation curve to the right. For an account of the significance of these changes for

O_2 transport, and more detailed discussion of effects of altered pH on cardiovascular function, the reader is referred to the recent review by Mitchell, Wildenthal, and Johnson (406a).

Metabolic processes within the cells are of course profoundly affected by all but small changes of the pH of extracellular fluid. Presumably it is the integration of vital cellular functions, particularly those whose stability depends upon stable values for the velocity constants of enzyme reactions, which is threatened in acidemia and alkalemia; and much of the disturbance within the cells is no doubt related to alterations of the acidic intensity of various compartments of cell water. But we know very little about the location and magnitude of pH changes developing in different areas within the cells in clinical acid-base disturbances, and interpretation of the metabolic significance of pH measurements on cell water compartments is at present highly speculative in molecular terms (Section A). Some gross metabolic consequences of acidosis and alkalosis for cellular metabolism have been recognized for decades because they are unambiguously reflected in extracellular fluid or urine; these include alterations of renal ammonia production (II-10B), which are inversely related to blood pH, and alterations of production of citrate by the kidney and tissues generally which are directly related to blood pH (438, 403). [Alteration of net production rates of organic acids by the brain (cf. Section A) in respiratory alkalosis and acidosis (396) may be related to arterial P_{CO_2} rather than pH.] The significance for the organism of these alterations of intermediary metabolism when plasma pH is altered is often clearer than the causal sequence which produces them (Section A). Sometimes, as with renal ammonia production, they serve a homeostatic purpose; sometimes they complicate the disorder. For example, decreases in citrate production and renal citrate excretion during acidemia contribute to the diathesis of calcium nephrolithiasis, particularly when the urine is alkaline and contains large amounts of calcium, as in the classical form of renal tubular acidosis (III-3B2).

Many other metabolic effects of altered pH of extracellular fluid are postulated (501a); experimental support depends for the most part upon *in vitro* studies whose relevance to the functioning of cellular enzymes *in vivo* is difficult to assess. For example, in spite of a very large literature of this kind relating the increased renal ammonia production in nonrespiratory acidosis to increased renal gluconeogenesis, it is not at all clear at present writing that nonrespiratory acidosis does in fact stimulate renal gluconeogenesis in the intact mammalian organism (96a, 528a).

Clinical manifestations ascribable to acidemia and alkalemia *per se* must be distinguished from those relating to changed concentration of CO_2 and HCO_3^-, and are commonly modified clinically by concomitant alterations of extracellular fluid volume and osmolality and by specific effects of altered extracellular concentration of particular electrolytes. There do not appear to be specific consequences of HCO_3^- depletion other than those of the associated acidemia; these include disordered cerebral function, especially mental dullness and apathy progressing to confusion, delirium and coma, hyperventilation and other brain-stem manifestations, cardiac dysrhythmias, and dehydration. In severe respiratory acidosis the elevated [CO_2] exerts a narcotic effect on both cerebrum and brain stem, and can cause fatal depression of the respiration when O_2 is administered.

Alkalosis may be associated with anxiety, confusion, increased neuromuscular irritability, hypotension, tachypnea, and seizures and other movement disorders (III-2B). Hypocapnia rather than alkalemia *per se* may be the principal cause of the marked reduction of cerebral blood flow seen in acute respiratory alkalosis, which is in turn probably a major contributing cause of the seizures. In general, clinical manifestations are more prominent in both acidemia and alkalemia when the conditions have developed acutely.

Hemorrhagic, cardiogenic, and septic shock produce complex abnormalities of acid-base regulation usually associated with hypoxemia, hyperventilation, and hypocapnia. Alkalemia is characteristic of the early stages; later, acidemia may supervene, and some degree of lactic acidosis is generally present. Because of the complexity of these states, with poorly defined alterations of the metabolism of heart, lung, brain, kidney, and peripheral tissues, the detailed pathogenesis of the associated acid-base abnormalities is often unclear (165).

NOTES

1. The tradition of analyzing blood, from the standpoint of its acid-base composition, as a physicochemical system in which strongly basic compounds like NaOH serve to "neutralize" the blood carbonic and nonvolatile acids goes back to the origins of acid-base chemistry (I-3; II-5), and has been continued by major contributors to the subject in the present century, including Van Slyke and Cullen (627), Gamble (182), Singer and Hastings (575), and Astrup and Siggaard-Andersen (18, 569). In this book, however—following Davenport (118b)—abnormalities of renal acid-base regulation are conceived as of resulting from accumulation of *excesses of nonvolatile base* or *excesses of nonvolatile acid*. (The latter expression has the same meaning as the "negative base excess" of Astrup and Siggaard-Andersen.)

From the standpoint of the student as well as the renal physiologist, it is natural to think of the kidney as regulating by *excreting* something; and so it is also considered here that the kidney *generates* base in extracellular fluid by excreting acid, and *generates* acid by excreting base (cf. II-4 and II-6). Sometimes, too, the net base or acid which the kidney must excrete to maintain acid-base balance has largely been generated extrarenally in extracellular fluid; for example, loss of gastric HCl in vomiting *generates* extracellular base which the kidney must excrete if balance is to be maintained. Normally, though, extrarenal losses of nonvolatile base or acid are trivial, and the net acid or base which the kidney must excrete to maintain acid-base balance results from the net *addition* of acidic and basic compounds to the body fluids by dietary intake and metabolism. Finally, we may consider that, when the net load is acidic, renal net acid excretion in the steady state is regenerating base which the invading acid is simultaneously destroying, while renal net base excretion regenerates acid destroyed by invading base.

2. We follow the usage of European authors (18, 569) in designating acidosis and alkalosis not due to altered arterial P_{CO_2} as "nonrespiratory" rather than "meta-

bolic." When spontaneous nonrespiratory acidosis is truly of metabolic origin, it results from metabolic overproduction of acidic compounds, and is not logically classifiable as renal: normal kidneys are simply overwhelmed by an abnormally large acid load. The general classification term "nonrespiratory" permits subclassification which draws a further needed distinction between acid-base disorders due to *renal dysfunction* and those due to an *excessive load* of nonvolatile acid or base; such loads may in turn be due (i) to generation of extracellular base by extrarenal acid loss, (ii) to generation of extracellular acid by extrarenal base loss, (iii) to metabolic overproduction of acidic compounds or metabolic generation of acid or base by metabolism of nonmineral cations or anions; or (iv) to rapid intake of alkaline salts or (rarely) mineral acids.

Loss of water in excess of solutes from extracellular fluid, with continued maintenance of arterial P_{CO_2} at 40 mm Hg, will give rise, without any alteration of acid-base balance, to rise in [HCO_3^-], and hence rise in pH due to the increased $HCO_3^-:CO_2$ ratio. To this entity the name "contraction alkalosis" has been applied (92); dilution of extracellular HCO_3^-, with normal arterial [CO_2], has been termed "dilution acidosis" (565, 663). Dilution acidemia and contraction alkalemia would seem to be the preferable terms, since there is an imbalance in respect to water rather than acid-base imbalance (165).

3. Compensation is in general incomplete, but (for example) severe nonrespiratory alkalosis may be found in a patient in whom a sustained prior elevation of arterial P_{CO_2} has been alleviated (III-2). What was originally a compensatory response of the kidney to respiratory acidosis then presents itself as nonrespiratory alkalosis.

4. In clinical jargon the total CO_2 of plasma or serum is called the "CO_2 content" and often is referred to simply as the "CO_2." This entity must of course not be confused with the dissolved CO_2 gas concentration in plasma or extracellular fluid, symbolized in this book and generally in the physiological literature as [CO_2].

5. The electrolyte structure of the plasma is in practice generally evaluated for reasons of convenience on samples of blood serum (defibrinated plasma). The electrolyte structure of plasma or serum is identical, but differs a little from that of extracellular fluid because of the presence of protein, which provides a net excess of anions (I-5D) which are restrained by the capillary wall from equilibrating with extracellular fluid. Differences in the concentrations of other ions, dictated by electroneutrality and osmolar equilibration, are defined by the Gibbs-Donnan distribution, resulting in small excesses of cationic equivalence in plasma over interstitial fluid, and in small reductions in the concentrations of the nonprotein anions in plasma as compared with interstitial fluid (193, 133, 241, 453, 182, 144, 65, 569).

6. Plasma phosphate is reckoned at an equivalence of 1.8 mEq/mole; its pK′ in plasma water is about 0.6 below plasma pH, and so it exists in plasma largely though not entirely as the divalent ion $HPO_4^=$. Most of the plasma organic acids are such strongly acidic compounds that they are essentially fully dissociated and cannot function as buffer anions in plasma. The cations of plasma water (Figures III-1.1 and III-1.2) are wholly dissociated and cannot participate in shifts of the acid-base equilibrium of the plasma. They can be thought of as derived from dietary alkaline salts like NaOH or $MgCO_3$ which have reacted with carbonic, mineral, and organic acids prior to or after ingestion (II-5; Appendix G).

7. The CO_2 gas dissolved in the plasma is obtained at 37° C by multiplying by the factor 0.03 mmoles liter^{-1} mm Hg^{-1} (Appendix D). If the extreme limits of arterial P_{CO_2} encountered clinically are taken as 15 to 150 mm Hg, plasma [CO_2] may range from 0.45 to 4.5.

two

NONRESPIRATORY ALKALOSIS

Table III-2.1 classifies the principal clinical causes of nonrespiratory alkalosis into two groups: excessive base load and renal alkalosis, i.e., accumulation of excess base in extracellular fluid due to faulty renal regulation. Obviously both factors may collaborate; renal disease in general diminishes the efficiency of the kidneys in maintaining acid-base homeostasis in the face of large net loads of acid or base, and in high-grade glomerulotubular failure (III-3) relatively small loads of either acid or base cause pathological alteration of the acid-base composition of extracellular fluid.

A. EXCESSIVE BASE LOAD (144, 65, 95, 467, 517, 255)

An excessive base load (II-5) is clinically most often occasioned by extrarenal loss of nonvolatile acid, specifically loss of HCl by vomiting or gastric suction. Metabolic overproduction of base is not known, but ingestion of base in quantities too large to be eliminated, without alkalosis, by kidneys whose function is reasonably well preserved is sometimes encountered clinically, principally in persons who continually take absorbable alkalis as remedy or preventive in peptic ulcer and other gastrointestinal disorders; naturally impaired renal function predisposes to alkalosis under these circumstances. Though it has also been thought that pure alkalosis may damage the kidneys, they seem in experimental situations to tolerate the condition well, provided that dehydration and K depletion are avoided (83).

Vomiting with or without diarrhea involves loss of K—mostly in the urine (83, 552a)—as well as of HCl, and conditions causing vomiting often prevent adequate K intake, so that alkalosis of the acid-loss type is frequently accompanied by K depletion. Loss of gastric and upper intestinal

TABLE III-2.1. *Classification of nonrespiratory alkalosis*

I. Extrarenal (excess base load)
   a. Loss of (nonvolatile) acid
      Vomiting; gastric suction
   b. Alkali excess
      Ingestion of absorbable alkali; overtreatment with HCO_3^-
II. Renal
   a. Mineralocorticoid excess
      Aldosteronism 1° and 2°; pseudoaldosteronism (licorice, drugs); adrenal steroid therapy
   b. K depletion
      Extrarenal losses (vomiting, diarrhea, laxative abuse, bowel drainage); aldosteronism; diuretics (thiazide, furosemide, ethacrynic acid, carbonic anhydrase inhibitors, mercurials); intrinsic renal disease; steroid therapy; inadequate K intake
   c. Abrupt reduction of established elevation of arterial P_{CO_2}

secretions in vomiting results in substantial deficits of extracellular Na and water. Contraction of extracellular fluid volume is itself a cause of alkalosis (III-1) and is therefore frequent in this type of alkalosis. It results in turn in "prerenal" depression of the kidney's regulatory efficiency including its normal ability to eliminate base as urinary HCO_3^- at a rapid rate, since the blood supply of the organ is reduced as a result of the diminished circulating blood volume. The dehydration normally stimulates aldosterone secretion, which, though it helps to conserve body Na stores and hence extracellular fluid volume, may further aggravate K depletion. In turn, K depletion aggravates alkalosis by producing inappropriate secretion of acid into the nephrons (Section B).

B. RENAL (144, 65, 467, 517, 487, 196, 642, 504, 268, 211)

A very dangerous type of alkalemia due to nonrespiratory alkalosis can appear with great suddenness in patients with emphysema and other types of pulmonary insufficiency treated by mechanical ventilation; this disorder was first identified by Refsum (491). In such patients, who have been chronically hypercapneic, renal compensation for respiratory acidosis has generally resulted in considerable accumulation of excess base in extracellular fluid, a process which ameliorates the acidemia. If then the arterial P_{CO_2} is abruptly reduced to 40 mm Hg or below during therapy, the underlying nonrespiratory alkalosis is unmasked, plasma pH rises rapidly to markedly abnormal values, and life-threatening complications appear (491, 514, 106, 169, 492, 493, 309, 399). But the alkalemia is not necessarily, or indeed usually, the exclusive result of earlier physiological compensation by the kidney for respiratory acidosis; there may be Cl depletion, and emphysematous patients often have sufficient K depletion to cause alkalosis, whether from prior treatment with diuretic drugs and adrenal cortical steroids, or both, or from insufficient K intake and losses via the gastrointestinal tract (491, 493, 399). Under these circumstances the alkalosis will not be abolished until the K and Cl depletions have been corrected; but the alkalemia can and must be alleviated at once by maintaining alveolar P_{CO_2} at levels higher than normal.

In addition to manifestations, especially cardiac arrhythmias, probably mainly due to the associated K depletion and hypokalemia, patients with this type of nonrespiratory alkalosis and alkalemia develop serious neurological disorders—anxiety, confusion, coma, tachypnea, asterixis and other tremors, myoclonus, and convulsions—as well as fever and hypotension. Hyperventilation alkalosis markedly reduces cerebral blood flow (308, 77), and it may be that cerebral ischemia secondary to rapid lowering of a chronically elevated arterial P_{CO_2} is instrumental in producing the neurological signs and symptoms in these patients. The syndrome is likely to prove rapidly fatal unless the alkalemia is promptly relieved. Patients with chronic lung disease on oxygen inhalation therapy are vulnerable also to acutely developing respiratory acidosis, which can also produce confusion and asterixis and other tremors. Characteristically, somnolence and coma are associated with hypercapnia, while hypotension, shock, and convulsions are absent; but the differential diagnosis between acidosis and alkalosis under these conditions is far from being an exact science, and frequent monitoring of the arterial pH is therefore imperative for proper guidance of therapy.

Etiologically, the commonest type of renal alkalosis is no doubt that produced by diuretic therapy. Thiazides (211) are probably the most commonly implicated diuretic drugs, but contraction of extracellular fluid volume and resulting loss of Na^+, Cl^-, and K^+ are common enough consequences of the use of most diuretics. Hypermineralocorticism contributes to the alkalosis of most such cases, and is the fundamental cause in Cushing's syndrome, primary hyperaldosteronism (whether due to adrenal cortical adenoma or hyperplasia), and aldosteronism secondary to many other conditions, including renal artery stenosis, malignant hypertension, and miscellaneous disorders associated with chronic depletion of extracellular fluid volume. The degree of alkalosis in these conditions is related not only to the degree of hormonal activity, but also to the amount of sodium reaching the distal part of the nephron, a finding which implicates distal exchange of secreted acid for Na reabsorbed as a major site of influence of these hormones (552b). Other disorders pathogenetically similar to hyperaldosteronism in this respect include genetic disorders of adrenal cortical steroidogenesis with overproduction of mineralocorticoid compounds other than aldosterone (congenital adrenal virilism and related conditions), pseudoaldosteronism due to licorice, and (apparently also rarely) nonhormonal, primary excessive distal exchange of protons for sodium (110, 552b).

Clinical and indeed experimental nonrespiratory alkalosis is, however, almost invariably influenced by a variety of interrelated physiological factors. Extensive experimental work has been reported over the past 15 years, focusing especially upon the behavior of the kidney in nonrespiratory alkalosis and attempting to account for it. Except for the alkalosis which can be produced in glomerulotubular failure by a base load of relatively modest proportions, all renal alkalosis (Table III-2.1) may be considered to be *renal tubular alkalosis* in which acid secretion into the nephrons is inappropriately high in relation to glomerular HCO_3^- filtration. The pathogenesis of the common types of renal tubular alkalosis is complex, and complete agreement on all aspects of the subject has not been attained. It is necessary to distinguish a number of related factors which may be involved in the pathogenesis of these conditions: (i) the depletion of K in renal tubular cell water, with the functional effect of stimulating secretion of acid into the nephrons; (ii) damage to kidney resulting from K depletion, producing kaliopenic nephropathy; (iii) direct effects of aldosterone (or any steroid possessing mineralocorticoid activity) upon renal tubular function; (iv) altered body content and distribution of Na and Cl, changes in extracellular volume, hemodynamic changes, etc., which may be in part the result of mineralocorticoid activity; and (v) the role of Cl in the correction of alkalosis.

The consequences of simple increase in the rate of acid secretion into the nephrons over the value expected at any level of plasma $[HCO_3^-]$ are directly predictable. A state of renal tubular alkalosis would result, during the genesis of which the urine would be inappropriately acid in relation to the acid-base load; base would therefore be retained by the kidney, and plasma $[HCO_3^-]$ would rise. This rise, however, by increasing the rate of glomerular HCO_3^- filtration at any acid-base load, would soon eventuate in a new steady state in which the acid-base composition of the urine would again become appropriate to the acid-base load. The internal environment would be inappropriate to it, however; nonrespiratory alkalosis would be present because of the excess base which had been retained during the transitional period. And indeed, because of the alkalosis, arterial P_{CO_2} would have risen, further increasing the inappropriately raised rate of renal acid secretion (III-1), so that, after attainment of the new, pathological steady state, the cumulative renal retention of base would be greater than it would be without pulmonary compensation. The alkalemia, however, would be moderated. In such a state of stabilized renal tubular alkalosis, the responses of the fundamental renal acid-base regulatory system to changing acid-base load would continue to be appropriate; plasma $[HCO_3^-]$ would be higher in the presence of a base load than during an acid load, though the value would be inappropriately high at all acid-base loads. The term "paradoxical aciduria" has been applied to excretion of acid urine in nonrespiratory alkalosis; in the steady state it is paradoxical only in relation to the internal environment, not to the acid-base load.

In practice, a pathophysiological state as simple as this is rarely encountered among the various types of nonrespiratory alkalosis encountered clinically. Aldosterone and other hormones with mineralocorticoid activity usually directly stimulate the renal tubular cells to secrete acid into the nephrons, but by their kaliuretic action they also give rise to K depletion, a condition generally associated with increased renal acid secretion, yet also at times (105, 268) with limitation of the ability of the secreting cells to lower the pH of tubular fluid. There are then two reasons why hypermineralocorticoidism should produce alkalosis: (i) directly by stimulation of renal acid secretion, and (ii) indirectly as a consequence of K depletion. A similar complexity results from exhibition of diuretic drugs like the thiazides which cause K depletion, or of agents like acetazolamide, which directly cause urinary loss of both K and HCO_3^- (211). By depleting the extracellular fluid volume, they activate the renin-aldosterone system, so that simple K depletion is compounded by hypermineralocorticism. To some extent the same complication develops when K is lost by the gastrointestinal route.

Moreover, a steady state in regard to body K content seldom exists in hypermineralocorticoidism, whether clinical or experimental; and is sometimes not even approximated. In experimental or clinical situations in which compounds possessing mineralocorticoid activity are given, K is progressively lost from the body over a prolonged period; an acid-base steady state does not exist during progressive K depletion insofar as the alkalosis is worsening. Accordingly the relative influence of K depletion and hormonal activity on renal acid secretion is quite variable. As an extreme example, after correction of hypermineralocorticoidism of any standing, as when hyperaldosteronism is surgically corrected, serious K depletion exists, whereas mineralocorticoid activity is no longer excessive, but rather subnormal, so that the hormonal stimulus to paradoxical aciduria is scarcely present. The pathological complexity of these various states is increased still further, and markedly, by the development of kaliopenic nephropathy (504, 268), a structural renal disorder which markedly influences renal medullary function. It is possible that the reduced ability to acidify the urine sometimes seen in K depletion is due to structural damage rather than being a simple functional consequence of reduced renal cellular or extracellular [K^+]. Finally, K depletion causes renal ammonia production to rise, so that urinary ammonia excretion may be inappropriately high for the urinary pH (268, pages 958–959; 28, page 297). Hyperammoniuria may contribute in variable degree to raised urinary pH in this condition, though not to net base excretion.

Steroid-induced K deficiency tends to be associated with higher urinary pH than are other forms of K depletion; Hollander and Blythe (268) seem inclined to ascribe this fact to a more severe K depletion due to the continued high renal K clearance at low plasma K levels in this condition. It may be so; but the subject deserves investigation. The urinary pH is not determined exclusively by the rate of renal excretion of acid or base; moreover, CO_2 loss after collection affects the pH of HCO_3^--containing urines to a variable extent depending on several factors (II-4C), and the pH of urine collected without loss of CO_2 has not been investigated in various types of K deficiency.

The suggestion of Berliner and his associates (54, 50) that there is competition for a secretory pathway between potassium and acid transport into the nephrons, each being linked with reabsorption of Na, is compatible with the reciprocal relation between K and acid secretion observed in the intact animal under a number of circumstances; but whether this concept can be implemented in either topographic or molecular terms is at present problematical, to say the least. It is true that stimulation of acid secretion, effected by hypercapnia (129, 135) or K deficiency (306, 73, 512, 199), is associated with reduced K^+ excretion and clearance; conversely, inhibition of acid secretion by hypocapnia (33, 591), administered K (355, 660, 53, 511, 436, 488), or acetazolamide (54, 211) promotes renal K^+ excretion. Secretion of acid and K^+ are also reciprocally affected by the circadian rhythm of urinary electrolyte excretion (400). However, the postulated competition between urine acidification and K^+ secretion would have to be localized to the distal nephron, since micropuncture studies disclose that K^+ is only reabsorbed, not secreted, in the proximal tubules and indeed in the first portion of the distal tubule (67, 383, 376, 647). Hypokalemia, however, appears from carefully controlled micropuncture studies (324) to have a marked effect *per se* upon the rate of proximal tubular acidification of glomerular filtrate; and some less conclusive evidence summarized by Rector (487, page 243) suggests that K excess may exert the opposite effect in the same location. Since there can be no competition between urine acidification and nonexistent proximal K^+ secretion, some other explanation is necessary to account at the molecular level for the proximal effect of hypokalemia upon acid secretion. In addition, parallel changes in acid and K^+ secretion are observed under the influence of mineralocorticoid hormones, as well as in clinical and experimental circumstances where the parallelism may or may not be entirely explicable on this hormonal basis (see end of the chapter).

Darrow and his colleagues (116, 118), and then Muntwyler and Griffin (417a), found K depletion in muscle cell water to be a routine complication of nonrespiratory alkalosis, including that produced by vomiting. This is unlikely to be a direct effect of the raised pH of extracellular fluid, for alkali and fluid therapy causes K to enter cells during the treatment of nonrespiratory acidosis. Probably, increased aldosterone activity secondary to depletion of extracellular volume contributes to K loss from the muscle cells; at any rate, adrenal cortical hormones possessing mineralocorticoid activity cause K^+ to leave, and Na^+ to enter, rabbit leukocytes (657). Balance data in man have suggested that both Na^+ and protons enter cell water in K depletion (148, 64); according to Darrow (116, 118), Muntwyler (417a, 417b), and Cooke (111) and their associates, balance data and muscle tissue analyses in rats indicate that two-thirds of the lost K^+ is replaced by Na^+, and one-third by protons. These conclusions were felt to be consonant with the decreased muscle cell pH reported earlier by Gardner, MacLachlan, and Berman in K-depleted rats (188), but subsequent reports have not confirmed this finding (552b, page 315).

There is little direct support for the idea that renal tubular cell water is also acidified in K deficiency, (9, 117). If it is, it may account for the increase in renal ammonia production associated with K deficiency (28), though the mechanism responsible for this supposititious relation would still remain conjectural. It is also often urged (e.g. 487, pages 243–244) that the increased rate of secretion of acid into the nephrons that is characteristic of K deficiency is ascribable to the increased acidity of the renal cell water during K depletion; these effects are apparently envisaged as physicochemical consequences of the altered cell-to-lumen electrochemical gradient. But it is the blood-to-lumen gradient, which would be unaffected by variations in the intercalated cell pH, which is the relevant thermodynamic factor that affects renal acid secretion, not the cell-to-lumen gradient. Secretion of base into the blood would be exactly as much hampered by renal cellular hyperacidity as acid secretion into the lumen would be favored; and since the two processes are obligatorily coupled, one would not expect on purely thermodynamic grounds that any effect would be exerted by cellular hyperacidity on the amount of overall secretory work involved in urine acidification.

Excessive secretion of acid into the nephrons in clinical nonrespiratory alkalosis is certainly not entirely the result of renal cellular K depletion. Gamble and his associates (216, 186, 523) originally directed attention to the part played by a specific influence of the adrenal cortical hormones in the genesis and maintenance of alkalosis, and Schwartz and his colleagues (19, 296, 298, 48, 550) have demonstrated that it is possible in man and dog to correct the alkalosis associated with moderate K deficiency, produced by prior aspiration of gastric contents or by dietary means, simply by treatment with NaCl, without any K replacement therapy. For this and other reasons (see below) these latter authors have placed more weight on Cl deficiency than on K depletion in the genesis and maintenance of clinical nonrespiratory alkalosis, in which both are generally present. However, it was not generally appreciated, at the time of the earlier studies by Schwartz and his colleagues, that expansion of extracellular fluid cell volume specifically suppresses Na and Cl reabsorption from glomerular filtrate and secretion of acid into it, while contracted volume has the reverse influence (II-7F). Correction of nonrespiratory alkalosis by administered NaCl without alleviation of K deficiency is therefore not necessarily relevant to the pathogenesis of the condition; one may be observing simply the concurrence of two abnormal influences on the rate of renal acid secretion which tend to cancel each other.

In the dog, alkalosis associated even with severe K depletion can regularly be completely "corrected" by NaCl therapy, but this appears to be the case neither in man (189, 297) nor in the rat (552b, 363). Administration of NaCl to patients with the alkalosis associated with hyperaldosteronism, hypercortisolism, or exogenous mineralocorticoid activity may make the alkalosis worse, rather than better, because the kaliuretic effect of these steroids depends quantitatively upon adequate amounts of Na reaching the distal portion of the distal nephrons and therefore becoming available for reabsorption in exchange for secreted K (503, 273, 552b). In addition, "saline-resistant alkalosis," of which the hallmark is continued loss of Cl^- into the urine when the condition is treated with NaCl, is of common enough occurrence in hyperadrenocorticism; and it is evidently now agreed that in some instances saline therapy will correct other types of alkalosis only after severe K depletion has been ameliorated (297). The factors responsible for saline correctability in this type of human alkalosis are not yet quantitatively defined under all circumstances, but the severity of the K depletion appears to be one, and the degree of mineralocorticoid activity another.

That mineralocorticoid activity stimulates renal acid secretion by a direct hormonal effect on the renal tubular cells is sometimes contested (487, page 242), but is usually considered established (196, pages 350–353; 448, 348, 175, 414); and mineralocorticoid deficiency may produce hyperchloremic acidosis in the absence of renal disease (411a). The immediate decline in urinary pH effected by administered mineralocorticoid (354a, 608a, 35, 401) cannot be due to K deficiency which has not had time to develop. J. L. Gamble, Jr. and his associates (216, 186, 523) originally drew attention to the importance of the direct effect of mineralocorticoid activity upon the secretory activity of the renal tubular cell for the genesis and maintenance of steroid-induced renal tubular alkalosis. They showed that alkalosis which they produced by steroid administration to dogs on a K-restricted diet was sustained only as long as the hormones continued to be administered (216, 186); alkalosis could also be produced by administration of cortisone to dogs which were prevented from losing K in the urine by dietary sodium restriction, though not as easily as when ample Na^+ reaches the urine (523). This specific effect of the adrenal steroids in sustaining established renal tubular alkalosis is also evident in the relative ease with which alkalosis can be produced by steroid administration as opposed to dietary measures and withdrawal of gastric HCl (487).

Effects of replacing NaCl intake by the sodium salts of various anions upon renal acid-base regulation have received much recent attention by Schwartz and his colleagues (48, 550). Repeated infusions of sodium nitrate can produce nonrespiratory alkalosis, because the nitrate ion is sufficiently reabsorbable from the nephrons to displace Cl^- in extracellular fluid; but NO_3^- is less reabsorbable than Cl^- and, when the infusions are stopped and no NaCl is given, the accumulated NO_3^- is poorly conserved by the kidney, creating a deficit of reabsorbable anion in extracellular fluid and glomerular filtrate. Sodium conservation is stimulated and (in the absence of sufficient non-HCO_3^- anion in glomerular filtrate) so are acid and K^+ secretion; alkalosis results. Sodium bromide does not do this, because the reabsorbed Br^- will be retained until NaCl is again furnished; the tubules seem hardly to discriminate between Cl^- and other halide ions (211). On the other hand, infusions of the Na salt of $SO_4^=$ do not produce alkalosis under similar conditions, because the sulfate is so poorly reabsorbable from the nephrons that it cannot replace Cl^- in extracellular fluid. In K^+ depletion, administration of K_2SO_4 without sodium promotes alkalosis by stimulating acid secretion; this is apparently linked to the stimulus to reabsorption of all available K^+ and Na^+ in the filtrate. Acid secretion is inappropriately high for acid-base balance, but favorable to restoration of the total alkali cation of extracellular fluid, i.e., maintenance of the volume of extracellular fluid.

Studies involving substitution of anions (other than HCO_3^-) for Cl^- in the intake, followed by the restriction of Na^+ intake (48, 550), are of theoretical interest in that they bring out the common dependence of acid-base and osmolality regulation upon the exchange of acid secretion for Na^+ reabsorption in the tubules. Though it has been suggested that there may be selective handling of Cl^- in some region of the nephron (108), and though active chloride transport out of one nephron segment (the thick portion of the ascending limb of Henle's loop) has now been demonstrated (81a), fractional reabsorption of Cl^- still appears to be largely determined passively, by Na^+ reabsorption. But under various abnormal circumstances, of which the most clinically relevant one is established deficit of extracellular fluid volume or total extracellular sodium, derangement of acid-base balance may ensue as a result of the strong stimulus to normalize total extracellular fluid Na^+ when Na^+, but not Cl^-, is available in the diet.

The dehydrated patient may not have access to table salt, or be unable to retain it, or his physician may restrict it for a variety of reasons. Under any such circumstances, contraction alkalemia may develop (III-1, Note 2); but in addition, two hormones which we know of probably influence acid-base balance in the course of regulating Na^+ balance under these conditions, and hemodynamic effects on renal function may further complicate the situation. Aldosterone secretion, stimulated by the contracted volume, tends to contribute to alkalosis by directly increasing renal acid secretion as well as by increasing urinary K loss; there may also be decreased natriuretic hormone activity, which would again tend toward increased acid secretion (642). Expanded extracellular fluid volume, conversely, may be associated with dilution acidemia (III-1, Note 2), while activity of natriuretic hormone and decreased aldosterone activity promote acidosis by suppressing acid secretion into the nephrons.

At the present time it seems quite impossible to reduce the relation between renal tubular K^+ secretion and reabsorption, on the one hand, and acid secretion on the other, to any simple and general formulation which will cover, in terms of molecular mechanism or topography, all the various situations encountered clinically and experimentally in alkalosis. Aldosterone activity promotes net K^+ secretion and acid secretion, but the link between the two functions is doubtless indirect, via separate links of each with Na^+ reabsorption. K depletion, which so often results from increased aldosterone activity, itself causes increased acid and decreased K^+ secretion, but, when severe, may perhaps cause increased cortisol secretion (alarm reaction) and consequent mineralocorticoid activity. On the other hand, kaliopenic nephropathy may interfere with urine acidification. Much remains to be learned about the effects of natriuretic hormone on acid and K^+ secretion, about the locus and mechanism of its renal action, and about the regulation of its secretion in relation to the usual clinical complex of nonrespiratory alkalosis, K depletion, and altered mineralocorticoid activity. The various links observed under various circumstances between renal K^+ excretion and acid secretion give every appearance of remaining intractably empirical, not readily reducible to a simple and general molecular formulation.

three

NONRESPIRATORY ACIDOSIS (182, 144, 65, 95, 467, 517, 165, 385)

Table III-3.1A presents a conventional physiological classification of this condition based upon a primary distinction between base loss, whether renal or extrarenal, and accumulation of acidic compounds in extracellular fluid. The acidosis of glomerulotubular insufficiency is not readily accommodated into this primary division.

Table III-3.1B gives an alternative classification focusing upon breakdown of the acid-base homeostatic function of the kidney in nonrespiratory acidosis; this classification, also physiological, is the basis of the ensuing discussion. Nonrespiratory acidosis is viewed as developing for one of two reasons (or both): (i) because a normal kidney is required to excrete acid more rapidly than is possible without development of acidosis; or (ii) because kidney damage interferes with renal excretory efficiency, i.e., reduces the rate at which acid can be excreted at values for arterial pH and plasma [HCO_3^-] within the normal range. From the standpoint of net renal acid excretion it is immaterial whether acid must be excreted because acidic compounds are being added rapidly to extracellular fluid or because acid is being generated rapidly in extracellular fluid by extrarenal bicarbonate loss: both processes produce an acid *load* for the kidneys to excrete (II-5A; III-1, Note 1).

Both the major types of nonrespiratory acidosis as classified in Table 1B may be divided into several subgroups. Abnormally great loads of nonvolatile acid may result from net loss of base from the organism (Section A-1) or they may be caused by rapid addition of nonvolatile acidic compounds to extracellular fluid, almost always because these compounds are being metabolically produced from neutral precursors (A-2). Acidosis due to

TABLE III-3.1A. *Classification of nonrespiratory acidosis based on base loss versus acid excess*

I. Bicarbonate loss with chloride acidosis
  A. Gastrointestinal (diarrhea, fistulas, intestinal drainage, ileostomy)
  B. Renal
    1. Renal tubular acidosis (Table III-3.4) with or without additional specific tubular transport abnormalities
    2. Acetazolamide and related diuretics, toxins; metabolic, immunological, and genetic disorders affecting renal acidification
    3. Renal disorders with conspicuously defective acidification of the urine
II. Accumulation of acidic compounds
  A. With chloride acidosis (NH_4Cl, etc.)
  B. Organic acidosis with increased Δ
    Ketoacidosis: lactic acidosis; drugs and toxins (methanol, salicylate, paraldehyde, ethylene glycol, borate)
III. Glomerulotubular insufficiency

TABLE III-3.1B. *Classification of nonrespiratory acidosis based on extrarenal versus renal etiology*

I. Excessive acid load
  A. Extrarenal base loss with chloride acidosis (diarrhea, fistulas, intestinal drainage, ileostomy)
  B. Rapid addition of acidic compounds to extracellular fluid
    1. Organic acids (Δ increased)
    2. With chloride acidosis (NH_4Cl, etc.)
II. Renal acidosis
  A. Selectively damaged renal acidification
    Renal tubular acidosis with or without additional selective tubular transport abnormalities; defective renal acidification due to metabolic, immunological, or genetic disorders affecting renal acidification; toxins, drugs, and diuretics
  B. Renal and systemic diseases (obstructive uropathy, chronic pyelonephritis, sickle-cell anemia, etc.) in which acidification defect is out of proportion to the degree of renal failure
  C. Glomerulotubular insufficiency

renal insufficiency results either because renal tubular acid secretion is more or less specifically impaired (B-1, B-2), or else because of glomerulotubular insufficiency (B-3). Renal acid-base excretory inefficiency in this latter condition can be viewed as resulting from glomerular insufficiency (decreased effective glomerular filtration rate), from the inability of specific tubular transport processes to function with high homeostatic efficiency in the presence of reduced glomerular filtration of particular substances and increased filtration rate per nephron, and from reduction of renal ammonia production.

A. EXCESSIVE ACID LOAD

1. Extrarenal Base Loss

Large daily volumes of alkaline secretions are normally produced by the intestines (182); these, and the alkaline pancreatic secretion, are normally entirely reabsorbed, but when they are lost by diarrhea, vomiting, or intestinal aspiration, or through a fistula or ileostomy stoma, a large net load of nonvolatile acid is likely to result (II-5). Patients with a permanent ileostomy or with chronic diarrhea may have an unusually large acid load which the kidney must excrete month after month; this load may be insufficiently large to result in systemic acidosis, yet large enough so that in order for acid-base balance to be maintained the kidney must continuously secrete a very acid urine. This unremitting urinary hyperacidity may be the cause of serious calculous disease of the urinary tract, leading ultimately to parenchymal renal disease (III-4). When the intestinal tract is acutely inflamed, the rate of production and loss of the intestinal secretions may be much increased, occasioning large base losses via the stool, and acute dehydration and acidosis may result. The dehydration decreases the renal blood flow and glomerular filtration rate, so that the regulatory efficiency of the kidney, including acid-base regulation, is compromised.

The most dramatic of all the acute diarrheal diseases leading to hyperacute dehydration and acidosis is epidemic Asiatic cholera, a disease which occupies a unique place in the history of clinical experience with acid-base disorders. Fluid is lost from the bowel in the acute diarrhea of severe cholera at a monstrous rate, and the patient's survival has in the past often depended primarily on defending extracellular fluid volume by prompt parenteral replenishment of the water and salts being poured out in the stools. Van Slyke (626) has reproduced in part some very remarkable observations reported to the *Lancet* during the cholera epidemic of 1831 in Great Britain. Dr. W. B. O'Shaughnessy concluded (437) from chemical examination of the blood from patients severely ill with cholera that water, "neutral saline ingredients," and carbonate of soda (really HCO_3^-) had been surrendered to the diarrhea fluid. On the basis of O'Shaughnessy's observations, Dr. T. Latta (330) treated a woman *in extremis* due to severe cholera, using repeated injections of several liters of "alkaline salt water"; the effect of the initial injection was described as "like magic" and recovery ensued.

Effective therapy of massive acute diarrhea and its physiological principles were rediscovered by Sellards, who treated cholera in Manila in 1910; later this author published a valuable monograph on acidosis (554). He observed the intense acidity of the urine of the patients, and improved their survival rate by treatment with intravenous bicarbonate solutions. Recently presented evidence (246) appears to indicate that the fluid loss from the bowel in cholera depends upon voluminous active secretion of fluid by the intestinal glands which, it is hoped, may prove controllable by pharmacological means.

2. Acid Excess

Diabetic ketoacidosis was the first condition recognized to be due to gross metabolic overproduction of acidic compounds (II-4A). It is still the commonest of these conditions; in all of them, of course, it is therapeutically urgent to put an end to the metabolic abnormality as soon as possible, as well as assisting the kidney to eliminate the excess acid rapidly by restoring blood and extracellular fluid volume. Huckabee's reports (276, 277) in 1961 made physicians aware that lactic acid accumulation is an important clinical cause of acidosis (II-4A) due to abnormal metabolic production of an acid. A number of drugs and poisons are metabolized to organic acids and their ingestion or exhibition can give rise to severe nonrespiratory acidosis (211); the presence of renal disease naturally increases the likelihood that commonly used drugs like aspirin may be taken in sufficient amounts to produce acidosis. Methanol, deteriorated paraldehyde (acetaldehyde), and polyethylene glycol (antifreeze) also produce acidosis due to accumulation of foreign acids. Chloride acidosis, which is typical of base-loss acidosis whether of primarily extrarenal or renal origin, is seldom encountered as a spontaneous occurrence as a result of acid excess; but physicians may produce chloride acidosis when testing patients with diminished kidney function for the ability of their kidneys to excrete acid. Usually the test acid load is given as ammonium chloride; this type of acid-excess acidosis represents, in effect, accumulation of HCl (I-3B). (The Cl^- accumulates in excess, while metabolic fixation of NH_3 leaves behind the proton of the NH_4Cl, and this proton is transferred to the buffers of extracellular fluid). Chloride acidosis used to be seen not uncommonly on an inadvertent basis when NH_4Cl was administered as a diuretic agent, usually with mercurials, especially when sufficient care was not taken to exhibit the medicament intermittently, usually for 3 or 4 successive days in the week. $CaCl_2$ and $MgSO_4$ have an acidifying effect similar to NH_4Cl (II-9A2). NH_4Cl acidosis is true metabolic acidosis, in that it results from the metabolism of NH_3 to urea. Ingested $CaCl_2$ and $MgSO_4$ are acidifying salts because less of the Ca^{++} and Mg^{++} are absorbed than of the accompanying anions; in effect HCl and H_2SO_4 are absorbed.

The differential diagnosis of diabetic acidosis is greatly assisted by simplified semiquantitative chemical tests for glycosuria, ketonuria, hyperglycemia, and hyperketonemia. Stupor and coma in the diabetic patient are not synonymous with ketoacidosis, but may be due to insulin-induced hypoglycemia, to extreme hyperglycemia with hyperosmolality of the body fluids, or to any concurrent cause of mental obtundation, such as a stroke.

It is therefore always important to establish the presence and severity of nonrespiratory acidosis in such patients independently of evaluation of the disturbance of sugar and fat metabolism. Nonketotic acidosis (lactic acidosis, acidosis secondary to glomerulotubular insufficiency, etc.) are not rare in the unregulated diabetic, and may coexist with ketoacidosis; pseudoketosis, without any acidosis, can occur due to intoxication with isopropanol (633).

During treatment it is wise to take full advantage of the kidney's ability to regulate acid-base balance by retaining administered Na^+ and excreting administered Cl^-. Provided the metabolic abnormality is being brought under control, and urine secretion is brisk and renal function adequate, alkali therapy is unnecessary after initial moderation of severe acidosis; and the common pitfalls of overtreatment with alkali, with development of iatrogenic alkalemia and increased risk of serious hypokalemia, are more easily avoided if the normal kidney is allowed to dictate the rapidity with which the last of the acid accumulation is excreted. Winters, Lowder, and Ordway (662) have shown that most patients recovering from ketoacidosis and other types of nonrespiratory acidosis pass through a period in which the blood pH is normal or raised in the presence of subnormal arterial P_{CO_2}; and repeated observations beginning with those of Stillman, Van Slyke, Cullen, and Fitz (593) and Peters (452a) have documented the persistence, during recovery from diabetic acidosis, of the hyperpnea which represents the normal pulmonary defense against acidemia. Too rapid restoration of normal or near-normal plasma $[HCO_3^-]$ during alkali therapy of diabetic and other types of nonrespiratory acidosis may result in alkalemia due to respiratory alkalosis; this complication is highly undesirable, especially because of the associated diminution of cerebral perfusion.

Studies during the past decade of the normal regulation of the pH of cerebrospinal fluid (CSF) and of the closely related pH of cerebral extracellular fluid have thrown considerable light on the mechanism of continued hyperpnea during the treatment of nonrespiratory acidosis. Owing to active stabilization of CSF pH (II-3), CSF $[HCO_3^-]$ does not decline as much as does plasma $[HCO_3^-]$ as nonrespiratory acidosis develops. Rapid correction of the acidosis by alkali therapy therefore has the paradoxical effect of producing abnormal acidification of CSF pH; this is because the tendency for compensatory hyperpnea to diminish as plasma $[HCO_3^-]$ rises is immediately reflected in rise of CSF P_{CO_2} as well as of blood P_{CO_2}. The acidifying effect of rising P_{CO_2} is marked in poorly buffered CSF and cerebral extracellular fluid (567a); and this acidification, by its local effect on the central chemoreceptors, perpetuates hyperventilation which has now become inappropriate to the acid-base status of the systemic blood. Restoration of blood $[HCO_3^-]$ to normal will inevitably result in alkalemia unless sufficient time has been allowed for CSF $[HCO_3^-]$ to rise to normal values by active ion transport between blood and CSF before restoring normal plasma HCO_3^- (431, 622).

Active regulation of plasma pH in response to systemic acid-base changes appears normally to be largely accomplished in considerably less than a 24-hour period, and it may be that other factors contribute to the continuing hyperpnea usually present during the treatment of clinical nonrespiratory acidosis. On purely theoretical grounds it seems highly probable that hyperammonemia, a strong respiratory stimulus, might play a role, and a more protracted one than that of the pH of the CSF. Return to normal levels of the elevated renal ammonia production of established acidosis lags for some days behind correction of experimental nonrespiratory acidosis (676), and if the urinary pH rises, as a result of correction of the acidosis, into the upper range of normal, hyperammonemia will necessarily result (II-9B). Whatever the exact mechanisms producing respiratory alkalosis, its occurrence shows that under these circumstances the kidneys, if their function is near normal and adequate water and salts are being provided, can usually regulate acid-base balance in more sensitive response to the patient's needs than can the physician.

Lactic acid is the reduction product of pyruvate, the end product of anaerobic glycolysis in the tissues generally. During abrupt violent exercise the metabolic rate of the skeletal muscles greatly and abruptly increases. By increasing glycolysis more than aerobic metabolism, and releasing substantial amounts of lactate and pyruvate into the blood, they cause the liver, with its ample blood supply, to share in the disposition of the pyruvate and lactate; accordingly, the muscles are not obliged to increase their oxygen supply in proportion to the increased metabolism. Some imbalance between muscle release of lactic acid, derived from glycogen stores and assimilated glucose, and the rate of removal of the compound by the liver is reflected by the rise in blood lactate during and for a while following vigorous muscular exercise. This admirable system gets out of hand in lactic acidosis, where a greater imbalance of the same type can result from unbridled anaerobiosis in muscles and other extrahepatic tissues, doubtless often compounded by reduced efficiency of the liver in assimilating circulating lactate and disposing of it either by gluconeogenesis or by metabolizing it to CO_2 and water.

Clinical elevation of the blood lactic acid concentration may accordingly be either physiological or pathological. Whenever carbohydrate metabolism is markedly accelerated, as when normal persons are given glucose rapidly, or diabetics insulin together with glucose, the rate of formation of pyruvate in the muscle cells is increased and additional lactate and pyruvate appear in the blood. This type of elevation of blood lactate is not uncommon in diabetics on therapy, and is characterized by a more or less proportional elevation of both lactate and pyruvate in the blood, the ratio of the con-

TABLE III-3.2. *Elevated blood lactate: classification based on the lactate-to-pyruvate (L:P) ratio*

I. Physiological (L:P normal)
  Exercise; rapid carbohydrate metabolism; insulin treatment
II. Pathological (L:P raised)
  A. Tissue hypoxia
    Shock, hypoxemia, vasopressor drugs
  B. Impaired cellular oxidation
    Hypocapnia, cyanide, phenformin
  C. Leukemia, neoplasia
  D. von Gierke's disease
  E. Idiopathic

centrations of these substances in blood being normally nearly 10:1. When, however, elevation of blood lactate is out of proportion to the elevation of blood pyruvate, the condition is apt to be one of the pathological types of lactic acidosis, even though the blood lactate level may be no higher than in some cases of physiological elevation.

Our awareness of the frequency with which lactic acidosis is encountered in medical practice owes much to two articles contributed in 1961 by Huckabee (276, 277). Clinical lactic acidosis (613) (Table III-3.2) is often associated with disorders obviously predisposing to tissue anoxia, whether due to tissue hypoperfusion (reduction of cardiac output or local ischemia), severe anemia or hypoxemia, or interference with the enzymatic machinery of cellular oxidation (cyanide, phenformin); sometimes it has a specific enzymatic basis, as in Gierke's disease or proliferation of malignant cells, whose metabolism is mainly anaerobic. Hypoxemia must be severe to produce lactic acidosis, which is seldom seen on this basis unless blood O_2 saturation is below 80%, and may be absent with oxygen saturations as low as 50%; lactic acidosis does not often complicate the majority of chronic hypoxemic states. It is, however, a not very uncommon complication of diabetic acidosis, probably principally because of associated hypotension; lactic acidosis should be suspected in this setting whenever the degree of acidosis seems out of proportion to the degree of ketonemia. Cyanide is the agent which most effectively poisons oxidation, and is a recognized cause of clinical lactic acidosis. Lactic acidosis associated with phenformin treatment of diabetes is also believed to be due to an interference by the drug with cellular respiration; Tranquada (613) thinks it likely that in most instances impaired renal function has allowed the accumulation of excessive amounts of phenformin when lactic acidosis results from its exhibition.

The prognosis is always guarded, but it is in the idiopathic form of lactic acidosis that a fatal outcome is the rule. Idiopathic lactic acidosis occurs in patients with severe illness of various kinds, but in whom an adequate reason for compromised oxidative metabolism is not apparent. It is associated with the greatest increases in the ratio of lactate to pyruvate, and with very high production rates of lactic acid. Therapy has left much to be desired, and in the case of the idiopathic form has been almost uniformly unsuccessful. Administration of sodium lactate is obviously irrational in a disease consisting of inability to clear the blood of lactic acid, and the ubiquitous possibility of undetected lactic acidosis is the principal reason why the treatment of nonrespiratory acidosis with lactate should be considered obsolete. Principal therapeutic reliance has traditionally been placed on the administration of sodium bicarbonate; drugs such as methylene blue have been tried in alkali-resistant cases in the hope that they might restore the function of cellular oxygen transport systems. Hemodialysis has been tried, and seems rational, especially in the severe form of the disease where the lactate cannot effectively be removed from the bloodstream by the kidneys; under these circumstances, any modest improvement in blood pH effected by bicarbonate is likely to be bought at the cost of increasing plasma hyperosmolality (cf. Section B2). Therapeutic application of hemodialysis has been reported (155), but there is yet no extensive series reported with this type of therapy.

B. RENAL ACIDOSIS

Because the basic processes employed by the kidney in regulating acid-base balance are so few, the primary physiological classification of renal acidosis is very simple: acidosis might result from intrinsic subnormality of acid secretion into the nephrons in relation to plasma [HCO_3^-], from inadequate delivery of buffer to the urine, or from both. As long as acid secretion into the nephrons remains inappropriately low in relation to glomerular HCO_3^- filtration, nonrespiratory acidosis will develop and progress. In some conditions, like the proximal form of renal tubular acidosis, there is a limitation on the total rate at which acid can ever be secreted into the nephrons; under these circumstances acidosis due to inadequate acid secretion may be present even when there is a large net load of base claiming excretion. A different type of defective renal acid secretion consists in a reduced ability of the cells of the terminal nephron segment to lower the urinary pH by establishing a transtubular electrochemical gradient; in this case acidosis will not develop as long as the load is sufficiently basic, so that only moderate lowering of urinary pH below plasma pH is called for. In addition, a contributory factor in the development of renal acidosis is often insufficient urinary buffer. Evidently, reduction of urinary pH to essentially normal minimal values may not suffice for elimination of any substantial acid load if urinary phosphate is reduced and especially if urinary ammonia excretion is lowered and cannot be substantially raised by developing acidosis; both these urinary buffers are in fact generally excreted at reduced rates in glomerulotubular renal failure (Section B-3). Relatively selective loss of the renal tubular capacity to produce ammonia appears to occur clinically (III-4) but causes unremitting urinary hyperacidity and its complications rather than acidosis as a rule.

The primary clinical classification of acidosis due to impaired kidney function is (i) *renal tubular acidosis* (RTA), a term signifying that an intrinsic disorder of the acid-secreting mechanism exists, with or without additional derangements of tubular transport functions, but without significant glomerular insufficiency unless this has developed secondarily or independently (Section B-2); and (ii) *glomerulotubular* insufficiency, which when advanced enough to produce chronic acidosis is usually accompanied by azotemia (Section B-3). The acidosis of glomerulotubular failure is associated with insufficiency of the total rate of acid secretion into the nephrons at normal concentrations of plasma [HCO_3^-]; this abnormality is reflected in some patients by continued wasting of urinary HCO_3^- as plasma [HCO_3^-] falls to low normal levels and on down to concentrations characteristic of frank nonrespiratory acidosis ("urinary HCO_3^- leak"). However, after the plasma HCO_3^-, and so the total effective HCO_3^- filtration, has declined to a greater or lesser extent, the damaged kidney is able to sustain acid secretion at a rate sufficient first to reclaim practically all the filtered HCO_3^-, and then to lower the urinary pH to approach normal minimal values and perhaps still further. The acidosis therefore becomes stabilized at some point, provided the acid load is only moderate; but then shortage of buffer for excess acid in the urine of these patients may be great enough, in relation to the acid load, that acidosis continues to develop after all HCO_3^- has disappeared from the urine and its pH lowered to 5.1 or less.

1. Tests of Renal Acidification

The sequence of events in the nephrons, as the glomerular filtrate is progressively acidified, is (*grosso modo*), *first*, mainly removal of HCO_3^- as (equilibrated) pH of luminal fluid falls to about 6.2; and, *second*, acidification of NH_3 and the nonvolatile buffers with further decline of a pH unit or more. In general, then, an acidification deficit may be suspected in patients on usual acid-ash diets and without evident extrarenal base loss whose urine pH remains persistently above 6.0, and strongly suspected if the pH cannot be further reduced by an acid load. Values up to 7.0 or even higher at some time during the day, especially midmorning, are quite usual and normal, however (II-1); also, high-pH urines are characteristically secreted by persons with nonrespiratory alkalosis due to hyperaldosteronism; this phenomenon has been discussed previously (III-2). Finally, urinary tract infection with strains of Proteus and other organisms which split urea may raise the urinary pH by means of the NH_3 added to it (410);

this artifactual cause of raised urinary pH is generally betrayed by the excess of total ammonia over that expected in relation to urine pH and flow. When evaluating clinically the normality of urinary ammonia excretion in relation to urinary pH and flow, the dependence of renal ammonia production upon renal blood flow should be borne in mind; it causes ammonia excretion to be positively correlated with glomerular filtration rate (GFR). The explanation appears to be (634a) that the glutamine extracted by the kidney is a rather constant proportion of the glutamine presented in the renal arterial inflow.

But when nonrespiratory acidosis is present, secretion of urine containing any consequential quantity of HCO_3^- is decisive evidence of insufficiency of renal tubular acid secretion, and in the absence of azotemia allows an immediate diagnosis of some form of *renal tubular acidosis*. Chloride acidosis of essentially the same pathogenesis sometimes accompanies azotemic glomerulotubular failure, especially in chronic pyelonephritis (329), medullary sponge kidney (124, 411), obstructive uropathy (56), and the nephropathy associated with implantation of the ureters into the colon (164).

Quite frequently, however, as for example in the diagnosis of unexplained osteomalacia, specific testing of the kidney's ability to excrete acid is undertaken in nonacidotic persons in order to disclose subclinical defects. The principle invoked is that of the tolerance test; it was first applied to renal acidification for diagnostic purposes by Albright and his associates (6, 7), who administered standard quantities of NH_4Cl orally over 5 days to observe the renal response of subjects on constant diets, after a suitable control period, to an acid load in terms of "the sum of the base sparing mechanisms" (II-4). NH_4Cl has been the preferred if not the only agent administered ever since to supply a standard excess of acid. It has an unpleasant taste and may be given in gelatin capsules; but enteric-coated NH_4Cl is unsatisfactory for test purposes as some of the dose given in this form often escapes absorption. Wrong and Davies (673) subsequently introduced a convenient 1-day ammonium chloride test, with simplified criteria of response; this procedure, commonly termed the "short" NH_4Cl test, is readily applicable to office or out patient practice, since the facilities of a metabolic ward are not required. Various modifications of these tests have been in widespread subsequent use, especially a 3-day test procedure (145).

Before the large effect of the urine flow rate upon the urinary pH was generally appreciated, reliance was sometimes placed upon determination of the urinary pH after standard test doses of NH_4Cl had been administered. Any patient with a high rate of urine flow for whatever reason, since his minimal urine pH is 0.6 higher than the lowest value he can achieve during antidiuresis (II-10), is likely to be assigned quite erroneously to the category of impaired or borderline urine acidification when the urinary pH is used as a criterion of response (602, 253). The proper criterion of response is the rate of renal net acid excretion (II-4) attained under specified conditions by the subject; in addition, the two constituents of the total acid excretion, NH_4^+ and the acid accommodated by the nonvolatile urinary buffers, can be individually appraised.

Many sources of unwanted variation may affect the subject's response to NH_4Cl when it is administered for 1 day, and to a lesser extent when it is given for 3 days, without prior control of diet and without efforts to maintain a steady state of acid-base balance during the testing. Evidently the test load is not determined entirely by the administered NH_4Cl, but in part by the diet as well. Prior acid-base status must affect the results unless a control period has created standard base-line conditions; and the 1-day test, which is really an 8-hour test, must be especially exposed to the influence of individual variations, characteristic of the subject, related to circadian rhythms of glomerular filtration rate and of other functions affecting the rate of net acid excretion. Posture (II-1) and water drinking (II-10) call especially for standardization during the test, otherwise observations may be made during a transitional stage during which the rate of acid excretion is substantially altered from steady-state conditions. Probably it is variability of these sorts which has sabotaged most efforts to improve the sharpness of interpretive criteria of the 1-day and 3-day tests, including the proposal of Elkinton, Huth, Webster and McCance (145) to relate renal net acid excretion to the plasma [HCO_3^-]. Again, urinary infection by urea-splitting organisms must be absent if the test is to be interpreted rationally.

Tannen (603) has reviewed normal variability of response in the ammonium chloride 8-hour test and has applied to it much greater uniformity of conditions than had been used by others. A 3-day period of standardized diet and control of the rate of urine flow does not eliminate large variation in response (6, 7), and unfortunately renders the test procedure almost as elaborate as the Albright 5-day test (6, 7) whose unsuitability for the clinic and the consulting room has stimulated most previous investigators to supplant it by a simpler approach. Individual variation of renal ammonia production was inferred to be the cause of a positive correlation of urine pH with net acid and ammonia excretion found after ingestion of NH_4Cl. Physicians desiring to use ammonium chloride testing, which remains a valuable means of evaluating the adequacy of the acid-secreting function of the nephrons, now have a variety of available plans to choose from (6, 7, 145, 603); criteria of normal response must of course be appropriate to the test procedure chosen.

When there is considerable reduction of the glomerular filtration rate d flow short of advanced azotemia, and indeed in severe ar failure itself (Section B-3), observation and tolerance ng is likely to disclose variable degrees of impaired acidification of the urine. We may be inclined to dub any such abnormality RTA when it is prominent, but these conditions are very different from highly selective defects of renal acid secretion such as those considered in the next section. One difficulty in classifying RTA stems from the fact that we cannot hope to set up, for renal disease generally, meaningful criteria for the degree of impairment of renal acid-base regulation appropriate to the degree of lowering of the GFR or of the renal blood flow. The anatomical and functional diversity of different renal diseases precludes so oversimplified a physiological analysis. Accordingly, one might prefer not to include among the varieties of RTA proper cases of impaired acidification secondary to conditions such as obstructive uropathy, chronic pyelonephritis, transplant kidney, medullary sponge kidney, nephrocalcinosis secondary to vitamin D intoxication or hyperparathyroidism, and sickle cell anemia (411a). Rather more specific impairment of urinary acidification has been observed on a functional basis, as in fasting (540, 310); secondary to chronic liver disease, especially Wilson's disease (180, 644); and in severe K deficiency. Whereas functional K deficiency increases renal acid secretion, there is also found in some cases, perhaps due to the structural changes of kaliopenic nephropathy, a decrease in the transtubular pH gradient against which acid can be secreted (105, 530). Hyperparathyroidism may produce the syndrome of RTA; hypercalcemia with or without nephrocalcinosis has been reported by some but not all authors to produce nonrespiratory alkalosis as well as nonrespiratory acidosis (504).

2. Renal tubular acidosis (352, 353, 84, 26, 7, 25, 505, 280, 402, 552b, 146, 529, 518, 411)

The first diagnoses of renal acidosis without glomerular insufficiency were made in young children by Lightwood and his associates (352, 353) and by Butler, Wilson, and Farber (84); only some years later was this kind of disorder recognized in adolescents and adults (26, 6, 7). It was first extensively studied from the physiological as well as the clinical standpoint by Albright and his colleagues (6, 7); their unwieldly term "tubular-insufficiency-without-glomerular-insufficiency" was condensed by Pines and Mudge (459) to the now conventional renal tubular acidosis or RTA. In all varieties of the disorder (except where glomerular insufficiency has supervened or is independently present) there is no anion gap, the depression of plasma [HCO_3^-] characteristically being equalled by an increase in plasma [Cl^-].

RTA is neither physiologically nor clinically a homogeneous disorder. Table III-3.3 presents a serviceable, if rough and ready, physiological

TABLE III-3.3.

	"Proximal"	"Distal"
HCO_3^- *Tm*	Reduced by 15% +	Normal
HCO_3^- "leak"	Disappears if acidosis severe enough	Always present
Urinary HCO_3^- when plasma $HCO_3^- > 24$ mEq/liter	Excessive	Appropriate
Alkali therapy	Very large amounts required to prevent acidosis	Alkaline-ash diet or about 1.5 mEq/kg alkali therapy per day suffices

classification; there is some divergence of opinion in regard to clinical classification (518, 411). Certain of these patients ("distal," Table III-3.3) are physiologically characterized by an absolute inability of the epithelium of the terminal nephron segment to establish transtubular pH gradients of normal maximal magnitude. Such patients cannot respond to an acid load with adequate reduction of the urinary pH, and this failure remains manifest even after severe acidosis has supervened. When the defect is marked, substantial amounts of HCO_3^- are lost in the urine under all conditions. On the other hand, renal acid secretion in these patients appears to be normal when the load is basic, indicating that the total acidification process can operate at high-normal capacity. More acid is secreted by normal kidneys into the nephrons as the base load increases (II-8D); and, since the apparent HCO_3^- *Tm* in these patients is not demonstrably abnormal, there is no indication that they fail to maintain a normal rate of acid secretion, provided only that the base load is sufficient to render unnecessary distal secretion against a large pH gradient. Because of the predominant contribution of the proximal segment to the total acid secretion, the findings are usually taken to mean that the malfunction is confined to the distal segment; it must be admitted, though, that proximal acid secretion makes up so large a proportion of the total as to preclude unequivocal demonstration that proximal acid secretion is wholly normal in these patients in the presence of net acid loads (411).

This variety of RTA is referred to either as "distal" or as "classical" RTA. Its pathogenesis is not understood and its etiology is multiform; it can be subclassified in several ways. The term "secondary" has been applied when the disease is associated with other conditions, of which a great many have been reported, including genetic and anomalous disorders, metabolic diseases and hepatic cirrhosis, intake of poisons, and exhibition of drugs. An association of the disease with abnormal and increased serum proteins is statistically quite striking. For more details on this subject, and conjectures as to the nature of the possible connection of these conditions with the defective renal acid secretion, the reader is referred to the comprehensive reviews listed above.

In the 'primary" form of distal RTA, which may appear at any age and favors female victims, all the manifestations appear to be consequences of the defect in urine acidification; the principal complications are urolithiasis, nephrocalcinosis, osteomalacia, K depletion, and dehydration. K depletion with hypokalemia results from hyperaldosteronism, which is incited by the tendency of extracellular fluid volume to shrink owing to the defective Na^+ reabsorption associated with the impaired acidification (202). The increased hormonal activity helps defend extracellular fluid volume, but at the cost of stimulating that portion of the tubular Na^+ reabsorption which is linked to K^+ secretion; K depletion ensues. The K deficiency is probably mainly responsible for the muscular weakness and the striking polyuria and inability to concentrate the urine found in these patients, and it may cause cardiac arrhythmias and grave and even fatal attacks of paralysis. Correction of acidosis and restitution of normal extracellular fluid volume by alkali therapy does not, however, always arrest K loss nor correct the hyperaldosteronism (411); perhaps some degree of autonomous adrenal cortical hyperfunction is involved.

Excessive urinary calcium excretion leads to both recurrent urolithiasis and severe clinical as well as chemical osteomalacia. Albright and his associates originally ascribed the osteomalacia and hypercalciuria to a direct effect of acidosis on the skeleton leaching out Ca^{++} from it, a view still in the field and still controversial (II-4; Appendix G). The calculi found in the urinary tract are entirely composed of calcium salts until infection supervenes, when other constituents may appear; stone formation is generally ascribed to the simultaneous alkalinity of the urine and its high calcium content, together with reduced urinary citrate excretion (553, 518). Glomerular filtration is usually somewhat reduced, sometimes profoundly so, as a result of long-standing chronic urinary infections and other complications of calculous urinary tract disease.

A majority of the cases of distal RTA are sporadic, but the disease has been reported in a number of kindreds; the principal mode of transmission is as an autosomal dominant. Recovery does not occur in the hereditary cases, and is uncommon in the sporadic cases, except in the infantile variety. The majority of these infantile cases were reported early, many of them from Great Britain, and it appears that some of them at least were, like the contemporaneous cases of idiopathic hypercalcemia, part of an epidemic traceable to excessive fortification of cow's milk with vitamin D (518). Infantile hypercalcemia, manifested by failure to thrive, anorexia, vomiting, constipation, wasting, and hypotonia, has in the past been thought of as a type of distal RTA often carrying a good prognosis for recovery and with a low incidence of nephrocalcinosis; but the evidence suggesting that many have been cases of secondary RTA, and uncertainty about the physiological type of their defective acid secretion (518), seem to justify open-mindedness as to their proper classification.

The quantity of alkali therapy needed to prevent acidosis in classical RTA is physiological (usually 1+ mEq daily/kg of body weight), since renal acid secretion is defective only when an acid load must be eliminated. Unfortunately, results of therapy often prove to be disappointing; the response of the osteomalacia to alkali therapy with or without vitamin D is likely to be unsatisfactory, and the complications of urolithiasis and chronic urinary tract infection are refractory. Persons are encountered who are essentially healthy save for the characteristic inability to lower urine pH normally; a particularly responsive increase in renal ammonia production to threatened acidosis has been credited with preventing overt disease in them (673). Some of these nonacidotic persons, however, develop nephrolithiasis and other complications of RTA (79a).

Another group of patients can elaborate maximally acid urine when they are sufficiently acidotic, but waste HCO_3^- at low-normal levels of plasma HCO_3^- and have massive bicarbonaturia at high-normal plasma HCO_3^-: the HCO_3^- *Tm* is lowered. This condition is referred to as "proximal" (or "HCO_3^--wasting" or "rate") RTA and is clearly the result of limitation on the absolute rate of acid secretion into the proximal tubules. When the rate of HCO_3^- filtration has been sufficiently reduced by acidosis, both proximal and distal tubular function is normal in the sense that all filtered HCO_3^- is reclaimed and the urine pH is lowered to the same extent seen in a normal subject during induced acidosis. The defect in these patients may be viewed as an inability of the proximal tubule—and perhaps also of the distal tubule (411)—to respond normally (II-8D) to increases in filtered HCO_3^- toward normal values by increasing the rate of acid secretion. Very large amounts of $NaHCO_3$, up to 10 mEq/kg daily, are necessary to prevent acidosis in these patients.

A "primary" form of this disease occurs in children (518); their growth is retarded, but the other features of distal RTA are generally absent. They appear to be protected by the acidity of their urine against nephrocalcinosis and urolithiasis, but the failure of the acidosis to cause osteomalacia and K depletion is somewhat puzzling. Perhaps a relatively short duration of this disease is responsible. Most commonly, proximal RTA is associated with other defects of proximal tubular transport, especially glycosuria and phosphaturia (Fanconi syndrome); these proximal tubular syndromes may be heredo-familial or due to a variety of toxic agents (553, 518, 411). Additional defects of acidification may be present in patients with proximal RTA, but it is not easy to evaluate them because of the quantitative predominance of proximal acidification. Determination of the HCO_3^- threshold is held by some authors (518) to suffice for distinguishing proximal from distal tubular RTA, but this criterion has been subjected to criticism (411).

3. Glomerulotubular Renal Failure

Nonrespiratory acidosis is the usual accompaniment of advanced azotemic renal failure, irrespective of etiology. Azotemia, in these conditions, is the consequence of reduced effective glomerular filtration, by which is meant plasma ultrafiltrate which, after passing through nephrons whose function remains to a greater or lesser extent unimpaired, makes its contribution to the volume of secreted urine. Obviously the patient's survival would not be compatible with manyfold increases in the concentration of plasma electrolytes comparable to the proportional increases in concentrations of urea and other nitrogenous plasma constituents seen in azotemic renal failure. Preservation of relatively normal plasma electrolyte concentrations in glomerular failure testifies to the efficiency with which tubular ionic transport processes are adapted to extreme reductions in GFR, even though considerable alteration of the internal environment is not prevented (182).

In view of the fact that the proportional removal of HCO_3^- from glomerular filtrate plays a central role in determining the rate of renal net acid or base excretion (II-6C; II-8D), the very existence of stabilized acidosis due to advanced glomerulotubular failure means that the rate of secretion of acid into the nephrons collectively has been reduced with great exactness and almost in exact proportion to the very marked reduction in the rate of glomerular bicarbonate filtration. The end result of this precise adjustment of tubular acid transport to the reduced glomerular HCO_3^- filtration is that the composition of the urine remains appropriate to the acid-base load, though of course the internal acid-base environment is abnormal. Glomerulotubular balance of this kind is obviously presupposed, in the case not only of filtered HCO_3^-, but in those of filtered Na^+, Cl^- and phosphate as well, ions whose rate of excretion is normally a small fraction of their filtration rate. Glomerulotubular balance is of relatively little significance in the case of a urinary constituent like K^+ or NH_4^+ where the excretion rate is normally little if at all beholden to glomerular filtration. Glomerulotubular balance is demonstrable also during lesser alterations in GFR produced under more or less physiological conditions, and has been extensively studied, especially in relation to Na^+, under such conditions by micropuncture methods (642, 659).

The concept that clinical azotemic renal failure is to be understood as resulting solely from reduction of the quantity of nephrons, the surviving nephrons remaining functionally normal (70, 69), is an oversimplification which cannot be sustained by reasoning originally put forward in its support (432, 63), and does not now enjoy general acceptance (551a, page 220).

Extensive investigations of the glomerulotubular balance associated with various types of renal damage producing glomerular insufficiency have recently been reviewed in detail by Seldin, Carter, and Rector (551a) and need not be considered comprehensively here. These investigations have been carried out principally in Bricker's laboratory, where attention has shifted in recent years from the original "intact nephron hypothesis" to observations of the effects of removal of a normal contralateral kidney upon the function of a remaining damaged kidney or renal remnant. The salient findings are that a variety of functional characteristics of the damaged kidney prior to removal of the normal kidney, and hence in the absence of uremia, are much closer to normal than after uremia and compensatory hypertrophy have developed as a result of removal of the normal kidney. These studies are held to demonstrate, in the words of Seldin *et al.* (551a), "that the chronically diseased kidney in a normal environment functions in an essentially normal fashion" and "that it is environmental factors rather than the diseased kidney *per se* which alter renal function." This seemingly circular formulation at first blush makes it a matter of some obscurity how the chemical derangements of gradually progressive glomerulotubular renal failure could ever develop in the first place. What seems to be intended is that the systemic consequences of reducing the total rate of glomerular filtration in the organism render tubular function abnormal; but it is by no means clear from the data offered that a uremic milieu, rather than functional changes accompanying compensatory hyperplasia, is responsible for the various functional abnormalities developing in the kidney remnant or diseased kidney after removal of the contralateral kidney (266a). Perhaps topographic descriptions of the functions of the several nephron segments of the kidneys of various types of renal failure, together with anatomical-functional analysis of the distorted renal architecture, will help to clarify the mechanisms that maintain glomerulotubular balance in these various conditions. At present, too little is known about mechanisms sustaining glomerulotubular balance of HCO_3^- in clinical glomerulotubular failure to warrant further comment on this subject. Disease processes attacking the kidney or any other organ are by their nature disorderly and various, and there are presumably differences, in respect to the kind and degree of heterogeneity among surviving functional nephrons, not only between, but also within, the various structural and etiological categories of clinical azotemic renal disease.

Azotemic renal acidosis can be analyzed into (i) a glomerular component and (ii) a tubular component, with due notice taken of the impossibility of disregarding the functional connection between altered tubular activity and reduced rates of glomerular filtration. The principal contribution of glomerular insufficiency to the acidosis of chronic glomerulotubular failure lies in the resulting decreased delivery of nonvolatile buffer to the urine.

Accumulation of the anions of sulfuric, phosphoric, and various organic acids in the plasma is the most conspicuous effect of glomerular failure upon the plasma electrolyte structure, but it does not necessarily entail any marked acidification of the plasma (see below). The accumulation of these anions does, however, reflect the consequences of glomerular insufficiency; their concentration is high because their filtration rate has fallen, and this fact—in spite of tubular compensatory activity in the case of phosphate—results in provision of less nonvolatile buffer to the urine. We shall examine these relations first in order to define the glomerular contribution to the acidosis of glomerulotubular insufficiency.

Sulfate is never a urinary buffer, being always completely dissociated; but an account of the renal handling of this ion will serve very well as a first illustration of the effect of glomerular failure on the concentrations of a number of anionic constituents of the plasma. At raised plasma concentrations of the ion, sulfate excretion in the urine is effected mostly by glomerular filtration, the rate of reabsorption being constant and representing a low proportion of the filtered material (467). For substances excreted in the urine essentially by filtration without regulatory reabsorption, as pointed out by Gamble (182), the following relation holds in the steady state where the load L equals the excretion rate E:

$$L = F \times P = E \tag{1}$$

where F is the glomerular filtration rate and P is the concentration of the substance in plasma water. From this relation it follows that P must increase hyperbolically as F declines at any load:

$$P = L/F \tag{2}$$

The concentration of $SO_4^=$ in the extracellular fluid accordingly increases particularly after F has been reduced below the normal value, and very markedly when F is less than 20% of normal.

Quite a few substances accumulate in extracellular fluid in high-grade glomerulotubular renal failure according to a relation between P, L, and F which resembles or approximates that of Equation 2; these include creatinine, urea, uric acid, and phosphate (182, Charts 22–24). Creatinine and urea clearances are essentially "glomerular clearances" like the clearance of $SO_4^=$ in glomerular failure; but Equation 2 does not hold exactly, because transtubular movement occurs in both cases. The value of the urea clearance in fact averages only about half the value calculated by Equation 2; but because the proportional reabsorption of filtered urea is stabilized at about 50% ± 10%, depending on the urine flow, the hyperbolic increase in P described by Equation 2 is closely approximated in the case of urea as well as creatinine.

Phosphate is a material whose tubular transport has a much more regulatory character than is true of urea or creatinine, especially (though not exclusively) as disease reduces F markedly. Like the clearances of Na^+ and Cl^- and K^+ and Ca^{++}, and unlike those of urea and creatinine, phosphate clearance is markedly and actively altered as load changes: plasma concentration is stabilized, as load changes, by tubular activity of homeostatic significance. Urinary phosphate normally accounts for only a small proportion of the filtered phosphate (15% or less); but when F is greatly reduced by disease the tubules must reject a much higher proportion of the filtered phosphate, and urine phosphate accordingly then becomes an important moiety of the filtered load. Under these circumstances, delivery of phosphate to the urine is appreciably reduced in direct consequence of the glomerular insufficiency; the high plasma phosphate resulting from glomerular insufficiency leads to diversion of some of the phosphate load to the stool. Tubular failure could be considered to play a role in this reduction of urinary phosphate only in the specialized sense that increased tubular rejection of filtered phosphate has failed to compensate entirely for reduced F. (We here disregard, as irrelevant to the principles being discussed, alterations of the filtrability of plasma phosphate in uremia.) Rise in the concentrations of another urinary buffer, uric acid (like that of some other substances, including K^+) in the plasma of uremic patients has its basis to a greater extent in tubular failure, because filtered uric acid is always extensively reabsorbed, the total urate excreted in urine being largely secreted.

Accumulation of extra $SO_4^=$, phosphate, and the ions of organic acids in uremic plasma must necessarily displace Cl^- or HCO_3^-, or both, if osmolality and volume regulation are accurate (Figure II-1.2). From the purely chemical standpoint, the kidney in high-grade glomerulotubular failure is not foreclosed from maintaining normal pH of extracellular fluid by accumulation of these ions in plasma: maintenance of $[HCO_3^-]$ within normal limits is achievable if the accumulating anions displace Cl^- instead of HCO_3^-. But in practice it is characteristically plasma HCO_3^- rather than Cl^- which is displaced in the acidosis of glomerulotubular insufficiency, a fact which means that tubular acidification (i.e., regeneration of base in peritubular blood) is defective.

Table III-3.4 illustrates the reflection, in the electrolyte structure of the plasma, of the progressive development of mild acidosis in a patient with high-grade glomerulotubular renal failure; it will be observed that the decrease in plasma $[HCO_3^-]$ is very closely paralleled by an associated decrease in $[Na^+ - Cl^-]$. The nearly equal reciprocal change in these parameters is the essential acid-base change in plasma electrolyte structure as acidosis develops because of withdrawal of alkali therapy in this condition. The equality means that Cl^- is replacing HCO_3^-; however, the

TABLE III-3.4. *Development of nonrespiratory acidosis in a patient with chronic glomerulonephritis and high-grade glomerular insufficiency**

Day	pH	$[HCO_3^-]$	$[Na^+] - [Cl^-]$	$\Delta[HCO_3^-]$	$\Delta [Na^+] - [Cl^-]$	$[Na^+]$
2	7.40	24	47			146
3	7.31	23	48	−1	+1	144
4	7.33	21	44	−3	−3	140
5	7.33	21	44	−3	−3	137
6	7.32	19	48	−5	+1	145
7	7.26	19	40	−5	−7	137
8	7.32	20	41	−4	−6	136
9	7.31	20	40	−4	−7	136
10	7.32	16	43	−8	−4	136
11	7.29	15	41	−9	−6	136
12	7.28	15	42	−9	−5	139
13	7.25	14	38	−10	−9	139
14	7.25	14	39	−10	−8	141
15	7.29	16	37	−8	−10	140
16	7.26	15	36	−9	−11	140

* Data obtained on blood serum are abstracted from Table II of Schwartz *et al.* (548). The patient had been previously treated with $NaHCO_3$, and was then observed on a constant dietary regimen without restriction of NaCl. Cl^- was not measured on the 1st day of observation, so $\Delta[HCO_3^-]$ and $\Delta[Na^+ - Cl^-]$ are reckoned from Day 2 as a base. In spite of provision of dietary NaCl, some loss of extracellular fluid probably occurred during the 1st week of observation, inasmuch as a loss of 1 kg of body weight occurred from Day 2 to Day 10 and the serum $[Na^+]$ declined. The patient's condition seems stabilized after Day 10 in respect to the pH, osmolality, and volume of extracellular fluid. The data are reproduced with the permission of the publisher.

equality between $\Delta [HCO_3^-]$ and $\Delta [Na^+ - Cl^-]$ is more constant than the equality between $\Delta [HCO_3^-]$ and $\Delta [Cl^-]$, as acidosis develops under these circumstances, because both the volume and the Na^+ concentration of extracellular fluid tend to decline. Accordingly it is change of $[Na^+ - Cl^-]$ which, when compared with $\Delta [HCO_3^-]$, most accurately reflects the regulation of acid-base balance in the renal tubules.

Decline of plasma $[HCO_3^-]$ to mildly acidotic levels as represented in Table III-3.4 reflects the failure of renal tubular acid secretion to conserve extracellular base reserves even though the acid load is modest; the efficiency of the acid-base homeostatic function of the patients' diseased kidneys has declined. Reclamation of filtered Na^+ must also be compromised insofar as acid secretion into the nephrons is linked to Na^+ reabsorption. Rise of $[Cl^-]$ in the plasma, as HCO_3^- declines, shows that the mechanism by which Na^+ is reabsorbed together with Cl^- is still able to respond in a regulatory manner to the imminence, or fact, of sodium depletion; if volume and osmolality control are accurate, as they may be if the patient is given, and can tolerate, sufficient dietary NaCl, then plasma Cl^- will rise reciprocally as plasma $[HCO_3^-]$ declines. On the other hand, if plasma $[Na^+]$ declines concurrently with the development of acidosis, then the increase in plasma $[Cl^-]$ will not be as great as the decline in $[HCO_3^-]$; but $\Delta\ [Na^+ - Cl^-]$ will still equal $\Delta\ [HCO_3^-]$.

While acidosis is developing in such a patient, the total rate of acid secretion into the nephrons is not appropriate to the $[HCO_3^-]$ filtration; but yet total renal acid secretion in these patients characteristically becomes adequate to allow acid-base balance to be approximated once the degree of acidosis has become stabilized (e.g., Table III-4). What is the explanation at the renal cellular level of the inadequate urinary acidification at first, and the subsequent stabilization of the acidosis by approximation to a state of nonvolatile acid balance?

One cause of the reduced capacity for renal acid excretion in these patients is undoubtedly inadequate delivery of buffer to the urine. That the kidneys of most patients with high-grade glomerulotubular failure retain to a remarkable extent the ability to lower the pH of the urine was noted already by Henderson and Palmer (244) and has often been documented in the later literature (673, 548). It needs to be specified, however, that in many of the patients such values are attained only when acidosis is well developed (see below). Under these conditions, urine pH of 5.1 is very regularly attainable by the kidneys even in far advanced glomerulotubular renal failure. This figure should not be compared with the overall lower limit of urinary pH of 4.5 in healthy persons, but rather with the lower limit of 5.1 in healthy persons during water diuresis (II-10B) because high-grade glomerulotubular failure is associated with obligatory polyuria due to solute diuresis (182, Chart 21). The ability of the cells of these patients to secrete acid against a gradient is remarkably close to normal; but the total quantity of buffer passing through the nephrons and being excreted in the urine is subnormal. Consequently, the rate of renal acid excretion at minimal urine pH is reduced. *The magnitude of the acid load which can be excreted without progression of acidosis is progressively reduced as urinary buffer for excess protons is reduced.*

Both volatile and nonvolatile buffer for excess acid are in short supply in glomerulotubular insufficiency. Glomerular filtration of the principal nonvolatile buffer, phosphate, is greatly reduced, and reduction of percentile tubular reabsorption of filtered phosphate, though moderating the resulting diminution of urinary phosphate excretion, cannot prevent it: in high-grade glomerulotubular failure the plasma phosphate is raised and the proportion of the dietary phosphorus eliminated in the stool is raised (6, 7). Other nonvolatile urinary buffers may similarly be diverted away from the urine as a result of glomerular failure. Interference with the metabolic activity of the renal parenchymal cells also contributes to reduction of urinary buffer: urinary ammonia excretion is markedly subnormal in advanced glomerulotubular failure (244, 673, 548), because the renal ammonia production is depressed (481). Reduced urinary ammonia excretion in relation to the urinary pH is to be expected, as a clinical rule of thumb, whenever the GFR declines much below 80 ml/minute in the adult, but becomes marked only with marked reductions in GFR (III-4); presumably it is the associated reduction of renal blood flow which is really responsible (Section B-1).

Though most if not all patients with severe glomerulotubular insufficiency can reduce urinary pH to close to 5.1, and some, during acidosis, still lower, there is considerable variation among them in the degree of acidosis which must develop before the urinary pH will be sufficiently lowered for acid-base balance to be approximated. Schwartz, Hall, Hays, and Relman (548) showed that some patients suffering from this condition, when rendered nonacidotic by alkaline therapy and then studied on constant dietary regimens yielding ordinary acid loads, may reduce their urinary pH rather promptly to a degree sufficient to establish acid-base balance, so that little or no acidosis supervenes; whereas others continue for a considerable period to elaborate urine of inappropriately high pH, permitting considerable acidosis to develop before their urine pH declines to a point compatible with maintenance of acid-base balance. The "bicarbonate leak" in these patients—excretion of urinary HCO_3^- after the plasma [HCO_3^-] has declined well below the value of about 25 mEq/liter that defines the normal threshold (411, 553)—, combined with an almost normal ability to reduce urine pH during acidosis, is reminiscent of the behavior of patients with "proximal" renal tubular acidosis.

What is the mechanism which permits the urinary pH finally to decline, in patients who have become acidotic, to a level where renal acid excretion is in balance, or nearly in balance, with a constant acid load? It is unnecessary to infer any increase in the rate of acid secretion into the nephrons, i.e., any specific influence of the acidosis upon function of the renal cells. Clearly, the same total rate of acid secretion into the nephrons which does not provide sufficient acid to react with all the bicarbonate filtered when plasma [HCO_3^-] is 22 mEq/liter may be more than sufficient when plasma [HCO_3^-] is 16 mEq/liter. There is no reason to doubt that, in glomerulotubular failure, as with normal kidneys, the rate of acid secretion into the nephrons actually lessens as the urine pH declines in response to declining plasma [HCO_3^-] (II-8D). Acidosis becomes stabilized in high-grade glomerulotubular failure primarily because the lowered rate of glomerular HCO_3^- filtration resulting from the acidosis lessens the rate at which the tubular cells must secrete acid into the nephrons to maintain acid-base balance. Specific effects of acidosis on the efficiency of acid secretion by the cell would appear to be of minor importance. The fact that the blood pH is depressed by several tenths of a pH unit presumably means that the same maximal transtubular pH gradient will be associated, during acidosis, with secretion of urine several tenths of a pH unit lower than under normal circumstances; the urine pH may fall as low as 4.8 during acidosis in these patients, lower than the physiological lower limit seen in moderate water diuresis (259). On the other hand, the hypocapnia compensatory to nonrespiratory acidosis might be expected to reduce the total rate of acid secretion (III-1).

Schwartz *et al.* concluded from their study (548) that "in general terms, all renal acidosis is 'tubular' acidosis"; and certainly when comparing patients of similar size who show the same degree of depression of GFR at similar acid-base loads, it is clear that if one patient achieves acid-base balance at a plasma [HCO_3^-] of 22 mEq/liter, and another only at 16 mEq/liter, the difference lies essentially in ability of the first patient to secrete more acid into the nephrons. But defective renal functions other than tubular transport also contribute to acidosis in glomerulotubular disease. Diminished ammonia production constitues renal tubular dysfunction only in a Pickwickian sense; it is a metabolic abnormality of the parenchymal cells. It would also be a major oversimplification to leave out of account the glomerular contributions to the acidosis of glomerulotubular renal failure. The most conspicuous one is the diminished total filtration of phosphate, which ensures that the rate of renal acid excretion is lessened, compared with normal persons excreting urine of comparable pH. Finally heterogeneity of nephrons in respect to compensatory hypertrophy, reduction and maldistribution of the renal blood supply, and distortion of renal architecture, with special reference to the integration of medullary and cortical function, are factors which must be supposed to affect renal function under pathophysiological circumstances and to account for some differences in the specifics of renal acid-base regulation among cases of glomerulotubular failure. In pure RTA, where the lesion is so highly selective functionally, and even in RTA associated with other defects of tubular transport, the pathophysiology is less heterogeneous, and differences between individual patients essentially express differences in the severity of the tubular transport disturbance which is responsible for the disease.

When all is said, however, the ruggedness of the renal design is manifest in severe glomerulotubular renal insufficiency, where responsiveness to changing acid-base and other changing loads is evident almost to the bitter end, though of course with far less sensitivity and capacity than in health. The therapist will take particular care, until such time as he decides that the function of the patient's kidneys must be replaced, to render support to extracellular fluid volume and plasma volume in order to sustain renal perfusion, to assure adequate blood oxygen carriage, to minimize the acid load, and to adjust the loads of other substances like water and salt in such a way that the kidney's regulatory performance is optimized. Even in its extremity the kidney remains a precision device within the greatly constricted limits imposed by disease.

four

DISORDERS RELATED TO URINARY pH

Briggs (71, 72), who did not accept the view that urinary ammonia serves renal maintenance of systemic acid-base balance, partially anticipated the doctrine of ammonia excretion by nonionic diffusion by advancing the idea that entry of NH_3 in the urine might function to prevent damage by urinary acidity to the nephrons, or the mucosa of the urinary tract, or both. Because urine pH never normally falls below 4.5 there are no experiments of nature in man or other mammals to disclose any protection which the urinary tract may owe to the fact that urinary acid excretion is effected by excretion of buffered acid, nor does Briggs's speculation seem to have led to any investigations in which the urinary tract has been perfused by solutions of acidic intensity higher than that normally attainable.

The urinary pH is, however, of known importance relative to disease essentially in two connections. One is during the treatment of bacterial urinary infections with mandelic acid and other compounds whose antibacterial efficacy is enhanced when the urine secreted is in the lowest pH range. Agents which acidify the urine are rather frequently prescribed, and acid-ash diets occasionally, in conjunction with exhibition of such antibacterial drugs. This subject is discussed in texts of therapeutics (211) and need not detain us here.

The urinary pH is important also as one of the factors causally related to the formation of several types in the urinary tract. The occurrence both of stones composed of calcium salts, the commonest of urinary calculi, and of uric acid stones, another common type, is powerfully influenced by the urinary pH, though as a rule urinary acidic intensity is ancillary to other factors—dietary, metabolic, and physiochemical—predisposing in the first place to the urolithiasis. High-pH urine favors the formation of calcium-containing calculi (579, 150) and is (for example) a major cause for the nephrolithiasis of classical renal tubular acidosis (III-B3). The pH range of the urine also influences the crystalline structure of calcium phosphate stones (150, 151): brushite stones are generally formed in urine of pH of about 5.9 to 6.6, whereas apatite or mixed apatite-struvite stones (containing also Mg^{++} and NH_4^+) form at a higher urine pH. Acidification of the urine is accordingly an appropriate part of the medical regimen aimed at controlling the formation of calcium phosphate stones; it is said to be useless when the calculi are composed of calcium oxalate (579), and may be unattainable in infections with urea-splitting bacteria (cf. page 287).

Victims of uric acid kidney stones fall mainly into two clinical categories: (i) those with hyperuricuria due to gout and other metabolic causes of increased purine catabolism, or associated with rapid cell turnover as in polycythemia vera, leukemia, or neoplasia; and (ii) those with unremitting urinary hyperacidity (579, 393). In uric acid urolithiasis with hyperuricuria, which will not be considered in detail here, formation of the calculi is undoubtedly due in large part to supersaturation of the urine with the increased urinary total urate. Marked reduction of the excessive production of urate is often attainable (211); and measures to increase the solubility in urine of the total urate which is excreted (see below) are also therapeutically appropriate.

Uric acid contains two dissociable protons, but only the first dissociation (at N-9) is of physiological importance, the second proton being transferred only at pH well above the physiological range. The singly charged urate ion, which we shall refer to as mono-urate,[1] is much more soluble in water than undissociated uric acid; Table III-4.1 presents an estimate of the solubilities of both members of this buffer pair, according to which a 20-fold gain in solubility of total urate appears to result from conversion of all undissociated uric acid to the mono-urate. Various values for pK of this buffer pair at 37° C are given, ranging from 5.4 to 5.75 (397); the latter value is accepted by Smith and Williams (579). The range of pK′ in human urine may be taken as approximately 5.3 to 5.6.

Most clinical conditions in which there is unremitting urinary hyperacidity are associated with a raised incidence of urinary uric acid calculi. This clinical association leaves no room for doubt that a major factor in the pathogenesis of the urolithiasis in these cases is the scanty solubility of total urate in water, specifically that of the undissociated uric acid. Figure III-4.1 shows that, whereas urate is present virtually exclusively as undissociated uric acid at the lowest attainable urine pH, practically quantitative conversion to the mono-urate results when urinary pH is raised to the range 6.5 to 7.0. The uniquely effective control of stone formation achieved in many cases of uric acid lithiasis by raising the urine pH, usually by administration of alkali medication, is due to (i) the much greater aqueous solubility of the various salts of the mono-urate ion, which is the conjugate

TABLE III-4.1. *Solubility of uric acid and of monosodium urate in water**

	Uric acid	Sodium urate
	mg/liter	
In H_2O at 25° C	60	1000
In H_2O at 100° C	300	8000
In H_2O at 37° C (estimate of interpolation)	100	2000

* Values at 25° C and at 100° C are taken from the Merck Manual (395); values at 37° C are interpolated. The figures are only approximate for urine, since solubility of the urate is affected by the presence of other solutes, and different salts of urate do not have the same solubility as sodium urate (the lithium salt is more soluble, ammonium urate less so).

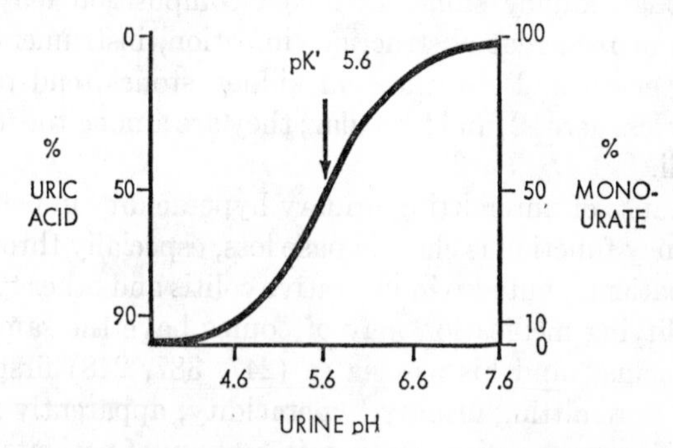

FIG. III-4.1. Percentage of total urate as uric acid (*left ordinate*) and mono-urate (*right ordinate*) as a function of the pH of dilute urine at 37° C. In high-flow urines pK′ is not much below pK.

base, and (ii) the uniquely favorable location of pK′ of this buffer pair in relation to the physiological range of urinary pH. Both factors are essential for the success of the alkali therapy; in the case of cystine, for example—another poorly soluble compound of C and N of which urinary calculi may be composed—the conjugate base is also much more soluble, but pK lies so high that even frank alkalinization of the urine, which can be maintained only with difficulty and carries the attendant risk of formation of calcium phosphate calculi, does not provide a very efficient means of increasing solubility (128).

The rate of urinary uric acid excretion varies directly with the purine content of the diet and also normally in relation to genetically determined differences in purine metabolism between individuals. In normal persons consuming diets that provide ample purine, urinary total urate excretion is generally in the neighborhood of 1 g daily, and on purine-free diets between 0.25 and a 0.5 g daily (579). At ordinary 24-hour urine volumes in the neighborhood of 1 liter daily, the urine is accordingly likely to be supersaturated with total urate even without hyperuricuria (Table III-4.1). The occurrence of uric acid urinary calculi in persons with unremitting urinary hyperacidity is evidently ascribable to loss of the normal rise in urinary pH to above 6.5 or even above 7.0 occurring at some time during the day, most commonly in the hours before noon (II-1). This daily period during which the solubility of total urinary urate is increased is apparently important in preventing development of stones from microcrystallization and seeding of the mucosa of the renal pelvis, for pure uric acid stones associated with urinary hyperacidity of hyperuricuria are true kidney stones, whereas large urate-containing calculi that form in the bladder are unrelated to these conditions. The stones formed recurrently in the renal pelvis in persons with unremitting urinary hyperacidity are characteristically composed of pure uric acid, though kidney stones of mixed composition may appear after complications of recurrent obstruction, infection, instrumentation, or surgery have supervened. Pure uric acid kidney stones tend to be multiple, small (1 cm or less across), and irregular; they are among the few radiolucent urinary calculi.

The usual cause of unremitting urinary hyperacidity in persons with well preserved kidney function is chronic base loss, especially through the stoma of ileostomy patients, but also in ulcerative colitis and other types of chronic diarrhea; acidifying medication may of course have the same effect (579). In 1958 Henneman and his associates (247, 387, 248) first reported the occurrence of unremitting urinary hyperacidity, apparently resulting from subnormal urinary ammonia excretion, in a group of patients suffering from uric acid urolithiasis. These patients appear to exemplify a single well defined syndrome, since referred to as "idiopathic." All of the patients originally reported were males of Mediterranean stock (Jewish or Italian); they they did not excrete excessive amounts of total urate in the urine nor did they have gout. The condition characteristically began around early middle life and worsened with the passage of time; multiple episodes of ureteral colic, urinary obstruction, and infection tended to develop and to cause progressive deterioration of renal function. Urinary calcium excretion was also low. Because of the predominance of persons of Mediterranean extraction among the victims, Henneman and his associates postulated that a mutant gene might be responsible, the genetic abnormality becoming expressed only in later life.

Whether highly selective deficiency of renal ammonia production is in fact a demonstrable clinical cause of urinary hyperacidity and uric acid urolithiasis is still *sub judice*. There is no doubt that unremitting urinary hyperacidity is very frequently associated with uric acid urolithiasis in adult male patients, especially Jews (579, 393, 247, 387, 248, 24, 21, 20, 36, 485, 219); the good response to alkali therapy shows that the urinary hyperacidity is a necessary contributing cause of the stone formation. Most students of the question have found urinary ammonia excretion to be low in many patients with uric acid urolithiasis, either absolutely or relative to urine pH according to the relation described by Wrong and Davies and others (Figure II-9.3). The majority view, however, seems at present to be that the lowered urinary ammonia excretion relative to urine pH in these patients (i.e., presumably lowered renal ammonia production) can probably be accounted for by the advanced age of the patients (36), by the presence of gout and related abnormalities of urate metabolism (219), or by renal damage including that secondary to the uric acid urolithiasis and its complications (397), or else by a combination of these factors.

Nevertheless, uric acid urolithiasis does appear sometimes to exist in what may be termed, for purposes of discussion, an "idiopathic" form (219), i.e. without hyperuricuria, with little renal damage, and often at a relatively early age. The striking congeries of maleness, Eastern Mediterranean origin, and maturity among the victims of this putative condition, and the frequency of gout in a similar group of patients, suggest some sort of a gene-related link between idiopathic uric acid nephrolithiasis and abnormal urate metabolism. Neither aging nor renal damage, however, is a process peculiar to any ethnic group. For this reason the suggestion of Gutman and Yü (219) that depressed renal ammonia production may be a metabolic consequence of gout (diversion of glutamic amide N to competing pathways) is attractive in relation to the etiology of idiopathic uric acid urolithiasis. In particular, a varying balance between depressed renal ammonia production and urinary urate excretion might explain the clinical occurrence of two independent causes of uric acid urolithiasis in a group of patients of distinctive ethnic origin; excess urinary urate might be responsible in some, unremitting urinary hyperacidity due to depressed renal ammonia production in others, while presumably both causes would cooperate in producing the disease in most patients with uric acid stones. On the other hand, Pollak and Mattenheimer (476) recorded normal urine ammonia secretion and renal glutaminase activity in four patients with gout.

It seems, then, that many gouty patients do not have gross reduction of renal ammonia production, while unremitting urinary hyperacidity due to depressed urinary ammonia excretion can be related in many gouty patients to age and especially to advancing renal damage. But these conclusions do not seem to rule out the possibility that more selective depression of renal ammonia production than is seen with aging and renal damage may sometimes occasion uric acid nephrolithiasis, nor to exclude a genetic linkage of some kind between idiopathic uric acid urolithiasis and the gouty diathesis.

It is extraordinarily difficult to arrive at a firm conclusion as to whether a primary and notably selective depression of urinary ammonia production does in fact sometimes cause idiopathic uric acid urolithiasis. The difficulty arises because the association of uric acid stones with depressed urinary ammonia excretion can clearly so often be explained on the basis of some combination of age, renal damage, and hyperuricuria; because the clinical facts as reported from different centers are so various; and because, as Metcalfe-Gibson and her colleagues have very justly emphasized (397), standards of normality of blood urate vary with differing methods used in different clinics. These authors directed attention especially to the importance of application of valid local standards for normal blood uric acid when reporting cases of idiopathic uric acid urolithiasis. Similarly, much greater precision than has yet been attained would be desirable in relating the deviation of urinary ammonia excretion to be expected to the kind and degree of renal damage present.

The author continues, on the basis of his own experience, to hold the belief, though not the conviction, that uric acid nephrolithiasis can sometimes be caused by a loss of the renal capacity to produce ammonia out of proportion to the effects which age and renal damage can reasonably be expected to explain (669). There follows a report of the most striking case among several such patients personally studied.

CASE REPORT

R. G., a male jeweler of Jewish extraction then 50 years of age, was seen in consultation for the first time in May, 1960, having been referred by his physician for investigation for possible metabolic abnormality underlying recurrent bouts of urolithiasis.

The patient's father had diabetes mellitus and his mother had died of cancer of the liver. He had two brothers and one sister who were in good health; there had been no urolithiasis in any other member of the family but a diagnosis (reliability unknown) of gout had been made in a maternal uncle. Physical examination disclosed a healthy-appearing, intelligent, and cooperative middle-aged male in no distress. He was 69 inches tall and weighed 175 pounds. The arterial blood pressure was 150/110 mm Hg. Pulse was 72/minute and regular. The ocular fundi showed moderate arteriolar narrowing; there were no hemorrhages or exudates. No costovertebral angle tenderness was elicited on either side; there was no cardiomegaly, and, except for atrophy of the muscles of the left leg, the physical examination was otherwise entirely unremarkable.

His urine gave a trace Sulkowitch reaction; serum calcium had been reported from elsewhere as 11.5 mg/100 ml, but was normal (11 mg/100 ml or less) in our laboratory on three occasions. An intravenous pyelogram made in 1956 appeared normal. The blood hemoglobin was 12 g/100 ml and the leukocyte count was normal. A urine sample showed pH 5.0 and specific gravity 1.021; no protein, sugar, or ketone was present and no formed elements were found in the sediment. Serum uric acid ranged from 5.1 to 5.4 mg/100 ml by the method of Carraway (94); upper limits of normal were reckoned in the clinical laboratory as 5.5 mg/100 ml without regard to sex. Serum urea N was 20 mgm/100 ml; total CO_2, 27 mEq/liter; phosphorus, 3.3 and 3.7 mg/100 ml; alkaline phosphatase, 2.7 Bodansky units; total protein, 8.37 g/100 ml; and albumin, 4.72 g/100 ml; the serum protein electrophoretic pattern was entirely normal. Renal clearance (mean of three periods) of endogenous creatinine was 79 ml/minute; of inulin, 80 ml/minute; of uric acid, 6.6 ml/minute. The proportional renal tubular reabsorption of filtered phosphorus, determined without special dietary preparation but in the fasting state at 8:30 AM, was 0.94. Urinary excretion of uric acid averaged 710 mg/24 hours on three successive determinations, and urinary calcium 73 mg/24 hours during dietary calcium restriction. Whole venous blood glutamine was well within normal limits (480) on three occasions.

An initial word-of-mouth report from a urologist who had arranged in 1957 for chemical analyses of two calculi which had been passed in that year stated that the stones had contained "calcium phosphate and oxalate." Subsequently, the original record was reexamined, and it showed that the report had in fact been "urates and oxalates." Meanwhile, the patient had disclosed that the first calculus passed had been analyzed, and had been found to be composed of pure uric acid. He had also saved a half-dozen of the calculi which he had passed over the years; they were brownish, irregular ovoids with average dimensions of approximately 1.0 × 0.5 × 0.5 cm. Three were submitted for chemical analysis and were found to consist entirely of uric acid, without any detectable admixture of calcium, phosphorus, oxalate, ammonium, or cystine.

As a screening procedure for urinary hyperacidity, the pH of freshly voided urine was determined on three occasions, twice during the morning and once during the afternoon; the patient was given Nitrazine paper, instructed in its use, and asked to test his urine pH 4 times daily for 4 days—midmorning, before lunch, before supper, and at bedtime—and to record the values. None of the values exceeded 5.5, and most were recorded as 5.0. He was then given ammonium chloride, 2 mEq/kg of body weight, daily in divided doses for 3 days; diet was uncontrolled. The procedure produced mild nonrespiratory acidosis; the increase in urinary titratable acidity on the 3rd day was within normal limits (145), but the ammonia excretion scarcely increased (Table III-4.2 and legend).

In order to minimize sodium intake in this hypertensive patient, an alkalinizing solution was compounded for the patient's use as follows:

Potassium citrate, 144 g
Sodium phosphate, dibasic, 53 g
Water, q.s., 1 liter

Ten milliliters of this solution yield 7.5 mEq of sodium, 3.7 mmoles of phos-

TABLE III-4.2. *Response of R. G. to 3-day ammonium chloride test**

Day	24-hour urine			Plasma
	Volume	Titratable acidity	NH_4^+	CO_2 content (end of 24-hour period)
	ml	*μEq/minute*		*mEq/liter*
Control 1	770	21.7	19.8	26.9
Control 2	695	22.9	27.0	
Mean		22.3	23.4	
3rd day of NH_4Cl	1025	36.3	28.0	18.4

* Two control 24-hour urine samples were collected under mineral oil and toluene, the aliquots being added to a collection bottle kept in the refrigerator. The patient was then given ammonium chloride, 2 mEq/kg of body weight/day, for 3 days, as a 10% solution in four divided daily oral doses. On the 3rd day of this regimen, a 24-hour urine was collected as before. In addition to the measurements on the 24-hour 3rd-day collection, two aliquots collected on the 3rd day were studied before refrigeration; one showed pH 5.1 and urinary ammonia excretion 10 μEq/minute, the other pH 5.2 and ammonia excretion 22 μEq/minute.

phate, and 13 mEq of potassium and citrate each. On June 7 the patient began to take 2 spoonfuls of the prescribed mixture diluted in water 3 times daily before meals. During the next week urine tests showed urinary pH to be 5.5 between 7:00 PM and 6:00 AM, and 6.0 between breakfast and supper. He had no ill effects from medication, and on June 15 he began to take 2 teaspoonfuls of the liquid medication at breakfast and lunch, 3 at supper and 1 at bedtime. The urine pH now averaged above 6.5; one value of 5.5 was obtained at 6:00 AM. Figure III-4.2 presents measurements of urinary ammonia excretion during treatment in relation to pH; in the majority of samples the ammonia excretion was more than 2 SD below the predicted value.

Over the ensuing 12 years, small adjustments have been made from time to time in the prescription; the objective has been to maintain the urinary pH near 6.5 at all times. The patient generally makes multiple observations of his urine pH, throughout a single day, once every few months in order to ensure that round-the-clock control is satisfactory. No more stones have been passed, and there have been no episodes of renal colic. The patient's general health remains excellent at present writing.

The principal points in favor of a specific reduction of renal ammonia production, as offered by this patient's medical history, may be summarized as follows: (i) blood and urine uric acid well within normal limits; (ii) marked depression of urinary ammonia excretion in relation to urinary pH at all times; (iii) hardly any response of urinary ammonia to mild nonrespiratory acidosis provoked by a 3-day acid load; (iv) no azotemia, and depression of

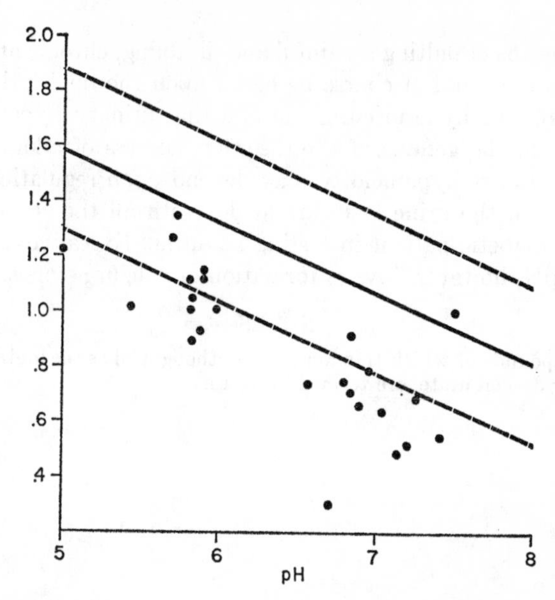

FIG. III-4.2. Logarithm to the base 10 of the urinary ammonia excretion of Patient R. G., in microequivalents per minute, plotted against urinary pH at 25° C. *Circles* represent patient's urine; 95% confidence limits for normal subjects (cf. Figure II-9.6) are designated by the *dotted lines*.

glomerular filtration modest even after 16 years of urolithiasis with recurrent infection and instrumentation and some years of arterial hypertension, hence presumably minimal when the uric acid urolithiasis began. In particular the degree of depression of urinary ammonia excretion seems far out of proportion to what would be expected from the glomerular filtration rate (397).

One general point in favor of the existence of a syndrome of idiopathic uric acid urolithiasis due to specific depression of renal ammonia production, similar to that described by Henneman and his colleagues, may be the striking lack of uric acid urolithiasis as a complication of advanced glomerulotubular renal failure, where urinary ammonia excretion is regularly markedly reduced and unremitting urinary hyperacidity is the rule. Whatever the reason for the failure of uric acid calculi to form in the urine of such patients—one may well be reduced urinary urate excretion—it does appear that a reduction of renal ammonia production disproportionate to the degree of generalized renal impairment is more favorable to uric acid urolithiasis than is general renal failure itself.

Whatever the outcome of this nosological problem may turn out to be, the physician has at his disposal a medicinal agent which, in certain cases of uric acid nephrolithiasis, can appear to the patient to provide an almost miraculous means of halting a painful and disabling, chronic and worsening illness. Therapy should of course be based upon thorough prior investigation of the role of hyperuricuria, unremitting urinary hyperacidity, and other factors in the genesis of the disorder; success of alkali therapy for unremitting urinary hyperacidity may depend upon regulation of urinary pH according to the urine test carried out with all the conscientiousness exacted of a diabetic patient in testing his urine. Excessive zeal in raising the urinary pH, however, invites formation of calcium stones.

NOTES

1. The compounds of which this ion may be thought of as the ionization product e.g., sodium-hydrogen urate, are termed biurates.

APPENDIX A

THE RELATION BETWEEN ARRHENIUS' α AND VAN'T HOFF'S i

NaCl in extracellular fluid behaves, in terms of colligative properties, as if it were not wholly dissociated into two ions, but rather as if every 1000 molecules gave rise to 1850 particles in solution. Since the osmotic activity of a solution of NaCl isotonic to plasma would be found to be 1.85 times the value expected, in the absence of dissociation, from its molality, van't Hoff's i would have the value

TABLE A-1.

	α	i	$1 + \alpha(n - 1)$
I. Nonconductors			
Ethanol	0.00	0.94	1.00
Mannitol	0.00	0.97	1.00
Sucrose	0.00	1.00	1.00
II. Acids			
HCl	0.90	1.98	1.90
H_3PO_4	0.08	2.32	1.24
H_2SO_3	0.14	1.03	1.28
III. Basic compounds			
NaOH	0.88	1.96	1.88
NH_3	0.01	1.03	1.01
IV. Salts			
NaCl	0.82	1.90	1.82
K_2CO_3	0.69	2.26	2.38

of 1.85, each mole of NaCl giving rise in solution to 0.15 mole of undissociated compound and 0.85 mole of each ion: $0.15 + 0.85 + 0.85 = 1.85$.

The general relation is then:

$$i = 1 - \alpha + n\alpha \quad (1)$$

The letter I has not been used to designate an Appendix in order to avoid possible confusion with Roman I.

where α is apparent proportional dissociation and n is the number of ions yielded in dissociation of the electrolyte molecule; or rearranging:

$$i = 1 + (n - 1)\alpha \qquad (2)$$

For example, assume a particular solution of $CaCl_2$ to behave, because of interionic attraction, as if it were 60% dissociated. Then α is 0.60, and i is $0.4 + (3 \times 0.6) = 2.2$ (Equation 1) or $1 + (2 \times 0.6) = 2.2$ (Equation 2).

Table A-1 reproduces sample paired values for α and i presented by Arrhenius in his paper of 1887 (15).

APPENDIX B

ETYMOLOGY OF *BUFFER*

The Oxford English Dictionary is certainly remiss in its account of the introduction into our language of the word *buffer* in its chemical sense: the first use recorded there is 2 years after the first edition (1920) of Clark's text (102), when the English word had been for some time in general use in physical chemistry. There is no doubt, however, that the chemical sense of *buffer* came into English, via Sørensen, from Hubert and Fernbach: "The word [tampon] was adopted by Sørensen (1909)," says Clark, "and in the German rendition of his paper it became 'Puffer' and thence the English 'buffer'." But this authoritative account (103) should not be taken to mean that *buffer*, in its chemical sense, was simply a transliteration from the German. The available facts appear on the contrary to suggest that *Puffer*, in the general sense of a cushion against change, and probably *tampon* in the same sense, derive from English usage.

The crucial question is: what was the image which Fernbach and Hubert (I-4A) had in mind when they applied the word *tampon* to the action of their phosphate buffer? It seems likely that they were thinking of the railwayman's use of the word, the first example of which is assigned by Robert (510) to 1864: "Plateau métallique vertical destiné à recevoir et à amortir les chocs. Tampons d'un locomotive . . . ." The lexicographic evidence suggests that the railroad buffer was originally a British device; in that case the French and German words would be translations of the English *buffer*, only the chemical sense being primarily French. The Oxford English Dictionary records (444) that equipment used by railways to cushion the shock of an impact ("a round plate or cushion usually supported by a strong spring fixed in pairs at the front and back of railway carriages or engines or on the face of a terminal wall") was termed a buffer at least as early as 1867, after an essentially identical engineering device already patented in Britain under the same name by 1835.

The device has given rise to additional metaphors. In referring to a Far Eastern country as a "buffer-state," the *Daily News* of July 23, 1883, asserted (444) that its use of the term "borrow(s) a simile from Indo-Afghan politics." (The politicians or journalists who coined "buffer-state" may originally have composed a simile; simile or metaphor, the image invoked must have come from railway parlance.) *Buffering* of oxidation-reduction systems is doubtless an extension from acid-base usage; *buffering* in electrical circuits may perhaps be a cognate of acid-base buffering, directly derived from the engineering sense (353).

It seems probable that the German *Puffer* is derived by transliteration from *buffer*, Sørensen having chosen it as the translation of *tampon* because it was already in use for a mechanical *buffer* and perhaps also already in derived senses like *Pufferstaat*. Trübner (616), after presenting older and unrelated meanings of *Puffer* (something that puffs; cakes of various kinds; etc.), goes on:

"In neuer Zeit sind Puffer auch federnde Vorrichtungen an Eisenbahnwagen und Prellböcken, die den Anprall lindern sollen. Ein solcher Puffer, der die Püffe, d.h. die Stösse auffängt, dient zu vergleichen. So von einem Bahnwagen selbst: 'Auch in den Nordstaaten ist der erste Wagen hinter der Lokomotive als Puffer der Weissen für die Schwarzen bestimmt' [from M. Eyth, *Im Strom unserer Zeit 1*:312, 1904, quoting a letter of 1867]—Vor allem ist *Pufferstaat* zum geopolitischen Begriff geworden. Verzeichnet ist das Wort erst 1905 (M. Heyne, Dt. Wb. *1*:1213). Damals wird als frühester Beleg für die Vorstellung eine Zeitungsstelle von 1898 beigebracht. Ein Streifen Landes wird als Puffer (zwischen zwei Staaten) betrachtet. Gleichzeitig 'Pufferstaat/bufferstate' (Thieme-Kellner, Handb. d. engl. u. dt. Spr. *2*:368, 1905)."

My suggestion that *tampon* acquired its mechanical sense as a translation of the English *buffer* is tentative, based as it is principally upon British poineering in railroad technology and upon the evidence of dictionaries. Also, Robert may be unpatriotically lax in tracing the railroader's *tampon* only to 1864, for he records the first application of the word to acid-base equilibria in the blood as in 1952 (!). He also traces "état tampon" only to 1935, whereas Professor Léonard R. Muller of Miami assures me that he recalls this expression as already current at the time either of the signing of the Treaty of Versailles or of the first meetings of the League of Nations.

APPENDIX C

INTERIONIC ATTRACTION

The deviations of extracellular fluid and urine from ideal behavior result essentially from the presence of their contained electrolytes. These solutions are sufficiently dilute that interaction between particles due to the relatively weak van der Waals forces of intermolecular attraction is negligible; but movements of ions in real solutions are influenced by the much stronger forces of interionic electrostatic attraction, which affects their mobility and hence their chemical potential whether or not an electromotive force is imposed across a volume of solution. As a result, the activity of an ion in solution differs from its concentration C to an extent which can be expressed mathematically as the *activity coefficient* γ ($= \boldsymbol{a}/C$). The value of γ for any ion depends, among other things, upon the concentrations and charges of all the various ions present in solution; the *ionic strength* μ which determines electrical environment of an ion in solution was shown in 1921 by Lewis and Randall (347) to be given as:

$$\mu = \tfrac{1}{2}\,(C_1z_1^2 + C_2z_2^2 + C_3z_3^2 + \ldots) \tag{1}$$

where C is the molar concentrations of the several ions present and z their respective valences.

The Debye-Hückel theory (I-3B1) is based on a mathematical argument which correlates the net electrostatic attraction exerted on an ion with the average distribution of charges around it. The general limiting equation of Debye and Hückel for strong electrolytes in dilute solution (123a) is:

$$\log \gamma = \frac{-Az_+z_-\sqrt{\mu}}{1 + Ba_i\sqrt{\mu}} \tag{2}$$

where γ is the activity coefficient of an electrolytic solute, A and B are constants depending on the dielectric properties of the solvent ($A = 0.5186$ and $B = 0.3314$ for water at 38° C), z_+z_- is the valence type of the solute in question—1 − 1 for NaCl, 2 − 1 for $CaCl_2$, etc.—and a_i is the average ionic diameter in Ångstroms.

The equation gives calculated conventional values for activity coefficients of individual ions (see below) which are in excellent accord with experiments for very dilute and moderately dilute solutions of strong electrolytes. In the case of weak acids and bases, Equation 2 provides us with a successful theoretical means of relating the apparent dissociation constant K', expressed in terms of actual concentrations, to the thermodynamic dissociation constant K, expressed in terms of activity; the latter takes interionic attraction into account:

$$K = \frac{a_{H^+}a_{H^-}}{a_{HA}} \tag{3}$$

For activity we can substitute the product of the concentration C and the activity coefficient γ, so that for any acid the thermodynamic dissociation constant K is given by:

$$K = \frac{(C_{H^+}\gamma_{H^+})(C_{A^-}\gamma_{A^-})}{(C_{HA}\gamma_{HA})} = \frac{C_{H^+}C_{A^-}}{C_{HA}} \times \frac{\gamma_{H^+}\gamma_{H^-}}{\gamma_{HA}} \tag{4}$$

Let α = proportional dissociation; then $C_{HA} = C_{(1-\alpha)}$, and the concentration of each ion $= C_\alpha$. Substituting in Equation 4:

$$K = \frac{(C_\alpha)^2}{C_{(1-\alpha)}} \times \frac{\gamma_H\gamma_{A^-}}{\gamma_{HA}} \tag{5}$$

$$K = K' \times \frac{\gamma_{H^+}\gamma_{A^-}}{\gamma_{HA}} \tag{6}$$

where K' is the apparent ionization constant calculated from the concentrations and from conductance data (Table II-4.1). The drift of the value of K' with rising concentrations is the result of neglecting the expression $\gamma_{H^+}\gamma_{A^-}/\gamma_{HA}$ in Equation 4. But we can evaluate this expression by a method which can be exemplified in terms of the two volatile buffer systems of particular interest in physiology. Putting Equation 3 into logarithmic form, we have for the H_2CO_3-HCO_3^- system:

$$\log K' = \log K - \log \frac{\gamma_{H^+}\gamma_{HCO_3^-}}{\gamma_{H_2CO_3}} \tag{7}$$

where K' is the apparent dissociation constant. According to contemporary acid-base theory (I-5) NH_4^+ is an acid and NH_3 its conjugate base; and so the logarithmic equation based on the (acidic) dissociation constant for this conjugate pair, similarly derived, is:

$$\log K' = \log K - \log \frac{\gamma_{H^+}\gamma_{NH_3}}{\gamma_{NH_4^+}} \tag{8}$$

H_2CO_3 and NH_3, being nonionic, are not affected by electrical fields, and hence in dilute solution have activity coefficients of unity. Activity coefficients of individual ions have been described as "convenient frictions" (422); since electroneutrality is preserved in solution we never have a situation in which the activity of an ion can be evaluated independently of that of some oppositely charged ion. Accordingly, what can really be determined is the product of the activity coefficients of ions of like charges, or the quotients of the coefficients of charges of unlike ions. If, however, a value for the activity coefficient γ is assigned by convention to any one ion, then the relative values of the coefficients of the other ions can be reckoned for any solution. By convention γ_{H^+} can be assigned a value of unity; in physiological solutions $[H_3O^+]$ is so low that it cannot in any case influence μ. Then for CO_2:

$$\log K' = \log K - \log \gamma_{HCO_3^-} \tag{9}$$

and for NH_4^+:

$$\log K' = \log K - \log \frac{1}{\gamma_{NH_4^+}} \tag{10}$$

$$\log K' = \log K + \log \gamma_{NH_4^+} \tag{11}$$

For sufficiently dilute solutions $Ba_i \sqrt{\mu}$ in Equation 1 is so small compared with unity that it may be neglected; and

$$\log \gamma = Az_+z_-\sqrt{\mu} \tag{12}$$

With valence types up to 3−1 this abbreviated Equation 1 is valid up to $\mu = 0.1$. Moreover, univalent ions so greatly predominate in extracellular fluid and urine that z_+z_- can be taken with little error as unity; where this is true, Equation 12 becomes:

$$\log \gamma = -0.52\sqrt{\mu} \tag{13}$$

For the same reason, μ in extracellular fluid and urine (Equation 2) is approximated by the sum of the molar concentrations of the principal cations: Na^+ and K^+ plus (in the case of urine) NH_4^+. Accordingly, substituting Equation 13 in Equations 9 and 11 we have for CO_2:

$$pK' = pK - 0.52\sqrt{[Na^+] + [K^+] + [NH_4^+]} \tag{14}$$

and for NH_3:

$$pK' = pK + 0.52\sqrt{[Na^+] + [K^+] + [NH_4^+]} \tag{15}$$

where the cation concentrations are expressed as moles per liter. The simplified Debye-Hückel equation has generally been applied to plasma and is in general satisfactory for urine also; values of pK′ so calculated ordinarily differ less than 0.03 from those obtained with the complete equation.

APPENDIX D

GAS SOLUBILITY AND pK′ OF VOLATILE BUFFERS IN URINE AND PLASMA

Blood plasma is not entirely homogeneous: ions are confined to plasma water, which averages 93% of normal postabsorptive human plasma, the remaining 7% of the volume being occupied by the plasma proteins. It is the electrolyte structure of plasma water, not of plasma, which is relevant to all physiological and medical considerations. Extreme deviation from normal of the protein content of pathological sera might result in alteration of up to 5% in plasma electrolyte concentrations; a larger potential source of error is increased fat content in the plasma of some patients with disorders of lipid metabolism. An erroneous diagnosis of hyponatremia may be entertained on the basis of a routine plasma sodium report in such patients unless it is realized that the analysis reflects "pseudohyponatremia" caused by decreased plasma water content, the sodium concentration in plasma water being normal.

The ion-to-gas ratios of the volatile buffers must be a little lower in plasma water than in plasma, because, whereas NH_4^+ and HCO_3^- are confined to the aqueous phase, the congener gases are soluble to some extent in the plasma lipids. It is likely that in lipemic plasmas the ratios may fall considerably, at any pH, below the ratios in plasma water; this question deserves more study than it has received. Other physiological fluids (such as cerebrospinal fluid and sundry secretions) which contain little fat or protein may be presumed to behave, like urine (I-4B; Appendix C), in conformity with the Debye-Hückel limiting equation.

Carbonic acid has two dissociable H's, but only the first dissociation occurs to an appreciable extent within the range of pH encountered in urine and extracellular fluid; therefore for brevity we refer throughout this book to pK and pK′ for CO_2, or for H_2CO_3, when strictly pK_1 and pK_1' are meant. pK for CO_2 was first evaluated by Hasselbalch (230); the value has been redetermined by many authors (447, 132, 646, 113, 114, 231, 137), and in physiology is conventionally taken to be 6.33 at 37° C (231). For ammonia, pK at 37° C is 8.89 (41).

Estimation of pK′ in plasma and urine is interrelated, in the case of both the volatile buffers, with evaluation of the solubility of the respective gas in these fluids; the solubility must be determined before the molar quantity of gas in solution can be reckoned from the gas pressure of the system, or vice versa. Symbolization and terminology of gas solvation among physiologists has been anything other than uniform, and, as a result, consultation of the relevant literature is often attended by formidable obstacles to comprehension. The difficulties have been much alleviated, if not entirely obviated, by a series of recommendations made by the Ad Hoc Committee on Terminology of the symposium "Current Concepts of Acid-Base Measurement" (418) sponsored by the New York Academy of Sciences.

Henry's law, formulated in 1805 by the English chemist William Henry (1775–1836), states that the (molar) quantity of a gas absorbed by a given volume of liquid at a given temperature is proportional to the gas pressure P. But since (according to Boyle's law) the volume of a gas is inversely proportional to its pressure, Henry's law also states that the *volume* of a gas absorbed at any temperature by a given volume of liquid is independent of the pressure (166). The extent to which a particular gas is absorbed by a liquid at a given temperature may accordingly be expressed in either of two ways: (i) as the solubility coefficient S, the volume of dry gas measured under the conditions of the experiment, dissolved in one volume of liquid, or (ii) in terms of α, Bunsen's absorption coefficient, defined as the volume of dry gas reduced to standard temperature and pressure (0° C, 1 atmosphere pressure) present in one volume of liquid. Since according to the law of Gay-Lussac the volume of a gas is inversely proportional to the absolute temperature T, it is apparent that for any gas obeying these laws:

$$\alpha = \frac{273\, S}{T}. \tag{1}$$

$$\text{or at } 37^\circ \text{ C} \qquad \alpha = 0.880\, S;$$

and

$$S = \frac{\alpha\, T}{273} \tag{2}$$

$$\text{or at } 37^\circ \text{ C} \qquad S = 1.135\alpha$$

The solubility of gases in liquids is depressed by solutes. In urine and plasma, as in other solutions, individual ions have their specific and additive effects on solubility and in each instance the effect varies as the concentration of the particular ion; nevertheless, gas solubility is not greatly affected by physiological concentrations of ions and other solutes. Solubility also decreases with rising temperature; in mammalian physiology we are of course concerned with temperatures differing greatly from 37° C only when studying fever or hypothermia, usually iatrogenic. Though many if not most of the reported physiological investigations of gas solubility have been conducted at 38° C, we follow here the recommendation of the Ad Hoc Committee on Measurement of the symposium "Current Concepts of Acid-Base Measurement" (418) that 37° C be made conventional for physiological work. Most commercial electrode assemblies for measurement of pH and P_{CO_2} are provided with constant temperature baths at 37° C.

1. CO_2

Van Slyke, Sendroy, Hastings, and Neill (630) obtained by tonometry values of α at 38° C for (total non-ionic) CO_2 in water, in various salt solutions, and in serum. No deviation from Henry's law was observed over the range of gas pressure studied The value obtained for water was 0.5455. Specific values were given for particular ions in terms of depression of α per unit molar concentration.

a. Urine

McGee and Hastings (390) applied the data of Van Slyke *et al.* (630) to biological fluids to obtain the approximation:

$$\alpha = 0.546 - 0.109\, \mu \tag{3}$$

where μ is the ionic strength (Appendix C). Details of the measurements by which the relation between α and μ was evaluated were not furnished. Their equation has generally been used to reckon α in urine, though Mainzer and Bruhn (372) reported somewhat lower estimates ranging from 0.441 to 0.514. The equation of McGee and Hastings leaves out of account, as negligible in biological fluids, the fact that the effects of ions upon gas solubility are specific, that is, not related directly to the contribution of the ion to μ.

One mole of CO_2 gas at standard temperature and pressure occupies 22.26 liters; hence 1 μmole occupies 22.26×10^{-3} ml and

$$[CO_2]\ \mu\text{moles/ml} = \frac{P_{CO_2}\ \text{mm Hg}}{760} \times \frac{\alpha}{22.26 \times 10^{-3}} \tag{4}$$

$$[CO_2]\ \mu\text{moles/ml} = 0.0591\, \alpha\, P_{CO_2}\ \text{mm Hg} \tag{5}$$

The factor 0.0591 α by which the P_{CO_2} in mm Hg must be multiplied to obtain $[CO_2]$ in μmoles ml^{-1} mm^{-1} was in the earlier physiological literature generally symbolized as *a*; this symbol, which is adopted here, may be termed the "molar CO_2 absorption factor." Van Slyke, Sendroy, Hastings, and Neill (630) and Hastings, Sendroy, and Van Slyke (232), both in 1928, used α as defined here, but referred to it as "Bunsen's solubility coefficient"; and Hastings and Sendroy (231) used the symbol *a* for Bunsen's absorption coefficient. Some more recent authors, or their editors, have written α when *a* was intended (278, 337, 252). The "salting-out coefficient" has unfortunately also been symbolized as α (137); this is a factor relating the solubility of CO_2 in water to its solubility in a salt solution of any molality, according to the relation

$$Q' = Q/\alpha m \tag{6}$$

where Q is moles of CO_2 at 1 atmosphere P_{CO_2} in water and Q' is the corresponding value for NaCl solution of molality m. The Ad Hoc Committee on Terminology of the symposium "Current Concepts of Acid-Base Measurement" advocated the use of S in the sense of "the [temperature-dependent] coefficient relating the sum of the concentrations of dissolved CO_2 and H_2CO_3 in mmol/l to P_{CO_2} in mm Hg." Though S has undeniably been used in this sense among physiologists (555), this recommendation of the Committee seems an unfortunate one in view of the long established sense given to S in the physicochemical literature.

Hastings and Sendroy (231) reported that their experimentally determined values for pK′ for CO_2 agreed well enough with an approximation of the Debye-Hückel equation:

$$pK' = 6.33 - 0.5\sqrt{\mu} \tag{7}$$

Portwood, Seldin, Rector, and Cade (479), using the abbreviated equation of Hastings and Sendroy, stated further that P_{CO_2} so calculated for urine agreed well with tonometer values. They took *a* equal to 0.0309, ignoring variation with urine flow; the relative error involved probably does not exceed 2% according to Equation 3. The added labor of computation using the detailed Debye-Hückel equation (Appendix C, Equation 1) yields values deviating by no more than 0.03 from those obtained with Equation 7.

The value 6.33 for pK of CO_2 at 37° C is a bit above that reported in the physicochemical literature (137). Preliminary studies (262) have suggested, however, that experimental data relating urinary P_{CO_2}, $[HCO_3]$, and pH accord with theory when conventional values for pK′ and the solubility of CO_2 for normal urine are applied. It seems clear also that the cerebrospinal fluid (407) and other essentially one-phase aqueous secretions (292) can be treated theoretically with little risk.

b. Plasma

The value in serum at 38° C for α_{CO_2} obtained by Van Slyke *et al.* (630) was 0.510, corresponding to $a = 0.0302$; slightly higher values have been obtained by others at 37° C (23, 292). Lack of observed deviation from Henry's law indicated that the colloids of normal sera take up little CO_2; but, since CO_2 is quite soluble in fat, application of this value to lipemic sera is probably subject to some error. Hastings *et al.* (232) restudied pK′ in serum and reviewed previous evaluations. They concluded that a variety of errors and misconceptions explained differing published values, and arrived at the figure 6.10 for pK′ in plasma and serum. These values for pK′ and pH are close to the corresponding values (0.0306 and 6.11, respectively) furnished by the Ad Hoc Committee on Methodology (418). The use of the symbol pK′ in connection with bicarbonate equilibria in plasma, though conventional, implies a degree of theoretical authority somewhat beyond what is strictly justified; for plasma, whether separated or true plasma (plasma equilibrated with the suspended blood cells), contains more than one phase. According to Siggaard-Andersen, plasma [HCO_3^-] determined gasometrically or titrimetrically is somewhat higher than the true value because the "HCO_3^-" includes small amounts of $CO_3^=$ and carbamino-CO_2; when these methods are used, there is some inverse variation in the apparent dissociation constant with pH. It is therefore recommended (568, 569) that the symbol pK_1'' be used in the Henderson-Hasselbalch equation when working with [HCO_3^-] determined titrimetrically and pK_1''' when [HCO_3^-] is determined gasometrically. If pH and temperature vary markedly from normal, such variation exerts an influence on the apparent dissociation constant of CO_2 which may have to be taken into account for accurate work (569, 556).

It has been assumed traditionally, on the basis of theoretical considerations as confirmed by studies of pathological sera reported by Robinson, Price, and Cullen (516), that the values for solubility and apparent dissociation of CO_2 established for normal plasma are applicable also to plasma from patients with pathological conditions other than lipemias. Recently the claim has been put forward (615, 294) that the Henderson-Hasselbalch equation is not applicable to some pathological sera without gross error. This view, or at any rate this way of stating the problem, is not readily acceptable. Sufficient deviation of the ionic strength of the aqueous phase of pathological sera from normal to affect pK′ appreciably could hardly occur during life. Certainly, abnormal sera may differ markedly from normal in the extent and type of lipid and protein present; but one may continue at the present time to entertain the possibility that abnormal constituents of the sera might affect the precision of the CO_2 electrode or other measurements or constants used to relate pH to CO_2 and HCO_3^- activity. The Henderson-Hasselbalch equation is not discredited if an erroneous relation between its terms results from inaccurate measurements of the pH or P_{CO_2} or from insertion of customary values for the relevant physicochemical constants in pathological plasmas where they may be inapplicable.

2. pK of H_2CO_3 at 37° C

About 1 part in 350 of dissolved CO_2 is present at 37° C in aqueous solution as H_2CO_3 (Table D-1). Roughton's method (525, 526, 527, 528) depends upon estimating the heats of the two reactions by which CO_2 is produced when HCl and $NaHCO_3$ are mixed in a flowing system: (i) formation of H_2CO_3 and (ii) its decomposition to CO_2 and H_2O. The former is practically instantaneous; the latter is relatively slow, the correspondingly slow heat liberation enabling the course of the decomposition of H_2CO_3 to be followed as the liquid flows along a tube, and so permitting calculation of K. Wissbrun, French, and Patterson (667) think that Roughton may have overestimated the heat of ionization, and consequently underestimated the heat of hydration, as his values for K do not agree closely with theirs. Their calculations are based on relaxation methods (high field conductance); they

TABLE D-1. *Estimates of pK and K for H_2CO_3 at 37° C**

Source	pK′	*K*	$CO_2:H_2CO_3$
1. Roughton (526)	3.47	3.388×10^{-4}	$(3.456/4.67) \times 10^3 = 738$
2. Roughton (527)	About 3.5	About 3.16×10^{-4}	About $(3.16/4.67) \times 10^3 = 682$
3. Wissbrun *et al.* (667)	3.810	1.55×10^{-4}	$(1.55/4.67) \times 10^3 = 340$

* Additional kinetic studies recently summarized by Malnic and Giebisch (375) suggest an equilibrium value for $CO_2:H_2CO_3$ at 37° C near 340.

find no consistent effect of temperature on *K* between 5° and 45° C and, estimate half-time for the ionic reaction as less than 10^{-7} seconds.

In Roughton's Harvey Lecture (526) the value pK = 3.470 at 38° C, based largely on earlier work (525), is given. In a chapter prepared for the Air Force manual of respiratory physiology (527) Roughton gives his own estimate that *K* is "about 3.5×10^{-4}" at 38° C and that about 0.15% of the dissolved CO_2 molecules are hydrated (1 in 680); he also furnishes the estimate of Wissbrun *et al.*

3. NH_3

a. pK′ in Urine and Plasma

Bank and Schwartz (30) studied the effect of sodium, potassium, chloride, sulfate, and thiocyanate ions on ammonia pK′. They measured the pH change effected by the conversion of ammonium salts to NH_3 by means of added alkali at 37° C; pK′ could be estimated from the known ratio $NH_4^+:NH_3$ and the measured pH. The specified ions were shown to exert no effect other than that predictable, according to the Debye-Hückel equation, from their contribution to ionic strength. Urea was found to have no effect. Accordingly, for urine at 37°:

$$pK' = 8.89 + 0.52\sqrt{[Na^+] + [K^+] + [NH_4^+]} \tag{8}$$

and for plasma

$$pK' = 8.93 \tag{9}$$

b. Solubility in Urine and Plasma

Jacquez, Poppel, and Jeltsch compared the solubility of NH_3 in plasma and plasma water (288). In their paper α, termed the "solubility coefficient," is conventionally defined as liters of dissolved dry gas at standard temperature and pressure per liter of solution at 1 mm Hg gas pressure. In order to obtain the molar solubility factor for NH_3, it is necessary to divide α by the volume in liters occupied by 1 mole of NH_3 under standard conditions, 22.09 at 37° C. From the available data (288, 327) *a* can be reckoned as 0.0281 for water and 0.0412 for plasma; it was suggested by Jacquez *et al.* that solvation of some NH_3 in plasma colloids might account for the difference between the two values. For plasma water at 37° C we can write:

$$NH_3\ \mu moles/liter = 28.1\ P_{NH_3}\ (\mu\ Hg) \tag{10}$$

and

$$P_{NH_3}\ (\mu\ Hg) = 35.6\ [NH_3]\ \mu moles/ml \quad (11)$$

No data are at present available on the effect of ions upon α_{NH_3}. If it is assumed that the "salting-out" of NH_3 is similar in magnitude to that for CO_2, values for urinary P_{NH_3} calculated according to Equation 11 average about 3% too low, the error being somewhat larger in concentrated and somewhat less in dilute urines.

c. [Ammonia] and [NH_3] of Blood, Plasma, and Plasma Water

P_{NH_3} of plasma and whole blood must evidently be identical. It is ordinarily desired to estimate plasma P_{NH_3} from measurements of the ammonia concentration of whole blood, because determination of whole blood ammonia is preferred for technical reasons to measurement of plasma ammonia. Spontaneous decomposition of nitrogenous blood constituents causing release of ammonia into shed blood results in artifactual elevation of ammonia levels; this process is minimized by immediate icing of the blood sample and collection of its free ammonia by vacuum distillation as promptly as possible after drawing the blood and without prior centrifugation, using the procedure of Archibald (13) or modifications of it such as that of Preuss, Bise, and Schreiner (480). Alkaline distillation methods (420), though normally satisfactory for determination of urinary ammonia, give values for blood ammonia averaging about 30% higher than those obtained by vacuum distillation, which are believed to be close to true blood ammonia.

Published values (74) for whole blood pK′ to the contrary notwithstanding, it is clear that there can be no apparent dissociation constant for a fluid which contains cell water within its suspended erythrocytes in variable proportions up to 40% or more of the total. Unless pH and ionic strength of erythrocyte water are known, the theoretical relation of NH_3 to NH_4^+ is calculable for plasma but not for whole blood. On the premise of NH_3 equilibration, however, the partition of NH_4^+ between erythrocytes and plasma can be ascertained empirically. Hills and Reid (258) investigated the distribution of ammonia between the phases of normal human blood water and reviewed investigations by others; they recommended taking normal whole plasma [ammonia] equal to blood ammonia ÷ 1.4, a factor which implies (assuming hematocrit 0.40) that erythrocyte [ammonia] is twice plasma [ammonia]. This estimate is compatible with higher pK′ and lower pH in erythrocyte water; it has been found to hold in alkalosis and in blood samples from patients with hepatic cirrhosis but is not necessarily applicable to other species. If samples of which the hematocrit (Hct) differs considerably from 0.40 must be analyzed, plasma [ammonia] p can be reckoned from the measured blood [ammonia] b on the assumption that the erythrocyte [ammonia] e is twice p. Then, since:

$$b = (2p) \times \text{Hct} + p\,(1 - \text{Hct}) \quad (12)$$

or

$$b = p \cdot \text{Hct} + p \quad (13)$$

$$p = b \div (\text{Hct} + 1) \quad (14)$$

The normal ratio $NH_4^+:NH_3$ in plasma water is given by a rearrangement of the Henderson-Hasselbalch equation:

$$NH_4^+:NH_3 = \text{antilog } (8.93-7.4) = 33.8 \quad (15)$$

i.e., 97% of the total ammonia is NH_4^+; and since there appears to be rather less than twice as much NH_3 in plasma as in plasma water (258), we may take it that approximately 98% of the total plasma ammonia is present in plasma water. Plasma P_{NH_3}, calculated as plasma water P_{NH_3}, can therefore be obtained from $[NH_3]$ of plasma water:

$$[NH_3]_{\text{plasma water}} = \frac{\text{plasma [ammonia]}}{\text{antilog } [8.93 = 7.4] + 1} = \frac{\text{plasma [ammonia]}}{34.8} \quad (16)$$

From Equation 11,

$$P_{NH_3}\ (\mu\ \text{Hg}) = 35.6\ [NH_3]\ \mu\text{moles/ml} \quad (17)$$

where P_{NH_3} is the ammonia pressure of blood, plasma, and plasma water, and $[NH_3]$ is the ammonia gas concentration in plasma water. As in urine, the estimated P_{NH_3} may be several percent too low owing to neglect of the probable salting-out effect of the solutes of plasma water.

d. P_{NH_3} and Ammonia Transport in Blood

Ammonia outflow from the kidney can be calculated from renal cortical P_{NH_3}, renal blood flow, and the hematocrit. $[NH_3]$ of plasma water is obtained from P_{NH_3} by Equation 10, and plasma [ammonia], according to Equation 16, is 34.8 times $[NH_3]$ of plasma water. Plasma carriage is renal plasma flow times plasma ammonia, and:

$$\text{Erythrocyte carriage} = 2 \times \text{plasma [ammonia]} \times (\text{renal blood flow} \times \text{Hct}) \quad (18)$$

Blood carriage is of course the sum of the two. It may be desirable to deduct 3% from the calculated figure to allow for the lower ammonia content of the effluent of the noneffective renal perfusion (258).

APPENDIX E

BUFFER STRENGTH OF HETEROGENEOUS EQUILIBRIA

Figure E-1 illustrates the behavior, under various conditions, of equilibria in which one member of the buffer pair is a gas. Four transitions from an initial to a final state of the CO_2 buffer system are represented; rise in pH of approximately 0.2 to 0.3 is related in each case to the quantity of alkali required to produce it. Ordinate and abscissa refer only to the large sigmoid curve which represents the

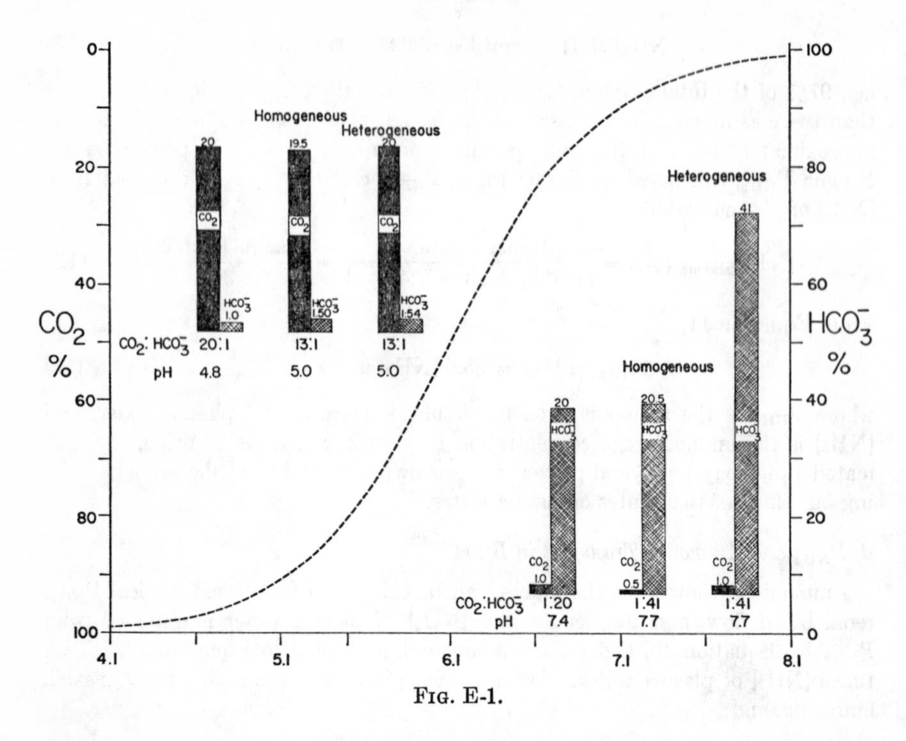

FIG. E-1.

percentage of total CO_2 present as dissolved CO_2 gas (*left ordinate*) and as HCO_3^- (*right ordinate*) in extracellular fluid at any pH, over the range shown on the abscissa. Solution pH in relation to this curve, and also for the *small inserted diagrams*, has been calculated according to the Henderson-Hasselbalch equation from the ion-to-gas ratio of the CO_2 buffer system, using the pK′ of 6.1 which is appropriate for any solution having the ionic strength of extracellular fluid.

Each of the two *small inserted diagrams*, each consisting of three *pairs of bars*, contrasts the effect of adding alkali to the liquid phase of (i) a system of homogeneous equilibrium and (ii) a system of heterogeneous equilibrium having a gaseous phase large enough to hold gas pressure essentially constant. The height of the *black bars* represents concentration of dissolved CO_2 in millimoles per liter, and the height of the *hatched bars* represents that of HCO_3^- in millimoles per liter; the CO_2:HCO_3^- ratio for each *pair of bars*, and the pH so defined, are listed below each *bar pair*.

Enough alkali is assumed to be added to each system in its initial state (designated by the *left-hand pair of bars* in each *small diagram*) in order to accomplish the same change in ratio of species in solution for both homogeneous and heterogeneous equilibrium (the two *right-hand bar pairs*), and hence to effect the same change in pH; initial and final gas-to-ion ratios and pH values are listed below the *bars*.

When in the initial state of the system the pH of the solution is considerably less than the value of pK′ for CO_2 (as illustrated by the *small six-bar diagram* inserted on the *left*), the molal concentration of dissolved gas is large relative to that of the ion. Under these circumstances, the buffer strength (ratio of alkali added to rise in pH effected) of the heterogeneous equilibrium does not greatly differ from the comparable homogeneous equilibrium. Conversion of a small amount of the gas to ion in both cases causes an identical proportional increase in the ion concentration which (because the original ion concentration is small) considerably increases the ion-to-gas ratio and the pH. The presence of the gaseous phase prevents change in gas concentration in the heterogeneous equilibrium; but the proportional change in gas concentration in the homogeneous equilibrium is small anyway and exerts very little effect on the change in ratio produced by adding alkali. Accordingly, there is not much difference in the magnitude of change in pH produced by adding the same quantity of alkali per liter to two types of equilibrium; buffer strength of both is relatively small.

When in the initial state of the system, solution pH is considerably greater than the value of pK′ for CO_2 (as illustrated by the *small inserted diagram* on the *right*), the molar concentration of gas in solution is small relative to that of the ion, and conversion of a small absolute quantity of gas to ion in a homogeneous equilibrium consequently entails a large proportional decrease in gas concentration, and hence a large decrease in the gas-to-ion ratio; consequently, a relatively very large increase in pH ensues when a relatively small quantity of alkali per liter is added to the system; buffer strength is very low. In the heterogeneous equilibrium, maintenance of the original gas concentration in solution during addition of alkali, through equilibration of solution gas pressure with a capacious gaseous phase, greatly increases buffer strength, since the increase in pH (i.e., of the ratio of ion to gas in solution) must now be brought about entirely by a change in ion concentration. To double the large initial $HCO_3^-:CO_2$ ratio requires $20.5 \div 0.5$, or 41 times more alkali in the case of the heterogeneous equilibrium as compared with the homogeneous equilibrium.

From these instances, it will be apparent (i) that buffer strength of a heterogeneous equilibrium with stabilized gas pressure is uniquely great provided the concentration of dissolved gas in solution is small relative to the concentration of ion; and (ii) that, given these conditions, buffer strength of this type of equilibrium increases indefinitely and exponentially as the ion-to-gas ratio increases (i.e., when the gas is the conjugate acid of the pair, as pK′ − pH increases).

APPENDIX F

EVALUATION OF THE URINARY TITRATABLE ACIDITY (*TA*)

1. Simple Titration

The principle is simply to determine the molar quantity of mineral alkali, generally KOH, which is needed to titrate the nonvolatile buffers of a timed urine collection back to the plasma pH (normally 7.4) at the time the urine was excreted, reversing in this manner the physiological process by which protons were transferred to the nonvolatile urinary buffers. The physiological datum of principal interest is of course the *rate of excretion* of titratable acid; it is commonly expressed in microequivalents $minute^{-1}$ and obtained as the product of the titration figure, expressed as microequivalents of alkali per milliliter of urine, and the urine flow rate in milliliters $minute^{-1}$.

The errors which may attach to estimation of *TA* by simple titration are of two kinds: (i) artifacts resulting either from titration of urinary NH_4^+ to NH_3 or from precipitation of calcium phosphate by the added alkali; and (ii) errors resulting from failure to take into account the difference in the value of pK_2' of phosphate between glomerular filtrate and the diluted urine sample at the end of the titration. (It is the second of the dissociations of three dissociable protons of phosphoric acid, i.e., shifts in equilibrium of the buffer pair $H_2PO_4^-$-$HPO_4^=$, which mainly concern us in physiology) (cf. I-5D.)

a. Calcium and Ammonia

These sources of possible error were first recognized by Folin (173) and have recently been reemphasized by Lemann and Lennon (333a). Titration of undiluted urine with alkali may result in precipitation of calcium phosphates, a process which increases the acidic intensity of the liquid phase; the ratio $H_2PO_4^-:HPO_4^=$ and the acid-to-base ratios of the other buffers present are artifactually increased. Additional alkali is consequently required to bring the urine to pH 7.4, and *TA* is overestimated. In addition, as the pH of the urine is raised by titration, some NH_4^+ must be converted to NH_3; and insofar as the titration figure comprehends titration of NH_4^+, the contribution of nonvolatile buffering is again overestimated. This error can never be large provided that NH_3 is not lost from the sample (see below); if NH_3 is lost, the error might become substantial and would be maximized when undiluted, very low-flow urine of low pH is titrated, especially when the urine is that of a patient with chronic acidosis, since ammonia concentration is greatest in such urines.

Dilution of the urine prior to titration obviates the first of these errors and renders the second negligible, but is likely to cause error by its effect on phosphate pK_2' (see below). Titration of diluted urine to pH 7.4 has been, at least until recently, the standard technic practiced since Henderson's original description of the method (239). Generally, the usual low-flow urines are diluted 1:10 or 2:25, which lowers their osmolality and ionic strength to that of samples collected during brisk water diuresis; alkalinization does not produce calcium phosphate precipitation at such dilutions even in samples with high calcium content.

Overestimation of nonvolatile buffering resulting from the inclusion of titrated NH_3 in estimated *TA* is small provided that NH_3 does not diffuse out of solution during the titration; this is because pK′ of ammonia is high enough so that NH_3 remains a small fraction of total ammonia at pH 7.4 and any lower pH, so that hardly any NH_4^+ is converted to NH_3 during the titration. Specifically, when the pH of the diluted urine has been raised to pH 7.4, NH_3 has risen only to 2½% of the total ammonia, and so overestimation of nonvolatile buffering for this reason could amount at most to around 2% of the total urinary ammonia. At the maximal normal rate of urinary ammonia excretion in healthy adults (80 μEq minute^{-1}), the error would average about 1% of total net acid excretion, or less than 1 μEq minute^{-1}; in chronic acidosis, where the rate of urinary ammonia excretion is abnormally high, it might perhaps reach 6 μEq minute^{-1}.

These estimates assume that there is no material evolution of NH_3 before and during the titration. When the conventional titration beaker contains 25 ml, whether the fluid represents a concentrated urine diluted 2:25 before titration, or 25 ml of a high-flow urine sample titrated without dilution, NH_3 evolution from the fluid in the beaker during titration is negligible because of the very low surface-to-volume ratio of the fluid. Diffusion is a highly effective means of transport over short distances (up to several score of microns), but it is extremely ineffective over the macroscopic distances such as are represented by the depth of a 25 ml-sample in the customary titration beaker, especially at the extremely low P_{NH_3} of the diluted or undiluted sample. Even at the pressures of CO_2 in acidified diluted urine, around 1000 times P_{NH_3}, vigorous bubbling of air through the solution for 10 minutes or more is required to remove most of the CO_2 dissolved prior to measurement of $TA - CO_2$ in a conventional titration beaker. Table F-1 shows that, even when an aliquot of acid urine diluted 2:25 and placed in a standard titration beaker is vigorously aerated after pH has been adjusted to 7.6, loss of NH_3 by diffusion is very slow. But it is likely that titration of undiluted low-flow

TABLE F-1. *Rate of NH_3 loss from diluted urine at pH 7.6**

Hours of aeration	pH	[Ammonia]
		µmoles/ml
0	7.60	30.9
¼	7.60	30.5
1	7.49	31.0
2	7.31	30.2
3	7.25	29.6
4	7.20	28.1

* A sample of urine of pH 5.5 was diluted 2:25 with water in a beaker, as in measurement of urinary titratable acidity, and adjusted with NaOH to pH 7.6. The total ammonia content per milliliter of urine and the pH were measured before and at intervals during continuous aeration; loss of NH_3 is reflected in decrease of both pH and total ammonia. Even after 4 hours of aeration, less than 10% of the ammonia has been lost.

urine may allow more NH_3 to escape, in which case overestimation of *TA* due to titration of NH_3 might be a serious source of error.

b. Neglect of Interionic Attraction (I-4B; Appendix C)

The conventional determination of urinary *TA* owes its attractive simplicity to the fact that no account is taken of which particular nonvolatile weak electrolytes may be present as urinary solutes; what is being investigated is simply the quantity of protons which have been accepted from the buffers of extracellular fluid by whatever nonvolatile buffers are present in the urine sample. However, it is often only approximately true that milliequivalents of protons which have been accepted per milliliter of urine by any given nonvolatile urinary buffer substance are identical with milliequivalents yielded by that same buffer during the back titration to plasma pH. The equality holds strictly only when the ionic strength μ of the solution being titrated is identical at the endpoint with the ionic strength of glomerular filtrate.

The error resulting from a difference in the value of μ between titration beaker at the end of the titration and glomerular filtrate does not exert a uniform effect upon the several classes of urinary nonvolatile buffer substances. In the case of the urinary organic acids, where the conjugate base possess a negative charge resulting from transfer of a proton by the uncharged acid to another compound, pK′ must be less than pK (I-4B1; Appendix C); for the same reason, pK′ for the second dissociation of a proton from phosphoric acid (converting $H_2PO_4^-$ to $HPO_4^=$) must be less than pK since the conjugate base has an extra negative charge compared with its conjugate acid. Moderately dilute solutions of any of these pure compounds, at a pH at which both buffer species are present, must exhibit rise in pH when water is added, for the effect of dilution is to raise the value of pK′ in the Henderson-Hasselbalch equation toward that of pK. But pK′ of the creatinine buffer system, in which the organic base accepts a proton in extracellular fluid so that the conjugate acid is the charged particle, is oppositely affected by changing ionic strength; its urinary pK′, like that of ammonia, exceeds pK, and dilution will cause the pK′ of a creatinine buffer solution to fall toward the value of pK, with decline in pH.

Only as the acidification process in the nephrons lowers pH below 6.0 can the creatinine and the ions of the dissociated salts of the organic acids of glomerular filtrate begin to accept appreciable quantities of protons; the pK values of these compounds are all too low (<5.0) to allow them to provide effective urinary buffering above pH 6.0. And in the pH range 4.5 to 6.0, the opposed effects of dilution upon the activity coefficients of the buffers present means that the titration of an aliquot over this pH range is hardly affected by sample dilution. But above pH 6.0 the situation is entirely different, since most of the nonvolatile buffer present is phosphate; consequently, dilution causes considerable rise in urine pH. In practice, the error of estimating *TA* caused by the decrease in μ resulting from dilution has been shown in both dogs (545) and humans (426) to be due virtually entirely to the dependence of the value of pK_2' for phosphate upon μ.

Nutbourne (426) found that comparable dilutions of urines obtained from hospital inpatients, and of phosphate buffer solutions of similar pH and phosphate concentrations, produced alterations in measured *TA* which were essentially indistinguishable. As deduced and demonstrated by Schwartz, Bank, and Cutler (545), alteration of phosphate pK_2' exerts such a sizable effect on *TA* because the urinary values for pK_2' (6.64 to 6.95) are relatively close to the normal pH of glomerular filtrate. This means that substantial apparent conversion of $H_2PO_4^-$ to $HPO_4^=$ is effected when concentrated urines are diluted. By contrast, pK′ is so low for all the urinary organic acids that their conversion to their conjugate bases has approached completion well below pH 7.4: the apparent quantities of conjugate acid and conjugate base at pH 7.4, and hence also the titration figure, are consequently little affected by variation of 0.2 or 0.3 in pK′. What little effect there is is counterpoised by small effects in the opposite sense on pK′ of creatinine. The error of estimating *TA* directly in a diluted sample is then practically entirely referable to urinary phosphate. The absolute error is not influenced by the original urine pH, being referable to a shift in the apparent ratio of species at pH 7.4 in the beaker as compared with glomerular filtrate. It follows that the percentile error is least in very acid urines where *TA* is high.

Estimation of *TA* on undiluted urine samples should be considered obsolete. Revival of the procedure in recent years has been inspired by the desire to avoid the errors caused by dilution; these errors, however, can be minimized in several ways (Section 2). If *TA* should nevertheless be estimated on undiluted urine samples, the error arising out of neglect of interionic attraction will be greatest in very concentrated urines, both because of their higher phosphate concentration and the large difference between μ of urine and glomerular filtrate. In aqueous phosphate solutions of 250 milliosmoles kg^{-1} there is no titration error (426); but, because of the contribution of solute (mainly urea) other than electrolytes to urine osmolality, urinary ionic strength cannot be deduced accurately from osmolality. Since urinary [urea] greatly exceeds glomerular filtrate [urea], urinary μ will be equal to μ in glomerular filtrate only in certain samples where osmolality exceeds considerably 250 milliosmoles kg^{-1}.

2. Indirect Methods

Should titration be carried out in such a way that the ionic strength μ of the fluid in the titration beaker is equal to that of plasma—0.167 according to Van Slyke and co-workers (627a)—no error due to alteration of pK′ could arise. However, aside from the inconvenience amounting to impracticality of attempting to adjust μ of the titration sample to 0.167 at the endpoint, other sources of error are likely to affect titration of such samples (Section 1), and ample dilution of the urine sample prior to titration is clearly preferable. Nutbourne (426) presents a variety of empirical methods which can be adapted to minimize the error due to the decrease in μ in the titration beaker resulting from sample dilution. For precise results, Schwartz *et al.* (545) recommend calculating the *TA* as follows: to the *TA* measured on dephosphorylated urine is added the "phosphate *TA*" calculated from (i) total urinary phosphate separately measured and (ii) the pH and phosphate pK_2'. The pK_2' is conventionally calculated (Appendix C) by taking μ equal to $[Na^+ + K^+ + NH_4^+]$ in the undiluted urine and pK equal to 7.18 at 37° C.

In investigations where *changes* in the rate of net renal acid or base excretion with changing urine flow are of interest, rather than the absolute rate, it is required only to ensure uniformity of μ for all samples at the endpoint of the titration, rather than ensuring equality of μ in urine and glomerular filtrate. Woeber, Reid, Kiem and Hills (668), for example, evaluated the effect of variation in urinary flow rate upon net acid excretion by compensating *in vitro* for the varying rate of water removal *in vivo* from the urine secreted per minute: pK_2' for phosphate was consequently always virtually the same in all samples (since the excretion rate of electrolytes other than HCO_3^- and NH_4^+ is not conspicuously nor consistently affected by varying urinary flow rate); and whatever the absolute error attaching to evaluation of *TA* may have been, it remained essentially constant as urine flow varied.

APPENDIX G

QUANTITATIVE APPRAISAL OF ACID-BASE BALANCE

1. Semantics

The recent literature on acid-base balance is greatly plagued by terminological nonuniformity. Though obeisance is now made in virtually all biomedical quarters to Brønsted-Lowry theory, Camien, Simmons, and Gonick (89) are nevertheless unquestionably correct that much continuing controversy in the literature about acid-base balance proper is traceable to incomplete or injudicious assimilation of contemporary chemical theory into physiological thought. What is sought in what follows, and in Chapters II-4 and II-5, is not so much reconciliation of prevalent styles of discourse on this subject—probably not a practical possibility in view of the controversial character of the current relevant literature—but rather unambiguous language, internally consistent and in agreement with the spirit and the letter of contemporary acid-base theory: communication with the reader rather than a synthesis of various parlances of writers in the field. It will nevertheless be apposite to consider here several of the more conspicuous divergences of opinion recorded on substantive issues in the literature together with related disagreements about terminology.

When is a proton not a proton? In the body fluids, the H nucleus acts as a proton only when it is a constituent of a chemical entity (i.e., an acid) from which it will (in part) instantly be *reversibly* detached by a decrease in the acidic intensity of the solution in a manner quantitatively deducible from the mass law (II-5). Whether the detachable H nucleus is still clearly associated with its electron, as in H_2CO_3, or has already in some sense been parted from it, as in NH_4^+, is immaterial.

According to an alternative position, however—one no doubt philosophically tenable—protons are present, and constitute "potential acid," wherever hydrogen atoms are present in a chemical compound. Just as a David could be said in some sense to have been already present in each of two marble blocks before the respective chisels of Michelangelo and of Bernini disclosed the fact, so it can be held that a glucose molecule (for example) contains numerous protons which can be revealed whenever a metabolic transformation occurs, producing one or more new chemical entities in which a formerly undetachable H nucleus will be separated from its attendant electron—e.g., in the course of the metabolic conversion of one glucose molecule to two of lactic acid. Such seems to be the stand of Camien, Simmons, and Gonick (89), who note that glucose is a Brønsted acid and reject the concept of "metabolic release of protons"—what is termed in this book metabolic production of an acid—on the ground that "the proton can neither be produced de novo nor released, in net amounts, from its combination with either one or another Brønsted base to constitute a Brønsted acid."[1]

Though Filley (165) and Elkinton (142, 143, 146) do not seem to concur with Camien *et al.* on this point, their views are related insofar as they tend to accord the metabolism of hydrogen a central position in the discussion of acid-base balance. While the relation, in health and disease, between protons transferred to the body fluids as the result of metabolic production of acidic compounds and the enormously larger number of H atoms in the food metabolized to H_2O and excreted as such must of course be understood in principle by the student of acid-base balance, the author prefers to consider H metabolism in health and disease a division of intermediary metabolism, and to remove exposition of this large and rather complex subject as far as possible from the center of the exposition of acid-base regulation. If the example of net lactic acid production and net utilization of administered lactate given in Chapter II-5 is understood (see Figure II-5.2), then the principles governing the contribution of metabolism to the net acid-base load should be clear, including the overproduction of organic acids in certain diseases. In any concrete setting in which the problem is to reckon the acid-base balance, the question which concerns us is not how much acid various dietary constituents might give rise to, but rather how much net nonvolatile acid is actually produced metabolically from them under the given conditions, and by what mechanisms and to what extent it is eliminated. Accordingly, "total *potentially* ionizable hydrogen," "turnover of actual and 'potential' hydrogen ion," "metabolic hydrogen turnover," and "hydrogen balance" (165, 142, 143, 146) are terms avoided in this book as sedulously as are the expressions "proton excretion" and "proton balance" (Section 2). This brings us to the two summary constituents of acid-base balance; load L and excretion E (II-5).

2. E and L

Brønsted-Lowry theory has been applied with excessive zeal in the case of the "acid excretion" by the kidney, an expression precisely defined physiologically several years before the theory was developed. Neither the rate of excretion of Brønsted acids in the urine, nor the relative quantities of urinary Brønsted acids and bases present, nor any function of these quantities, bears any consistent relation to renal acid excretion in the conventionally physiological sense (II-4C). Unfortunate consequences entailed by the belief that *E*, the renal net excretion of acid, is definable in purely chemical terms without reference to the extracellular fluid are illustrated in an article by Hunt (279). This investigator estimated the renal acid excretion as titratable acid (*TA*) reckoned by titrating to pH 7.0 (neutrality at room temperature), plus NH_4^+ "which is an acid in the sense that it is a hydrogen donor." The resulting data naturally defy meaningful analysis. Not realizing that net excretion of base may occur when the urine is chemically acidic (279), Hunt interpreted his observation that his alkaline-ash diets failed to alkalinize the urine (i.e., to raise its pH above 7.0 at 25° C) as an indication that dietary ash is not, in health, the principal (or even an important) determinant of the rate of net renal excretion of base of nonvolatile acid. And yet it has been clear since Liebig's paper of 1844 (II-4; II-5) that the dietary ash is a principal determinant of variation in the rate of renal excretion of nonvolatile acid or base in the urine of healthy persons, an observation reconfirmed whenever the ash of a diet is prescribed or an alkalinizing or acidifying salt administered.

Closely related to the misconception that renal acid excretion means excretion of Brønsted acids is the impression that "the phenomena of acid-base balance . . . primarily involve the production, consumption, and excretion of protons" (500). Hills (255) pointed out that anyone who excretes 1 liter of water daily eliminates very close to 1 Eq of protons daily in the Brønsted acid H_2O, protons which no one would include when reckoning the urinary acid excretion; however, just as glucose is too weak an acid ever to function in extracellular fluid as a proton donor of any quantitative pretentions, so too, quantitatively speaking, is H_2O too weak an acid to function as such physiologically. But $H_2PO_4^-$ functions as a physiological proton donor, yet not all the protons of urinary $H_2PO_4^-$ which this ion can part with to form $HPO_4^=$ can be counted toward urinary acid excretion (II-4B and II-4C). Renal acid excretion can not be reckoned by counting the dissociable protons of the urinary buffer acids.

3. Recent Applications of the Metabolic Balance Technic

Views representative of those championed by the two principal groups recently active in research in this area are conveniently accessible in summary in articles by Relman (500, 499), Lemann and Lennon (333b), and Camien, Simmons, and Gonick (89).

Relman and his associates (336, 502, 340, 334), using liquid soy protein diets to minimize stool formation and errors attaching to chemical analysis of the intake, originally concerned themselves primarily with defining net acid *production* and had concluded by 1964 that this term was expressible in their experiments as the net metabolic production of organic acids plus mineral acid arising from the metabolism of dietary S and from phosphoester residues, whether by oxidation of phosphodiesters or by oxidation of organic cations neutralizing phosphomonoesters. Relman *et al.* (502), for example, furnished data represented as demonstrating mean equality of renal net acid excretion in volunteers with production of acid derived from the sources enumerated above; this article warrants brief quantitative review.

If acid production equaled net renal acid excretion during a control period of observation in which the 10 subjects studied were in acid balance (to a near approximation), as was concluded in this article, then net balance of Na + K + Ca +

Mg must also (to a near approximation) have equaled net balance of Cl. It appears from the sample experiment (their Table I) reported in detail that data were collected permitting evaluation of all mineral balances except that of Mg, at least in this experiment; but these data were not related to acid balance. The willingness of Relman *et al.* (502) to neglect the balance of the other dietary minerals as a factor contributory to acid-base balance appears to have stemmed from their endorsement of Hunt's doubts as to the contribution of the net dietary mineral ash to L_A (Section 2).

Mean acid balance was reckoned by these authors (their Table II) to have varied from −8 to +13 mEq/day in their 10 subjects during a control period varying from 3 to 8 days, during which observations were made after a precontrol stabilization period of 2 days. Apparent positive control-period balances of Na and Cl (14 mEq daily in the illustrative experiment) were uniformly found; they were assumed by the authors to reflect skin losses. On this assumption the negative control-period balances of K and Ca of 5 and 14 mEq, respectively, reported in their Table I, would indicate a mineral contribution of 19 mEq daily to the net acid load; presumably, Mg losses also occurred on the nearly Mg-free diet, so acid generation ascribable to the mineral balances must have been (insofar as the sample experiment was representative) considerably greater than the greatest of the negative acid balances reported.

On the other hand, since values for TA were not corrected, a systematic error may have attached to evaluation of E_A in all subjects (II-4; Appendix F); if E_A was underestimated, as seems probable, recorded acid balances near zero may have rested in part on underestimation of both L_A and E_A. A recurrent problem in interpreting the absolute balance in acid-base metabolic balance studies using solid diets is that of evaluating the magnitude of the several analytical errors attaching to endogenous acid production, and sometimes also to E_A. It will hardly be doubted that normal adults do come into approximate acid-base balance when they receive maintenance calories supplied by any ordinary diet; but coincidence of L_A and E_A is of course compatible with substantial errors of estimate of both—notably where analytical methods for one or more of the constituents of these terms are suboptimal. Another source of imprecision in the liquid-diet experiments of Relman *et al.* (502) is identifiable in the fact that phosphorus balance was not achieved, at least according to data presented in the sample experiment. The authors chose to ascribe these imbalances to the inherent difficulties of obtaining *pro rata* stool collections in metabolic studies of relatively short duration; in effect they assumed zero P balance. Evidently, however, there was scope here for considerable error in evaluating the contribution of dietary P to L_A.

In subsequent reports by this school (210, 339, 335, 228, 333b), similar methods were applied to the study of patients with chronic renal disease as well as of normal persons rendered acidotic by administration of NH_4Cl. The balance data were interpreted to mean that substantial amounts of acid are retained on a continuing basis both in patients with acidosis due to renal disease and in normal subjects rendered acidotic; plasma $[HCO_3^-]$ reflected no net accumulation of nonvolatile acid in the body, however. Alkali therapy raising plasma HCO_3^- to the high-normal range generally restored calculated b_A to zero in normal subjects and patients. It was suggested that during chronic nonrespiratory acidosis dissolution of calcium salts of the skeleton probably provides alkali to neutralize acid accumulating in the body and may accompany extra acid excreted in the stool. This contention, originally put forward by Albright and his associates (6, 7), has subsequently been both supported and contravened (499, 127, 590, 589, 333b).

Insofar as the evidence on net acid balance provided by the balance studies is concerned, methodological queries previously mentioned arise again, this time in respect to Δb as well as b. The Van Slyke-Palmer method (628) does not estimate urinary organic acids quantitatively; the method is also rather nonspecific and, among other deficiencies, fails to detect much of the citrate present. But it is well known (424, 89) that urinary citrate excretion rises sharply when the body fluids are alkalinized and falls when they are acidified (403); and our knowledge of the identity and quantity of individual urinary organic acids and of their response to acid-base alterations is sufficiently incomplete (424) that the possibility of increased excretion of additional organic acids (such as ketoglutarate) in the urine during alkali therapy is a real one. To the extent that alkalinization may have resulted in increased organic acid production and excretion not adequately reflected in the analyses, Δb_A would erroneously suggest a shift in balance in the direction of base retention. It is difficult to convince oneself, in reviewing these studies of stable acidosis, that the positive acid balance, corrected by alkali therapy, which the experimenters believed their data to have demonstrated might not have represented in actual fact zero balance during acidosis, with an artifactual rather than a real decrease in b_A with alkali therapy.

More generally, the element of circularity, pointed out by Camien, Simmons, and Gonick (89), attaching to the reckoning of b_A is unavoidable (255): the contribution of metabolically produced organic acids to L_A can be recognized only insofar as their anions do in fact appear in the urine and stool (II-5B). Studies from the laboratory of these latter investigators (89, 88, 90, 87, 208, 209) have made use of a simplified titrimetric procedure (the "ash-TA") for determination of the collective net ash of diet, urine, or feces (88, 90) without estimation of the individual mineral elements present; values obtained with this method agree well with those calculated from the amounts of the several minerals present. Camien, Simmons, and Gonick (89) have recently recommended that the intake-production term of acid balance be evaluated under two headings. The mineral balances exert an influence on L_A which is variable in magnitude, but which is often the dominant factor, and this contribution to L_A can be estimated without circularity from the "ash-TA," taken in milliequivalents as (millimoles Na + K) + 2 (millimoles Ca + Mg) − (Cl + 2S + 1.8 P in millimoles). Subtraction of the separately determined urinary organic S and P from the last term is comprised in a residual term R which also adds to L_A net urinary organic acid (i.e., urinary conjugate bases of organic acids less urinary conjugate acids of organic bases). This approach has the considerable advantage of isolating the unavoidably circular aspect of all estimations of b_A in the R term. (Acid retention manifested by a positive balance of mineral acid would presumably permit one to suspect similar retention of organic acid not detectable in the balance data.)

Lennon, Lemann, and Litzow, working with a wider range of diets than they had previously used, and taking fecal losses into account, have presented acid-base balance data obtained from normal subjects (339, 335) and patients with acidosis due to chronic renal disease with and without alkali therapy (353a). The contribution to *L* of the difference in equivalents between dietary Cl and the dietary Na, K, Ca, and Mg is now for the first time included by this group, in the conventional manner, in their evaluation of acid-base balance. A large measure of agreement therefore appears to have been reached among contemporary workers in the field on the theoretical issues which had been the subject of dispute; there remain small differences in technic between different laboratories. Lennon and his colleagues now use an improved method for urinary organic acids and a calculated value for urinary *TA*: they continue to find a positive value for b_A in the acidosis of chronic renal disease, correctable by alkali therapy (228). Recently attention has been directed to the possibility that the gastrointestinal tract may make an important contribution toward regulating net acid balance in some circumstances (309a, 333b).

NOTES

1. Glucose is, to be sure, classifiable as a conjugate acid in the broadest sense, but it is so very weak (pK 12 to 13) that, practically speaking, it never donates protons reversibly at any physiological pH (76, page 254). There appear to be a number of reasons why the misconception that glucose is an acidic compound in physiology keeps cropping up in the clinical literature. The unfortunate term "glucose titration study" (II-7F) may suggest it, and the acidic intensity of aqueous glucose solutions marketed for parenteral use has done more than suggest it to quite a few physicians.

The pH of glucose solutions (2.5 to 50%) commonly supplied for parenteral administration ranges between 3.6 and 6.5; generally it is 4.0 to 5.0. Tap water and distilled water usually contain enough CO_2 to account for acidic intensity of this degree; were the acidity to be due entirely to the presence of mineral acid, which is sometimes added in trace quantity, there would still be present at pH 4.0 only 0.1 mEq/liter. An absorbing correspondence (456a) in the *New England Journal of Medicine* was initiated by a letter drawing attention to the acidity of commercial glucose solutions. Anxiety was subsequently expressed that acidosis might result from infusion of these acid solutions; after it was brought out by another correspondent that the quantity of acid administered could never be significant, a third correspondent drew the moral that pH is a very misleading measure of acidity, to which "$[H^+]$" is to be preferred.

The original letter tentatively ascribed damage to the zona pellucida of rat ova observed *in vitro* to the acidity of the commercial 5% dextrose solution in which they had been bathed; this inference was topped by the ascription by a group of university surgeons of the phlebitis that they observed within 24 hours in 30 to 50% of infants and 25 to 30% of adults given glucose intravenously to the acidity of the commercial glucose solutions infused. These surgeons supplied a bibliography in support of their finding that "neutralization" of the solutions with 7.5 g of bicarbonate solution/100 ml (!) offered protection against the alleged acid-induced phlebitis. The entire correspondence deserves attentive study by everyone interested in the place of acid-base chemistry and physiology in the medical curriculum and in postgraduate medical education.

APPENDIX H

EQUILIBRATION OF P_{CO_2} AND P_{NH_3} IN PAPILLARY TISSUE AND URINE

The presumption is now strong that nonionic diffusion of NH_3 within the kidney proceeds so rapidly that P_{NH_3} of blood and luminal fluid remain equilibrated throughout the kidney (and *a fortiori* at the papilla) as urine flow and pH vary (668, 258, 596, 261, 469, 235). The only appreciably NH_3 pressure gradient believed to exist in the kidney is the axially directed one which causes papillary P_{NH_3} to exceed cortical P_{NH_3} except in high-flow, markedly acid urine (596). The much higher values for P_{CO_2} as well as P_{NH_3} in bladder urine as compared with papillary urine are produced, as urine moves beyond the papilla, by continuing dehydration of H_2CO_3 present in papillary urine at a concentration above the equilibrium value with CO_2 (495, 258, 261).

Equilibration of P_{NH_3} at the papilla has important consequences in that it permits quantitative rationalization of the relation obtaining during water diuresis between renal ammonia supply (production plus inflow), P_{NH_3} of cortex and papilla, and the rate of outflow of ammonia into urine and renal venous blood as the urine pH is varied. In antidiuresis the simplicity of the system is compromised by the superimposed effects of water reabsorption from the distal nephron, but during water diuresis the theoretical values for urinary ammonia excretion as a function of pH are in good accord with experimental observations (II-9). One of the minor ironies of the recent literature on renal physiology is that the 17-year-old account by Orloff and Berliner (434) of the mechanism of urinary ammonia excretion, which is squarely based upon an alleged failure of the P_{NH_3} of most urines to attain diffusion equilibrium in the medullary collecting ducts, continues to be cited as authoritative by writers who have abandoned the premises on which it is based (467, 469, 28). The treatment of Orloff and Berliner, and the more elaborate discussion of Milne, Scribner, and Crawford (403), are applicable to certain dissociable organic bases of larger molecular weight, and hence of much lower diffusivity than NH_3 (I-6), but not to NH_3. The entire basis for its application to NH_3 resided in the claim of Orloff and Berliner that brisk water diuresis increases the rate of urinary ammonia excretion only in higher-pH urines, and not when the pH is less than 6.0 to 6.25. This conclusion was clearly faulty, however. Moreover, the theoretical treatment of urinary ammonia excretion originally based upon it is conceptually unsound even if the premise should be granted.

The experiments of Orloff and Berliner were not well designed to disclose the relation of urinary ammonia excretion to urine flow, nor was sufficient care taken to avoid simultaneous imposition of conditions in addition to the water load, which might affect urine acidification. "Water diuresis" was produced at the lower urine pH values by infusing sodium sulfate solutions, and the rate at which sulfate was administered varied widely; also, when osmotic diuresis was studied, an infusion of sodium sulfate with or without mannitol was given. $SO_4^{=}$ is a poorly reabsorbable, doubly charged anion; i.e., its properties are optimal for increasing the transtubular electrochemical gradient, and hence for increasing acid secretion into the nephrons (403). In addition, from the very scanty data offered by Orloff and Berliner on water diuresis, it would appear that almost all their observations were of hypotonic urines —the only data given on "water diuresis" consist of observations of urines of a mongrel dog with flow rates of 1.5 to 8.0 ml/minute—whereas the principal effect of urine flow on ammonia excretion is seen when flow falls toward minimal values and hyperosmolol urines are being secreted (I-10). The authors gave no evidence of being aware that the well designed and carefully conducted experiments of Hubbard and Munford (275) and of Hubbard (274) had already shown that water diuresis always increases the ammonia excretion in the urine, irrespective of its pH, as has regularly been found by subsequent investigators using sound experimental designs of several types (668, 611, 119, 386, 602). There are few better-documented observations in physiology than the uniform increase in urinary ammonia excretion ensuing when low flows of urine of any pH are raised by brisk physiological water diuresis.

In addition, serious conceptual confusion was embodied in the concept of "production limitation" introduced by Orloff and Berliner, adopted by Pitts and his associates (465, 126, 658), and incorporated by Pitts into both editions of his textbook (464, 467). (The assertion, however, that the ammonia excretion rate in urines of pH 6.0 or less is unaffected by changing urine flow has been deleted from the second edition.) "Production-limited" ammonia excretion can have either of two meanings.

Sense 1: Outflow of ammonia from the kidney is production-limited in the sense that in the steady state the ammonia outflow from the kidney is restricted by the ammonia supply, i.e., ammonia inflow plus ammonia production (I-9B). However, passive outflow of ammonia by any channel is always production-limited in this sense.

Sense 2: Ammonia production might fail to increase upon imposition of some new condition, e.g., water diuresis. Orloff and Berliner thought this was the case in the more acid urines, and interpreted it as reflecting "production limitation," in contrast to a possible diuresis-induced increase in renal ammonia production. However, urinary ammonia excretion would necessarily increase during water diuresis without any increase in ammonia production because the urine would be favored by an increased flow in its competition with the renal venous blood for the ammonia supply (I-9). Only if the urine were the only channel of ammonia outflow from the kidney could urinary ammonia excretion fail to increase with water diuresis because of limitation of renal ammonia production in this sense.

In the "Summary and Conclusions" of their article, Orloff and Berliner (434) wrote: "In the acid range of urine pH accumulation is limited by the maximal rate of intracellular production in the case of ammonia. . . . Consequently, alterations in urine flow do not affect the rate of excretion. In more alkaline urines, however, urine flow influences excretion rate so that diffusion . . . between cells and tubule urine presumably approaches equilibrium." The second and third sentences quoted are conceptually faulty. The second seems to take "production limitation" in Sense 2, and to presuppose a single outflow channel, the urine, for the renal ammonia supply. The fact that diuresis increases ammonia excretion is inevitable with or without increased renal ammonia production as long as ammonia outflow represents a competition between two outflow channels (II-9B). The third quoted sentence is also incorrect; diuresis-induced increase in urinary ammonia excretion does not in the least imply diffusion equilibrium, but must occur, as long as non-ionic diffusion is granted, whether or not an appreciable cell-to-lumen P_{NH_3} gradient develops.

Pitts, reviewing urinary ammonia excretion in 1964 (465, page 762), wrote as follows: " . . . ammonia in renal tubular cells is essentially in diffusion equilibrium with that in renal venous blood. . . . The addition of ammonia to renal venous blood under usual conditions . . . is therefore flow limited. It may also be concluded that the rate of urinary excretion in acidosis is in large part production-limited, since an increase in apparent production (provision of preformed ammonia in arterial blood) results in increased excretion with only minor increase in cellular P_{NH_3}. This suggests . . . that diffusion equilibrium may exist between tubular cells and tubular urine; at least permeability to free-base ammonia is not the major factor limiting the overall rate of excretion."

The second sentence is tautologous and its inference misleading, since on the premise that ammonia diffuses into the renal venous [better: capillary] blood, whether or not to equilibrium, the rate of addition of ammonia to the renal venous blood must be flow-limited whether or not the blood and tissue are equilibrated with respect to P_{NH_3}. In the third sentence, "production-limited" is used in Sense 1 and indeed is so defined; but urinary ammonia excretion must always necessarily be production-limited in this sense, under any circumstances and not simply in acidosis. The inference drawn in the fourth sentence is unwarranted: increasing the ammonia supply to the kidney must increase urinary ammonia excretion whether or not the P_{NH_3} of tissue and urine are fully equilibrated.

Windhager (658, page 50) summarizes the conclusions of Pitts and his associates as drawn from the data of Denis, Preuss, and Pitts (126) as follows: "Enrichment (Anreicherung) of venous blood with ammonia is not [normally] limited by . . . production rate but only by blood flow. But urinary ammonia excretion in the acidotic dog appears to be production-limited, since raising arterial ammonia leads to increased excretion without essential alteration of cell P_{NH_3}." It is impossible to make sense of this. On the universally granted premise of passive partitioning of the renal ammonia supply between urine and renal venous blood, the rates of ammonia outflow via both channels are always restricted by the ammonia supply, and must (other things being equal) increase as the supply increases.

The idea of "production-limited" urinary ammonia excretion has been a fertile source of semantic confusion and should have no further place in discussions of the subject.

APPENDIX J

KINETIC ANALYSIS OF URINARY AMMONIA EXCRETION AS A FUNCTION OF URINE pH IN DIURETIC MAN

It can hardly be doubted that the P_{NH_3} of urine is equilibrated with that of renal tissue at the papilla, and that the effective renal blood flow is equilibrated with the uniform NH_3 pressure of the renal cortex (II-9C; 11-10; Appendix H). If it is then postulated as an initial oversimplification that the tissue ammonia pressure is uniform throughout the kidney, i.e., that the concentration c of dissolved NH_3 gas in papillary urine remains equal to that in renal venous blood as the pH of high-flow urine is varied, the value of c must obviously decline as the urinary pH is lowered, and c and the rates of ammonia outflow via urine and renal venous blood will be related to each other as continuous functions of the urinary pH by the following expression:

$$S = c\,(k_1 + K_2) \tag{1}$$

where k_3 and K_2 are the velocity constants for the ammonia outflows in urine and renal venous blood, and S the rate at which ammonia is supplied to the system. Capitals indicate invariant values; K_2 and S can be treated as invariant because both the renal blood flow, and the rate at which ammonia is supplied to the system by renal ammonia production plus the arterial inflow, remain constant, to a near approximation, as the urinary pH varies physiologically. It is assumed also for purposes of kinetic analysis that the function of the two healthy kidneys is so similar that they may be treated for present purposes as a functional unit.

S, K_2, and k_1 have been evaluated by Hills and Reid (258) for healthy men. S was originally assigned by them a value of 160 μmoles/minute on the basis of simultaneous measurements carried out by them and by Owen and Robinson (442) of the rate of urinary outflow from both kidneys as the urinary pH of high-flow urine varied in healthy persons. Both pairs of investigators, however, estimated the ammonia content of their blood samples by alkaline distillation of blood filtrates, and it is generally agreed that some artifactual ammonia is included when this method is applied to blood. In our laboratory, estimates of blood ammonia obtained by the method of Nathan and Rodkey (420) appear, on the average, to exceed by 15 to 20% the true value, taken as synonymous with the vacuum distillation methods of Archibald (13) or Preuss, Bise, and Schreiner (480). Recomputation of S from the original data by application of this correction to the values for renal venous blood ammonia outflow yielded a revised estimate of S as 135 μmoles/minute (497).

No significant variation of S has ever been demonstrated in experiments lasting up to 4 to 5 hours in man or other mammalian species as the urinary pH is caused to vary by physiological variation of the acid-base load (II-9). Small amounts of NH_4Cl or $NaHCO_3$, administered orally during and for up to 16 hours before the experiments in order to ensure observation of the full physiological range of urinary pH, can have little effect on S; even the acute production of moderate to severe experimental acidosis or alkalosis affects renal ammonia production only gradually, and under these circumstances 5 to 6 days are required for the maximal increase in ammonia production to occur. It seems clear that S can be considered, practically speaking, to remain invariant for purposes of analyzing kinetically the physiological control of urinary ammonia excretion as urinary pH fluctuates in the normal manner in response to diet and circadian influences.

The values of the first-order outflow velocity constants k_1 and K_2 in Equation 1 are determined entirely by the pH and the flow rate of the papillary urine and the renal venous blood, respectively. The detailed derivation of the values for high-flow human urine, including a small correction for noneffective renal blood flow, are presented elsewhere (258). K_2 has to be reckoned empirically from experimental determination of the ratio of whole blood [ammonia] to plasma [ammonia], for in whole blood, a nonhomogeneous fluid, NH_3 has no pK′. [This fact has not prevented publication of values for it (74) nor textbook citation of these values (28, page 259), symbolized for some reason as pK_{ab}.] Hills and Reid (258) assigned to the blood-to-plasma [ammonia] ratio in normal man the value 1.4 on the basis of their own data and a review of published values. Evidently the value would decline in anemia in approximate proportion to the erythrocyte count or hematocrit. K_2 has been assigned the invariant value 77.0 liter minute^{-1}; k_1 varies during water diuresis from very low values in highest-pH urine to values moderately exceeding K_2 in lowest-pH urine (258).

Figure J-1 compares the relation between urinary ammonia excretion and urine pH experimentally observed in diuretic man (cf. Figure II-9.1) with a theoretical relation excretion reckoned from Equation 1, using the experimentally determined values for S, k_1, and K_2. In this computation no change in urinary pH beyond the papilla has been allowed for; agreement of theory and experiment is unsatisfactory, though there cannot be any substantial error attaching to S and K_2. A nearer

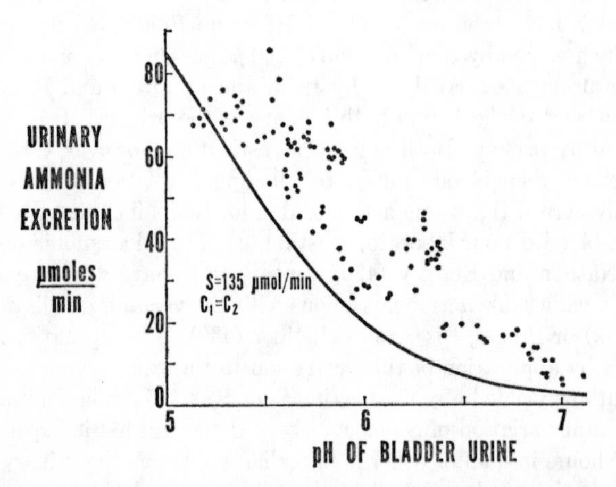

FIG. J-1. Urinary ammonia excretion: experimental values plotted against pH of bladder urine, compared with theoretical function calculated from Equation 1. The figure is identical with Figure II-9.7 except that the theoretical function is not projected into a low range of urinary pH unattainable physiologically.

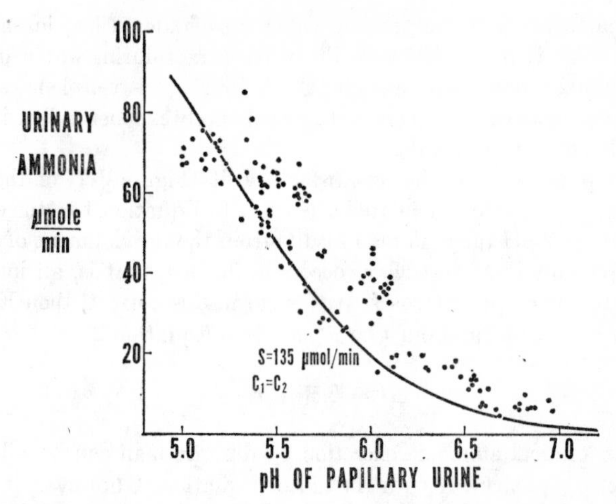

FIG. J-2. Urinary ammonia excretion as a function of estimated papillary urine pH. The pH of papillary urine has been estimated for the experimental samples by deducting the rise in urine pH estimated to occur beyond the papilla from the measured pH of the bladder urine ΔpH. The basis of estimating ΔpH is presented in Appendix K. The theoretical function is naturally unaltered; the effect of the correction is to move most of the experimental values somewhat to the *left*, toward the corresponding theoretical value. ΔpH ranges from 0.3 in highest-pH urine to 0 in lowest-pH urine.

approximation is obtained (Figure J-2) when the theoretical function is computed from the pH of papillary urine, reckond by deducting from the measured pH of the collected urine the estimated rise in pH occurring beyond the papilla as a result of continued dehydration of H_2CO_3 (II-10A). Agreement of experiment and theory is now fair for the most acid urines, but as the papillary pH rises, the theoretical function increasingly underestimates the values found.

Such a discrepancy is to be expected in that it is likely that a papilla-to-cortex NH_3 pressure gradient may develop (II-10A), particularly when higher-pH (ammonia-poor) urine is being secreted and practically no ammonia is being withdrawn into the urine from the region of the hairpin bends of Henle's loop and the vasa recta (261). Such a gradient would of course have the effects of raising the urinary ammonia outflow at any given value for S and for urine pH, and of decreasing to the same extent the renal blood ammonia outflow. And in fact a papilla-to-cortex NH_3 pressure gradient is observed at higher urinary pH in diuretic dogs infused with carbonic anhydrase intravenously, an experimental procedure which can have little effect upon the system governing the distribution of the ammonia outflow between urine and renal venous blood (258).

In normal man (258) as in the dog (261) papillary P_{NH_3} approximates cortical P_{NH_3} in the most acid high-flow urines, whose pH is not affected beyond the papilla by delayed dehydration of H_2CO_3. Figure II-10.3 shows the P_{NH_3} of renal venous blood and urine observed in three greyhounds during intravenous carbonic anhydrase infusion, plotted as functions of the pH of high-flow bladder urine; at high urine flows P_{NH_3} and pH of papillary urine are synonymous in the enzyme-infused animal with the corresponding values in bladder urine, inasmuch as CO_2 and NH_3 are scarcely reabsorbed from the urinary tract during water diuresis (261). Urinary and renal venous P_{NH_3}, and also the difference between them, which reflects the papilla-to-cortex gradient, increased approximately exponentially with increasing urinary pH in these experiments.

It appears probable that the unsatisfactory fit (Figure J-2) of the theoretical urinary ammonia excretion calculated according to Equation 1 to the experimental data for the higher-pH range in man results from the development of a papilla-to-cortex NH_3 pressure gradient such as occurs in the dog, that is, an increasing c_1-c_2 ratio as the papillary pH increases. If this surmise is correct, then Equation 1 is not precisely applicable, and must be replaced by Equation 2:

$$S = k_1c_1 + K_2c_2 \tag{2}$$

Before the theoretical ammonia excretion by diuretic man can be calculated as a function of the papillary pH with the aid of Equation 2, however, it is necessary that a value be assigned to the ratio $c_1:c_2$ as a function of pH of papillary urine. The only currently available data on this function are those of Hills and Reid in the dog (Figure II-10.3). On the assumption that c_1/c_2 does rise exponentially as the papillary pH rises, in man as in the dog, from a value of unity at pH 5.0, Equation 2 can be implemented by expressing c_1 in Equation 2 as a function of c_2, $f(c_2)$:

$$S = k_1 f(c_2) + K_2c_2 \tag{3}$$

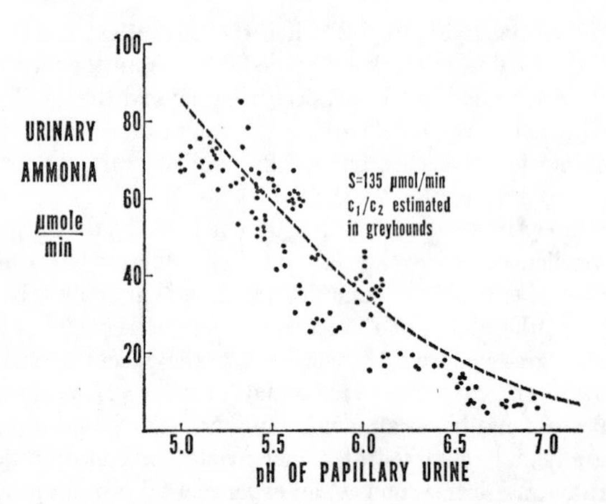

FIG. J-3. Urinary ammonia excretion as a function of papillary pH in diuretic man. The experimental data are plotted as in Figure J-2; the theoretical function is calculated from Equation 3 on the assumption that a papillary-to-cortex NH_3 pressure gradient develops progressively as an experimental function of urinary pH, reaching a value of 6:1 when highest-pH urine is being secreted. This estimated ratio of 6 is the smallest value which can reasonably be obtained from lines hand-drawn to fit the data presented for renal venous and urinary P_{NH_3} in greyhounds being infused with carbonic anhydrase (261).

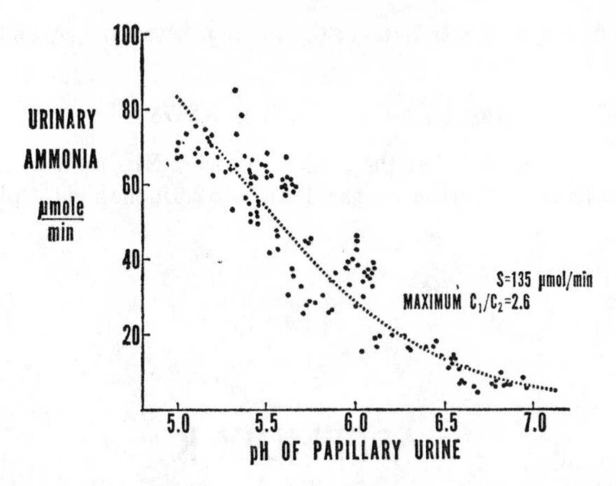

FIG. J-4. Urinary ammonia excretion as a function of papillary pH in diuretic man. Experimental data plotted as in Figures J-2 and J-3. The theoretical function was calculated according to Equation 3 and was selected for best fit to the data from a family of curves obtained by substituting various values from 2 to 4 for c_1/c_2 in highest-pH urine. In each curve of the family it was assumed that $c_1 = c_2$ at pH 5.0 and that the ratio increased exponentially with rising urine pH.

However, estimation of c_1/c_2 as an exponential function of urine pH based upon Figure II-10.3 leads to overcorrection (Figure J-3) of the discrepancy between theory and observation seen in Figure J-2.

To account for this overcorrection, it seems unnecessary to assume that a larger papilla-to-cortex NH_3 pressure gradient develops in the dog than in man at high flow of high-pH urine, though it may be so. Rather, it seems probable that the data obtained by Hills and Reid in carbonic anhydrase-infused greyhounds (261) overestimate to some extent the papilla-to-cortex NH_3 pressure gradient which would be associated with brisk water diuresis in this animal. It is impossible to obtain maximal water diuresis in normal dogs in experiments of the type represented in Figure II-10.3 because of the release of endogenous vasopressin provoked by anesthesia and surgery; these animals received some mannitol in their infusions to keep the urine flow rate at levels corresponding to brisk water diuresis, but vasopressin secretion cannot be supposed to have been inhibited to the same extent as in the conscious and cooperative human volunteers during maximal water diuresis. Consequently, the enhancing effects of this hormone upon the efficiency of the countercurrent system (609, 621) were probably a factor increasing c_1/c_2 in these experiments.

Figure J-4 reverses the approach of Figure J-3 and evaluates the papilla-to-cortex gradient in man secreting high-flow, high-pH urine by finding the papilla-to-cortex NH_3 pressure gradient at high flow which gives the best fit of theory and experiment. It selects that theoretical line given by Equation 3, among those obtainable by assuming that $c_1{:}c_2$ increases exponentially from a value of unity at a papillary pH of 5.0 to some unspecified higher value at pH 7.0, which best fits the data. This value of c_1/c_2 for highest-pH, high-flow urine turns out to be 2.6. Agreement of theory and experiment are very satisfactory; the relation between c_1/c_2 and papillary pH upon which it is based is:

$$\log (c_1/c_2) = -1.261 + 0.2473 \text{ pH} \quad (4)$$

It is accordingly suggested that the papilla-to-cortex NH_3 pressure gradient varies n diuretic man from unity in the most acid urines to 2.6 in the highest-pH urines.

APPENDIX K

ESTIMATION OF POSTPAPILLARY RISE IN URINE pH (ΔpH)

Continuing dehydration of H_2CO_3 beyond the papilla causes rise in both urinary P_{CO_2} and pH as CO_2 is generated by reaction of nonvolatile buffer and HCO_3^- (I-6C; II-10B); and the relative magnitudes of the changes in P_{CO_2} and pH are determined by differences in the nonvolatile buffer strength. Estimates of the pH of high-flow papillary urines used in constructing Figure II-9.9 and Figures J-2 to J-4 were derived by subtracting estimated postpapillary Δ pH from the measured pH of the bladder urines; Δ pH was reckoned as a function of urine pH from the mean nonvolatile buffer strength of urines and from estimations of the rise in P_{CO_2} in high-flow human urine beyond the papilla.

Figure K-1 shows our best estimate of the P_{CO_2} of bladder urine and papillary urine as functions of the pH of high-flow human bladder urine; the difference between the two functions gives postpapillary P_{CO_2} as a function of the pH of the bladder urine. The equation for the difference in elevation of the two lines is:

$$P_{CO_2} \text{ (mm Hg)} = -10.77 + 3.28 \text{ pH} \quad (1)$$

The line representing bladder urine P_{CO_2} as a function of bladder urine pH is the regression best fitting the paired values for P_{CO_2} and pH measured in 104 samples of the urine of normal volunteers during water diuresis; all flows were 10 ml/minute or greater. The value of papillary P_{CO_2} is inferential in man, since intravenous infusion of carbonic anhydrase in human subjects has not been thought a licit procedure; moreover, it is likely that papillary P_{CO_2} may be somewhat lower than normal when carbonic anhydrase is being infused, since then less CO_2 must be reabsorbed from the medullary collecting duct because H_2CO_3 newly synthesized in the distal tubules and cortical collecting ducts is there dehydrated and the CO_2 is locally reabsorbed (261).

The *lower line* of Figure K-1 was accordingly derived by connecting by a straight line an estimate of papillary P_{CO_2} in normal man at lowest-pH and at highest-pH, high-flow urine. When the most acid urine is being secreted, urinary P_{CO_2} is raised

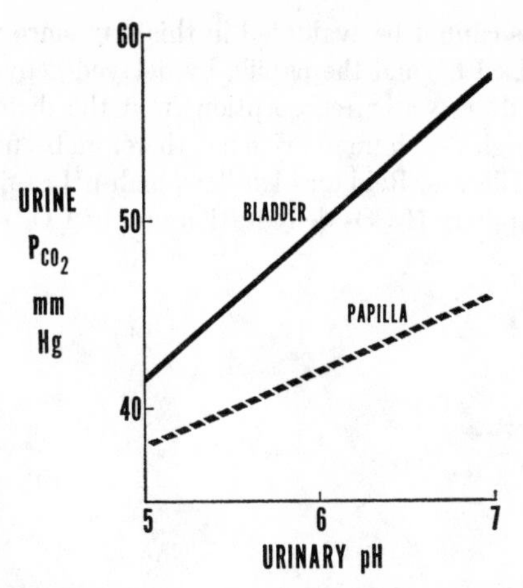

FIG. K-1. P_{CO_2} of bladder urine and estimated P_{CO_2} of papillary urine as functions of the pH of bladder urine in man during water diuresis. See text for basis of estimation of papillary P_{CO_2}.

by water diuresis from minimal values in the isosmolar range of urine flow; this rise appears to be due largely if not entirely to postpapillary dehydration of H_2CO_3 (261). Papillary P_{CO_2} in the most acid urines has accordingly been reckoned as the average P_{CO_2} at urine flows between 3 and 5 ml/minute in four experiments in volunteers secreting the most acid urines (Figure K-2). The value of papillary P_{CO_2} in

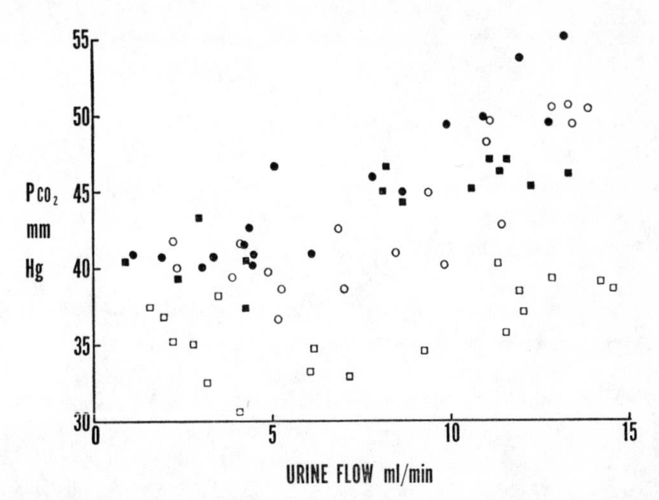

FIG. K-2. Effect of water diuresis on P_{CO_2} of very acid urine. Four experiments are represented, each type of *symbol* (*open circles*, *open squares*, *closed circles*, *closed squares*) referring to the data from a single experiment. Results of three of the experiments have been published previously; they were the three in the lowest pH range of 15 such experiments relating urinary P_{CO_2} to urine flow rate of healthy volunteers. *Open squares* represent values obtained in a similar experiment in a volunteer with borderline nonrespiratory acidosis produced by prior administration of NH_4Cl.

the highest-pH urines cannot be evaluated in this way since the P_{CO_2} of isosmotic urine is markedly raised beyond the papilla by delayed dehydration of the H_2CO_3 synthesized as a result of water reabsorption from the distal nephron. Papillary P_{CO_2} in highest-pH, high-flow human urine has therefore been assigned the value 47 mm, as recorded by Hills and Reid for high-flow, high-pH canine urines on the basis of estimates of postpapillary H_2CO_3 dehydration and of CO_2 reabsorption from the urinary tract (261).

REFERENCES

1. Addae, S. K., and Lotspeich, W. D.: Relation between glutamine utilization and production in metabolic acidosis. Am. J. Physiol. *215:* 269–277, 1968.
2. Addae, S. K., and Lotspeich, W. D.: Glutamine balance in metabolic acidosis as studied with the artificial kidney. Am. J. Physiol. *215:* 278–281, 1968.
3. Adler, S.: The simultaneous determination of muscle cell pH using a weak acid and weak base. J. Clin. Invest. *51:* 256–265, 1972.
4. Albert, M. S., Dell, R. B., and Winters, R. W.: Quantitative displacement of acid-base equilibrium in metabolic acidosis. Ann. Intern. Med. *66:* 312–322, 1967.
5. Albright, F., Bauer, W., Ropes, M., and Aub, J. C.: Studies of calcium and phosphorus metabolism. IV. The effect of the parathyroid hormone. J. Clin. Invest. *7:* 139–181, 1929.
6. Albright, F., Burnett, C. H., Parson W., Reifenstein, E. C., Jr., and Roos, A.: Osteomalacia and late rickets. The various etiologies met in the United States with emphasis on that resulting from a specific form of renal acidosis, the therapeutic indications for each etiological sub-group, and the relationship between osteomalacia and Milkman's syndrome. Medicine (Baltimore) *25:* 399–479, 1946.
7. Albright, F., and Reifenstein, E. C.: *The Parathyroid Glands and Metabolic Bone Disease: Selected Studies.* Williams & Wilkins Co., Baltimore, 1948.
8. Alving, A. S., and Gordon, W.: Studies of urea, creatinine, and ammonia excretion in dogs in acidosis. J. Biol. Chem. *120:* 103–113, 1937.
9. Anderson, H. M., and Mudge, G. H.: The effect of potassium on intracellular bicarbonate in slices of kidney cortex. J. Clin. Invest. *34:* 1691–1696, 1955.

9a. Anrep, G. V., and Cannan, R. K.: The concentration of lactic acid in the blood in experimental alkalaemia and acidaemia. J. Physiol. (London) *58:* 244–258, 1923.

The order in which the references are listed at any location in the text corresponds usually to the chronological order of the appearance of the original work cited. In the case of reviews the listing may be in order of usefulness to the general reader.

10. Araki, T.: Über die Bildung von Milchsäure und Glycose im Organismus bei Sauerstoffmangel. Z. Physiol. Chem. *15:* 335–370, 1891.
11. Araki, T.: Über die Bildung von Milchsäure und Glycose im Organismus bei Sauerstoffmangel. Zweite Mittheilung. Über die Wirkung von Morphium, Amylnitrit, Coca'n. Z. Physiol. Chem. *15:* 546–561, 1891.
12. Araki, T.: Über die chemischen Änderungen der Lebensprocesse in Folge von Sauerstoffmangel. Z. Physiol. Chem. *19:* 422–475, 1894.

12a. Arbus, G. S., Hebert, L. A., Levesque, P. R., Ersten, B. E., and Schwartz, W. B.: Characterization and clinical application of the "significance band" for acute respiratory alkalosis. New Eng. J. Med. *280:* 117–123, 1969.

13. Archibald, R. M.: Quantitative microdetermination of ammonia in the presence of glutamine and other labile substances. J. Biol. Chem. *151:* 141–148, 1943.
14. Army Malaria Research Unit: Factors affecting the excretion of Mepacrine in the urine. Ann. Trop. Med. Parasitol. *39:* 53–60, 1945.
15. Arrhenius, S.: Über die Dissociation der in Wasser gelösten Stoffe. Z. Physik. Chem. *1:* 631–648, 1887.
16. Arrhenius, S.: The theory of electrolytic dissociation (Faraday Lecture). J. Chem. Soc. *105:* 1414–1426, 1917.
17. Arrhenius, S. A.: *Theories of Solutions.* Yale Univsity Press, New Haven, Conn., 1913.
18. Astrup, P., Engel, K., Jørgensen, K., and Siggaard-Andersen, O.: Definitions and terminology in blood acid-base chemistry. In *Current Concepts of Acid-Base Measurement,* pp. 59–65. Edited by G. G. Nahas. Ann. N.Y. Acad. Sci. *133:* 1–274, 1966.
19. Atkins, E. L., and Schwartz, W. B.: Factors governing correction of the alkalosis associated with potassium deficiency: the critical role of chloride in the recovery process. J. Clin. Invest. *41:* 218–229, 1962.
20. Atsmon, A., deVries, A., and Frank, M.: *Uric Acid Lithiasis.* Elsevier, Amsterdam, 1963.

21. Atsmon, A., Frank, M., Lazebnik, J., Kochwa, S., and deVries, A.: Uric acid stones: a study of 58 patients. J. Urol. *84:* 167–176, 1960.
22. Auerbach, A.: Über die Säurewirkung der Fleischnahrung. Virchows Arch. *98:* 512–526, 1884.
23. Austin, W. H., Lacombe, E., Rand, P. W., and Chatterjee, M.: Solubility of carbon dioxide in serum from 15 to 38° C. J. Appl. Physiol. *18:* 301–304, 1963.
24. Badenoch, A. W.: Uric acid stone formation. Br. J. Urol. *32:* 374–382, 1960.
25. Baines, A. D.: Effect of extracellular fluid volume expansion on maximum glucose reabsorption rate and glomerular tubular balance in single rat nephrons. J. Clin. Invest. *50:* 2414–2425, 1971.
26. Baines, G. H., Barclay, J. A., and Cooke, W. T.: Nephrocalcinosis associated with hyperchloraemia and low plasma-bicarbonate. Q. J. Med. *14–15(NS):* 113–122, 1945–46.
27. Balagura, S., and Pitts, R. F.: Excretion of ammonia injected into renal artery. Am. J. Physiol. *203:* 11–14, 1962.
28. Balagura-Baruch, S.: Renal metabolism and transfer of ammonia. In *The Kidney: Morphology, Biochemistry, Physiology*, Vol. 3, Ch. 5, pp. 253–327. Edited by C. Rouiller and A. F. Muller. Academic Press, New York, 1971.
28a. Baldamus, C. A., and Uhlich, E.: Countercurrent considerations relating to renal medullary P_{CO_2} and bicarbonate concentrations. In *Renal Transport and Diuretics*, pp. 187–193. Edited by K. Thurau and H. Jahrmfirker. International Symposium, Feldäfing, June 21–23, 1968. Springer-Verlag, Berlin, 1969.
28b. Baldwin, E.: *Dynamic Aspects of Biochemistry*, Ed. 4. Cambridge University Press, New York, 1963.
29. Bang, I.: Untersuchungen über den Reststickstoff des Blutes. I. Mittheilung. Biochem. Z. *72:* 104–118, 1916.
30. Bank, N., and Schwartz, W. B.: Influence of certain urinary solutes on acidic dissociation constants of ammonium ion at 37° C. J. Appl. Physiol. *15:* 125–127, 1960.
31. Bank, N., and Schwartz, W. B.: The influences of ion penetrating ability on urinary acidification and the excretion of titratable acid. J. Clin. Invest. *39:* 1516–1525, 1960.
32. Barclay, J. A., Cooke, W. T., Kenney, R. A., and Nutt, M. E.: The effects of water diuresis and exercise on the volume and composition of the urine. Am. J. Physiol. *148:* 327–337, 1947.
33. Barker, E. S., Singer, R. B., Elkinton, J. R., and Clark, J. K.: The renal response in man to acute experimental respiratory acidosis and alkalosis. J. Clin. Invest. *36:* 515–529, 1957.
33a. Barnett, G. D., and Addis, T.: Urea as a source of blood ammonia. J. Biol. Chem. *30:* 41–46, 1917.
34. Bartels, H., and Wrbitzky, R.: Bestimmung des CO_2-Absorptionskoeffizienten zwischen 15 and 38° C in Wasser und Plasma. Pfluegers Arch. Gesamte Physiol. *271:* 162–168, 1960.
35. Bartter, F. C., and Fourman, P.: The different effects of aldosterone-like steroids and hydrocortisone-like steroids on urinary excretion of potassium and acid. Metabolism *11:* 6–20, 1962.
36. Barzel, U. S., Sperling, O., Frank, M., and deVries, A.: Renal ammonium excretion and urinary pH in idiopathic uric acid lithiasis. J. Urol. *92:* 1–5, 1964.
37. Bates, D. V., Mackiem, P. T., and Christie, R. V.: *Respiratory Function in Disease.* W. B. Saunders Co., Philadelphia, 1971.
38. Bates, R. G.: *Determination of pH: Theory and Practice*, Ch. 7. John Wiley and Sons, New York, 1965.
39. Bates, R. G.: *Determination of pH: Theory and Practice*, Ch. 1–4. John Wiley and Sons, New York, 1965.
40. Bates, R. G.: Acids, bases, and buffers. In *Current Concepts of Acid-Base Measurement*, pp. 25–33. Edited by G. G. Nahas. Ann. N.Y. Acad. Sci. *133:* 1–274, 1966.
41. Bates, R. G., and Pinching, G. D.: Acidic dissociation constant of ammonium ion at 0° to 50° C, and the base strength of ammonia. J. Res. Natl. Bur. Stand. *42:* 419–430, 1949.
42. Bazett, H. C., Thurlow, S., Crowell, C., and Stewart, W.: Studies on the effects of baths on man. II. The diuresis caused by warm baths, together with some observations on urinary tides. Am. J. Physiol. *70:* 430–452, 1924.
43. Becquerel, A., and Rodier, A.: *Traité de Chimie Pathologique Appliquée a la Médicine Pratique.* Germer-Baillière, Paris, 1854.
44. Bell, R. P.: *The Proton in Chemistry.* Cornell University Press, Ithaca, N. Y., 1959.
45. Benedict, F. G.: A study of prolonged fasting. Publication 203, Carnegie Institution, Washington, D. C., 1915.
46. Bennett, C. M., Brenner, B. M., and Berliner, R. W.: Micropuncture study of nephron function in the rhesus monkey. J. Clin. Invest. *47:* 203–216, 1968.
47. Bennett, T. P., and Frieden, E.: Metamorphosis and biochemical adaptation in amphibia. In *Comparative Biochemistry: A Comprehensive Treatment*, Vol. 4B, Ch. 11. Edited by M. Florkin and H. S. Mason. Academic Press, New York, 1962.
47a. Bergstrom, W. H., and Wallace, W. M.: Bone as a sodium and potassium reservoir. J. Clin. Invest. *33:* 867–873, 1954.
48. Berkman, P. M., van Ypersele de Strihou, C., Needle, M. A., Gulyassy, P. F., and

Schwartz, W. B.: Factors which determine whether infusion of the sodium salt of an anion will induce metabolic alkalosis in dogs. Clin. Sci. *33:* 517–525, 1967.
49. Berl, S., Takagaki, G., Clarke, D. D., and Waelsch, H.: Metabolic compartments *in vivo*. Ammonia and glutamic acid metabolism in brain and liver. J. Biol. Chem. *237:* 2562–2569, 1962.
50. Berliner, R. W.: Renal secretion of potassium and hydrogen ions. Fed. Proc. *11:* 695–700, 1952.
51. Berliner, R. W.: Outline of renal physiology. In *Diseases of the Kidney*, Ed. 1, Ch. 2, pp. 30–79. Edited by M. B. Strauss and L. G. Welt. Little, Brown and Co., Boston, 1963.
52. Berliner, R. W.: Outline of renal physiology. In *Diseases of the Kidney*, Ed. 2, Vol. 1, Ch. 2, pp. 31–85, Edited by M. B. Strauss and L. G. Welt. Little, Brown and Co., Boston, 1971.
53. Berliner, R. W., Kennedy, T. J., Jr., and Hilton, J. G.: Renal mechanisms of excretion of potassium. Am. J. Physiol. *162:* 348–367, 1950.
54. Berliner, R. W., Kennedy, T., and Orloff, J.: Relationship between acidification of the urine and potassium metabolism. Am. J. Med. *11:* 274–282, 1951.
55. Berliner, R. W., and Orloff, J.: Carbonic anhydrase inhibitors. Pharmacol. Rev. *8:* 137–174, 1956.
56. Berlyne, G. M.: Distal tubular function in chronic hydronephrosis. Q. J. Med. *30(NS):* 339–355, 1961.
57. Bernal, J. D., and Fowler, R. H.: A theory of water and ionic solution with particular reference to hydrogen and hydroxyl ions. J. Chem. Phys. *1:* 515–548, 1933.
58. Bernard, C.: *Leçons sur la Chaleur Animale, sur les Effets de la Chaleur, et sur la Fièvre.* J. B. Ballière et Fils, Paris, 1876.
59. Bernard, C.: *Leçons sur les Phénomenès de le Vie Communs aux Animaux et aux Végétaux*, Vol. 1. J. B. Balliere et Fils, Paris, 1878.
60. Bernstein, B. A., and Clapp, J. R.: Micropuncture study of bicarbonate reabsorption by the dog nephron. Am. J. Physiol *214:* 251–257, 1968.
61. Berthelot, P. E. M., and St.-Gilles, L. P. de: Recherches sur les affinités. Ann. Chim. Phys. (3e Ser.) *65:* 385–422, 1862; *66:* 5–110, 1862; *68:* 225–359, 1863. Ostwald's Klassiker no. 173.
62. Bessman, S. P., and Bradley, J. E.: Uptake of ammonia by muscle; its implications in ammoniagenic coma. New Eng. J. Med. *253:* 1143–1147, 1955.
63. Biber, T. U. L., Mylle, M., Baines, A. D., and Gottschalk, C. W.: A study of micropuncture and microdissection of acute renal damage in rats. Am. J. Med. *44:* 664–705, 1968.
64. Black, D. A. K., and Milne, M. D.: Experimental potassium depletion in man. Clin. Sci. *11:* 397–415, 1952.
65. Bland, J. H.: *Clinical Metabolism of Body Water and Electrolytes.* W. B. Saunders Co., Philadelphia, 1963.
66. Bloch, W.: Gay-Lussac und Thenard. In *Das Buch der Grossen Chemiker*, Vol. 1, pp. 386–404. Edited by G. Bunge. Verlag-Chemie, Weinheim/Bergstrasse, 1929; reprinted 1965.
67. Bloomer, H. A., Rector, F. C., Jr., and Seldin, D. W.: The mechanism of potassium reabsorption in the proximal tubule of the rat. J. Clin. Invest. *42:* 277–285, 1963.
67a. Bock, A. V., Dill, D. B., and Edwards, H. T.: Lactic acid in the blood of resting man. J. Clin. Invest. *11:* 775–788, 1932.
68. Bockris, J. O'M., and Reddy, A. K. N.: *Modern Electrochemistry: An Introduction to an Intradisciplinary Area.* Plenum Press, New York, 1970.
68a. Brazeau, P., and Gilman, A.: Effect of plasma CO_2 tension on renal tubular reabsorption of bicarbonate. Am. J. Physiol *175:* 33–38, 1953.
69. Bricker, N. S., Klahr, S., Lubowitz, H., and Rieselbach, R. E.: Renal function in chronic renal disease. Medicine (Baltimore) *44:* 263–288, 1965.
70. Bricker, N. S., Morrin, P. A. F., and Kime, S. W., Jr.: The pathologic physiology of chronic Bright's disease. Am. J. Med. *28:* 77–98, 1960.
71. Briggs, A. P.: The acidosis of nephritis. Arch. Intern. Med. *49:* 56–76, 1932.
72. Briggs, A. P.: The significance of the urinary ammonia. J. Lab. Clin. Med. *28:* 174–179, 1942.
73. Broch, O. J.: Low potassium alkalosis with acid urine in ulcerative colitis. Scand. J. Clin. Lab. Invest. *2:* 113–119, 1950.
74. Bromberg, P. A., Robin, E. D., and Forkner, C. E., Jr.: The existence of ammonia in blood *in vivo* with observations on the significance of the NH_4^+-NH_3 system. J. Clin. Invest. *39:* 332–341, 1960.
75. Brönsted, J. N.: Einige Bemerkungen über den Begriff der Säuren und Basen. Réceuil Travaux Chim. Pays-Bas *42:* 718–728, 1923.
76. Brønsted, J. N.: Acid and basic catalysis. Chem. Rev. *5:* 231–338, 1928.
77. Brown, E. B., Jr.: Physiological effects of hyperventilation. Physiol. Rev. *33:* 445–471, 1953.
78. Brown, F. A., Jr., Hastings, J. W., and Palmers, J. D.: *The Biological Clock: Two Views.* Academic Press, New York, 1970.
79. Brunton, C. E.: The acid output of the kidney and the so-called alkaline tide. Physiol. Rev. *13:* 372–399, 1933.
79a. Buckalew, V. M., McCurdy, D. A., Ludwig, G. D., Chaykin, L. B.; and Elkinton, J. R.: Incomplete renal tubular acidosis. Physi-

ological studies in three patients with a defect in lowering urine pH. Am. J. Med. *45:* 32–42, 1968.

80. Bunge, G.: *Textbook of Physiological and Pathological Chemistry*, 2nd English Ed. Translated from the 4th German edition by F. A. Starling and E. H. Starling. P. Blakiston's Son and Co., Philadelphia, 1902.
81. Bünning, E.: *The Physiological Clock*, Ed. 2, revised. Springer-Verlag, Berlin, 1967.

81a. Burg, M.: Function of the thick ascending limb of Henle's loop. Abstracts of Plenary Sessions and Symposia, Fifth International Congress of Nephrology, Mexico City, 1972, p. 45.

82. Burnell, J. M.: Changes in bone sodium and carbonate in metabolic acidosis and alkalosis in the dog. J. Clin. Invest. *50:* 327–331, 1971.
83. Burnett, C. H., Burrows, B. A., and Commons, R. R.: Studies of alkalosis. I. Renal function during and following alkalosis resulting from pyloric obstruction. J. Clin. Invest. *29:* 169–174, 1950.
84. Butler, A. M., Wilson, J. L., and Farber, S.: Dehydration and acidosis with calcification of renal tubules. J. Pediatr. *8:* 489–499, 1936.
85. Cahill, G. F., Jr., and Owen, O. E.: The role of the kidney in the regulation of protein metabolism. In *Mammalian Protein Metabolism*, Vol. 4, Ch. 39. Edited by H. N. Munro. Academic Press, New York, 1970.

85a. Caldwell, P. C.: Intracellular pH. Int. Rev. Cytol. *5:* 229–277, 1956.

86. Camerer, W., Jr.: Beobachtungen und Versuche über die Ammoniakausscheidung im menschlichen Urin, mit Berücksichtigung nach weiterer stickstoffhaltiger Urinbestandteile und Bestimmung der Acidität nach Lieblein. Z. Biol. *43:* 14–45, 1902.
87. Camien, M. N., and Gonick, H. C.: Relationship of renal "net acid" excretion to titratable ash-acidity (Ash-TA) in diet and feces. Proc. Soc. Exp. Biol. Med. *126:* 45–51, 1967.
88. Camien, M. N., and Reilly, T. J.: Determination of titratable ash-acidity (ash-TA). Proc. Soc. Exp. Biol. Med. *126:* 51–55 1967.
89. Camien, M. H., Simmons, D. H., and Gonick, H. C.: A critical reappraisal of "acid-base" balance. Am. J. Clin. Nutr. *22:* 786–793, 1969.
90. Camien, M. N., Smith, L. M., Reilly, T. J., and Simmons, D. H.: Determination of total cation-forming mineral elements in feces and urine and its relation to renal "net acid" excretion. Proc. Soc. Exp. Biol. Med. *123:* 686–691, 1966.
91. Campbell, J. A., and Webster, T. A.: Day and night urine during complete rest, laboratory routine, light muscular work and oxygen administration. Biochem. J. *15:* 660–664, 1921.
92. Cannon, P. J., Heinemann, H. O., Albert, M. S., Laragh, J. H., and Winters, E. W.: "Contraction" alkalosis after diuresis of edematous patients with ethacrynic acid. Ann. Intern. Med. *62:* 979–990, 1965.
93. Carlsten, A., Hallgren, B., Jagenburg, R., Svanborg, A., and Werkö, L.: Arteriohepatic venous differences of free fatty acids and amino acids. Studies in patients with diabetes or essential hypercholesterolemia, and in healthy individuals. Acta Med. Scand. *181:* 199–207, 1967.
94. Carraway, W. T.: Determination of uric acid in serum by a carbonate method. Am. J. Clin. Pathol. *25:* 840–845, 1955.

94a. Carter, N.: Intracellular pH. Kidney Int. *1:* 341–346, 1972.

94b. Christensen, H. N.: *Body Fluids and Their Neutrality*. Oxford University Press, New York, 1963.

95. Christensen, H. N.: *Body Fluids and the Acid-Base Balance*. W. B. Saunders Co., Philadelphia, 1965.
96. Christensen, H. N.: In *Current Concepts of Acid-Base Measurement*, p. 97. Edited by G. G. Nahas. Ann. N. Y. Acad. Sci. *133:* 1–274, 1966.

96a. Churchill, P. C., and Malvin, R. L.: Relation of renal gluconeogenesis to ammonia production in the dog. Am. J. Physiol. *218:* 241–245, 1970.

97. Clapp, J. R.: The effect of alterations of acid-base balance on bicarbonate reabsorption by the dog proximal tubule. Clin. Res. *13:* 302, 1965.
98. Clapp, J. R., Owen, E. E., and Robinson, R. R.: Contribution of the proximal tubule to urinary ammonia excretion. Am. J. Physiol. *209:* 269–272, 1965.
99. Clapp, J. R., Watson, J. F., and Berliner, R. W.: Osmolality, bicarbonate concentration, and water reabsorption in proximal tubule of dog nephron. Am. J. Physiol. *205:* 273–280, 1963.
100. Clare, W.: Experimenta de excretione acidi sulfurici per urinam. Dissertation. Dorpat, 1854.
101. Clark, G. M., and Eisenman, B.: Studies in ammonia metabolism. IV. Biochemical changes in brain tissue of dogs during ammonia-induced coma. New Eng. J. Med. *259:* 178–180, 1958.
102. Clark, W. M.: *The Determination of Hydrogen Ions*, Ed. 1. Williams & Wilkins Co., Baltimore, 1920.
103. Clark, W. M.: *The Determination of Hydrogen Ions*, Ed. 2, Williams & Wilkins Co., Baltimore, 1922.
104. Clark, W. M.: *Topics in Physical Chemistry: A Supplementary Text for Students of Medicine*. Williams & Wilkins Co., Baltimore, 1948.

105. Clarke, E., Evans, B. M., MacIntyre, I., and Milne, M. D.: Acidosis in experimental electrolyte depletion. Clin. Sci. *14:* 421–440, 1955.
106. Cochran, R. T., Jr.: Pulmonary insufficiency and hypercapnia complicated by potassium-responsive alkalosis. New Eng. J. Med. *268:* 521–525, 1963.
107. Cohen, E.: J. H. van't Hoff: In *Das Buch der Grossen Chemiker*, Vol. 2, pp. 391–407. Verlag-Chemie, Weinheim/Bergstrasse, 1929; reprinted 1965.
108. Cohen, J. J.: Selective Cl retention in repair of metabolic alkalosis without increasing filtered load. Am. J. Physiol. *218:* 165–170, 1970.
109. Comroe, J. H.: *Physiology of Respiration: An Introductory Text*. Year Book Medical Publishers, Chicago, 1965.
110. Conn, J. W., Rovner, D. R., and Cohen, E. L.: Licorice-induced pseudoaldosteronism. J.A.M.A. *205:* 492–496, 1968.
111. Cooke, R. E., Segar, W. E., Cheek, D. B., Coville, F. E., and Darrow, D. C.: The extrarenal correction of alkalosis associated with potassium deficiency. J. Clin. Invest. *31:* 798–805, 1952.
112. Coranda: Über das Verhalten des Ammoniaks im menschlichen Organismus. Arch. Exp. Pathol. Pharmakol. *80:* 76–96, 1879.
113. Cullen, G. E.: Studies of acidosis. III. The electrometric titration of plasma as a measure of its alkaline reserve. J. Biol. Chem *30:* 369–388, 1917.
114. Cullen G. E.: The pK' of the Henderson-Hasselbalch equation for hydrion concentration of serum. J. Biol. Chem. *66:* 301–322, 1925.
115. Damian, A. C., and Pitts, R. F.: Rates of glutaminase I and glutamine synthetase reactions in rat kidney *in vivo*. Am. J. Physiol. *218:* 1249–1255, 1970.
116. Darrow, D. C.: Changes in muscle composition in alkalosis. J. Clin. Invest. *25:* 324–330, 1946.
117. Darrow, D. C., Cooke, R. E., and Coville, F. E.: Kidney electrolyte in rats with alkalosis associated with potassium deficiency. Am. J. Physiol. *172:* 55–59, 1953.
118. Darrow, D. C., Schwartz, R., Iannucci, J. F., and Coville, F.: The relation of serum bicarbonate concentration to muscle composition. J. Clin. Invest. *27:* 198–208, 1948.
118a. Dastur, D. K., Seshadri, R., and Talageri, V. R.: Liver-brain relationships in hepatic coma with special reference to ammonia and keto acid metabolism. Arch. Intern. Med. *112:* 899–916, 1963.
118b. Davenport, H. W.: *The A B C of Acid-Base Chemistry*, Ed. 4. University of Chicago Press, Chicago, 1958.
119. Davies, H. E. F.: Rise in urine pH and in ammonium excretion during a water diuresis. J. Physiol. (London) *194:* 79–80P, 1968.
120. Davies, H. W., Haldane, J. S., and Kennaway, E. L.: Experiments on the regulation of the blood's alkalinity. I. J. Physiol (London) *54:* 32–45, 1920.
121. Davies, R. E.: Hydrochloric acid production by isolated gastric mucosa. Biochem. J. *42:* 609–621, 1948.
122. Davis, K. S., and Day, J. A.: *Water, the Mirror of Science*. Doubleday and Co., Garden City, N. Y., 1961.
123. Davis, R. P.: Logland: a Gibbsian view of acid-base balance (Editorial). Am. J. Med. *42:* 159–162, 1967.
123a. Debye, P., and Hückel, E.: Zur Theorie der Electrolyte. I. Gefrierpunktsniedrigung und verwandte Erscheinungen. Physik. Z. *24:* 185–206, 1923.
124. Deck, M. D. F.: Medullary sponge kidney with renal tubular acidosis: a report of 3 cases. J. Urol. *94:* 330–335, 1965.
125. Deetjen, P.: Tubulärer Transport schwacher organischer Säuren und Basen. In *Normale und Pathologische Funktionen des Nierentubulus*, pp. 19–36. Edited by K. J. Ullrich and K. Hierholzer. Dritter Symposion der Gesellschaft fûr Nephrologie, September 11–13, 1964, Berlin. Hans Huber, Bern and Stuttgart, 1965.
126. Denis, G., Preuss, H., and Pitts, R.: The P_{NH_3} of renal tubular cells. J. Clin. Invest. *43:* 571–582, 1964.
127. Dent, C. E., Harper, C. M., and Philpot, G. R.: The treatment of renal-glomerular osteodystrophy. Q. J. Med. *30(NS):* 1–31, 1961.
128. Dent, C. E., and Senior, B.: Studies on the treatment of cystinuria. Br. J. Urol. *27:* 317–332, 1955.
129. Denton, D. A., Maxwell, M., McDonald, I. R., Munro, J., and Williams, W.: Renal regulation of the extracellular fluid in acute respiratory acidaemia. Aust. J. Exp. Biol. Med. Sci. *30:* 489–510, 1952.
130. deVries, H.: Osmotische Versuche mit lebenden Membranen. Z. Physik. Chem. *2:* 413–433, 1888.
131. de Wardener, H. E., Mills, I. H., Clapham, W. F., and Hayter, C. J.: Studies on the efferent mechanism of sodium diuresis which follows the administration of intravenous saline in the dog. Clin. Sci. *21:* 249–258, 1961.
131a. Dick, D. A. T.: *Cell Water*. Appleton-Century-Crofts, New York, 1966.
132. Donegan, J. F., and Parsons, T. R.: Some further observations on blood reaction. J. Physiol. (London) *52:* 315–327, 1919.
133. Donnan, F. G.: Theorie der Membrangleichgewichte und Membranpotentiale bei Vorhandensein von nicht dialysierenden Elektrolyten. Ein Beitrag zur physikalisch-chemischen Physiologie. Z. Elektrochem. *17:* 572–581, 1911.

134. Donnan, F. G.: The theory of membrane equilibria. Chem. Rev. *1:* 73–90, 1924–1925.
135. Dorman, P. J., Sullivan, W. J., and Pitts, R. F.: The renal response to acute respiratory acidosis. J. Clin. Invest. *33:* 82–90, 1954.
135a. Duda, G. P., and Handler, P.: Kinetics of ammonia metabolism *in vivo*. J. Biol. Chem. *232:* 303–314, 1958.
136. Dunlop, J. C.: On the action of large doses of dilute mineral acids on metabolism. J. Physiol. (London) *20:* 82–96, 1896.
137. Edsall, J. T., and Wyman, J.: *Biophysical Chemistry*, Vol. 1: *Thermodynamics, Electrostatics, and the Biological Significance of the Properties of Matter*. Academic Press, New York, 1966.
138. Eggleton, M. G.: Urine acidity in alcohol diuresis in man. J. Physiol. (London) *104:* 312–320, 1946.
139. Eggleton, M. G.: Some factors affecting the acidity of urine in man. J. Physiol. (London) *106:* 456–465, 1947.
140. Eigen, M.: Protonenübertragung. Säure-Base-Katalyse und enzymatische Hydrolyse. Teil I: Elementarvorgänge. Angew. Chem. [Engl.] *75:* 489–508, 1963.
141. Eisenberg, D., and Kauzmann, W.: *The Structure and Properties of Water*. Oxford University Press, New York, 1969.
142. Elkinton, J. R.: Whole body buffers in the regulation of acid-base equilibrium. Yale J. Biol. Med. *29:* 191–210, 1956–1957.
143. Elkinton, J. R.: Hydrogen—ionized and otherwise. Ann. Intern. Med. *57:* 687–688, 1962.
144. Elkinton, J. R., and Danowski, T. S.: *The Body Fluids: Basic Physiology and Practical Therapeutics*. Williams & Wilkins Co., Baltimore, 1955.
145. Elkinton, J. R., Huth, E. J., Webster, G. D., and McCance, R. A.: The renal excretion of hydrogen ion in renal tubular acidosis. I. Quantitative assessment of the response to ammonium chloride as an acid load. Am. J. Med. *29:* 554–575, 1960.
146. Elkinton, J. R., McCurdy, D. K., and Buckalew, V. M., Jr.: Hydrogen ion and the kidney. In *Renal Disease*, pp. 110–135. Edited by D. A. K. Black. F. A. Davis Co., Philadelphia, 1967.
47. Elkinton, J. R., Singer, R. B., Barker, E. S., and Clark, J. K.: Effects in man of acute experimental respiratory alkalosis and acidosis on ionic transfers in the total body fluids. J. Clin. Invest. *34:* 1671–1690, 1955.
48. Elkinton, J. R., Squires, R. D., and Crosley, A. P., Jr.: Intracellular cation exchanges in metabolic alkalosis. J. Clin. Invest. *30:* 369–380, 1951.
149. Ellinger, P. The site of acidification of urine in the frog's and rat's kidney. Q. J. Exp. Physiol. *30:* 205–218, 1940.
150. Elliott, J. S., Sharp, R. F., and Lewis, L.: Urinary pH. J. Urol. *81:* 339–343, 1959.
151. Elliott, J. S., Quaide, W. L., Sharp, R. F., and Lewis, L.: Mineralogical studies of urine: the relationship of apatite, brushite, and struvite to urinary pH. J. Urol. *80:* 269–271, 1958.
152. Emerson, K., and Dole, V. P.: The renal excretion of atabrine. Malaria Report no. 246, Board of the Coordination of Malaria Studies, 1943 (cited in Reference 14).
153. Endres, G.: Über Gesetzmassigkeiten in der Beziehung zwischen der wahren Harnreaktion und der alveolaren CO_2-Spannung. Biochem. Z.: *132:* 220–241, 1922.
154. Epstein, F. H., Goodyer, A. V. N., Lawrason, F. D., and Relman, A. S.: Studies of the antidiuresis of quiet standing: the importance of changes in plasma volume and glomerular filtration rate. J. Clin. Invest. *30:* 63–72, 1951.
155. Ewy, G. A., Pabico, R. C., Maher, J. F., and Mintz, D. H.: Lactate acidosis associated with phenformin therapy and localized tissue hypoxia. Report of a case treated by hemodialysis. Ann. Intern. Med. *59:* 878–883, 1963.
156. Eylandt, T.: De acidorum sumptorum vi in urinae acorem. Dissertation. Dorpat, 1854.
157. Färber, E.: Boyle. In *Das Buch der Grossen Chemiker*, Vol. 1, pp. 173–191. Edited by G. Bunge. Verlag-Chemie, Weinheim/Bergstrasse, 1929; reprinted 1965.
157a. Farkas, G.: Über die Concentration der Hydroxylionen im Blutserum. Pfluegers Arch. Gesamte Physiol. *98:* 551–576, 1903.
158. Faurholt, C.: Études sur les solutions aqueuses d'anhydride carbonique et d'acide carbonique. J. Chim. Phys. *21:* 400–455, 1923–1924.
159. Feder, L.: Über die Ausscheidung des Salmiaks im Harn. Z. Biol. *13:* 256–298, 1877.
160. Feldman, J. M., and Lebovitz, H. E.: Ammonium ion, a modulator of insulin secretion. Am. J. Physiol. *221:* 1027–1032, 1971.
160a. Fenn, W. O.: Carbon dioxide and intracellular homeostasis. Ann. N.Y. Acad. Sci. *92:* 547–558, 1961.
161. Ferguson, E. B., Jr.: A study of the regulation of the rate of urinary ammonia excretion in the rat. J. Physiol. (London) *112:* 420–425, 1951.
162. Fernandez, L. B., Gonzalez, E., Marzi, A., and Ledesma de Paolo, M. I.: Fecal acidorrhea. New Eng. J. Med. *284:* 295–298, 1971.
163. Fernbach, A., and Hubert, L.: De l'influence des phosphates et de quelques autres matières minérales sur la diastase protéolytique du malt. Compt. Rend. *131:* 293–295, 1900.
164. Ferris, D. O., and Odel, H. M.: Electrolyte pattern of the blood after bilateral uretero-

sigmoidostomy. J.A.M.A. *142:* 634–641, 1950.

165. Filley, G. F.: *Acid-Base and Blood Gas Regulation: For Medical Students before and after Graduation.* Lea and Febiger, Philadelphia, 1971.
166. Findlay, A.: *Introduction to Physical Chemistry,* Ed. 3. Revised by H. W. Melville. Longmans, Green and Co., New York, 1953.
167. Findlay, A.: *A Hundred Years of Chemistry,* Ed. 3. Revised by T. I. Williams. Gerald Duckworth and Co., London, 1965.
168. Fleck, G. M.: *Equilibria in Solution.* Holt, Rhinehart, and Winston, New York, 1966.
169. Flemma, R. J., and Young, W. G., Jr.: The metabolic effects of mechanical ventilation and respiratory alkalosis in postoperative patients. Surgery *56:* 36–43, 1964.
170. Flenley, D. C.: Another non-logarithmic acid-base diagram? Lancet *1:* 961–965, 1971.
171. Flock, E. V., Block, M. A., Grindlay, J. H., Mann, F. C., and Bollman, J. L.: Changes in free amino acids of brain and muscle after total hepatectomy. J. Biol. Chem. *200:* 529–536, 1953.
172. Flock, E. V., Mann, F. C., and Bollman, J. L.: Free amino acids in plasma and muscle following total removal of the liver. J. Biol. Chem. *192:* 293–300, 1951.
173. Folin, O.: On the acidity of urine. Am. J. Physiol. *9:* 265–278, 1903.
174. Folin, O.: Laws governing the chemical composition of urine. Am. J. Physiol. *13:* 66–115, 1905.
175. Forsham, P. H.: The adrenal cortex. In *Textbook of Endocrinology,* Ed. 4. Edited by R. H. Williams. Ch. 5, Part 1: P. H. Forsham and K. L. Melman, The Adrenals. W. B. Saunders Co., Philadelphia, 1968.
176. Forster, J.: Versuche über die Bedeutung der Aschebestandtheile in der Nahrung. Z. Biol. *9:* 297–380, 1873.

176a. Fraenckel, O.: Eine neue Methode zur Bestimmung der Reaktion des Blutes. Pfluegers Arch. Gesamte Physiol. *96:* 601–623, 1903.

177. Friedenthal, H.: Über die Reaktion des Blutserums der Wirbelthiere und der Reaktion der lebendegen Substanz im allgemeinen. Z. Allg. Physiol. *1:* 56–66, 1902.
178. Frömter, E., and Hegel, U.: Transtubuläre Potenzialdifferenze an proximalen und distalen Tubuli der Rattenniere. Pfluegers Arch. Gesamte Physiol. *291:* 107–120, 1966.
179. Fülgraff, G., and Pitts, R. F.: Kinetics of ammonia production and excretion in the acidotic dog. Am. J. Physiol. *209:* 1206–1212, 1965.
180. Fulop, M., Sternlieb, I., and Scheinberg, I. H.: Defective urinary acidification in Wilson's disease. Ann. Intern. Med. *68:* 770–777, 1968.
181. Gamble, J. L.: Carbonic acid and bicarbonate in urine. J. Biol. Chem. *51:* 295–310, 1922.
182. Gamble, J. L.: *Chemical Anatomy, Physiology and Pathology of Extracellular Fluid: A Lecture Syllabus,* Ed. 6. Harvard University Press, Cambridge, Mass., 1964.
183. Gamble, J. L., Blackfan, K. D., and Hamilton, B.: The diuretic action of acid producing salts. J. Clin. Invest. *1:* 359–388, 1925.
184. Gamble, J. L., Ross, G. S., and Tisdall, F. F.: The metabolism of fixed base during fasting. J. Biol. Chem. *57:* 633–695, 1923.
185. Gamble, J. L., Jr.: Regulation of the acidity of the extracellular fluids: teaching syllabus. Johns Hopkins University School of Medicine, Baltimore, 1966.
186. Gamble, J. L., Jr., and Cooke, R. E.: Potassium deficiency and adrenocortical hormones in the etiology of metabolic alkalosis. J. Pediatr. *55:* 296–302, 1959.
187. Gamble, J. L., Jr., Zuromskis, P. J., Bettice J. A., and Ginsberg, R. L.: Intracellular buffering in the dog at varying CO_2 tensions. Clin. Sci. *42:* 311–324, 1972.
188. Gardner, L. I., MacLachlan, E. A., and Berman, H.: Effect of potassium deficiency on carbon dioxide, cation, and phosphate content of muscle, with a note on the carbon dioxide content of human muscle. J. Gen. Physiol. *36:* 153–159, 1952.
189. Garella, S., Chazan, J. A., and Cohen, J. J.: Saline-resistant metabolic alkalosis or "chloride-wasting nephropathy." Ann. Intern. Med. *73:* 31–38, 1970.
190. Gäthgens, C.: Zur Frage der Ausscheidung freier Säuren durch den Harn. Vorläufige Mittheilung. Centralblatt Med. Wissenschaften *53:* 833–835, 1872.
191. Gäthgens, C.: Über Ammoniakausscheidung. Z. Physiol. Chem. *4:* 36–54, 1880.
192. Gerez, C., and Kirsten, R.: Untersuchungen über Ammoniakbildung bei der Muskelarbeit. Biochem. Z. *341:* 534–542, 1965.
193. Gibbs, J. W.: On the equilibrium of heterogeneous substances. Trans. Conn. Acad. Arts. Sci. *3:* 108–248 and 343–524, 1874–1878.
194. Giebisch, G.: The contribution of measurements of electrical phenomena to our knowledge of renal electrolyte transport. Progr. Cardiovasc. Dis. *3:* 463–482, 1961.
195. Giebisch, G.: Functional organization of proximal and distal tubular electrolyte transport. Nephron *6:* 260–281, 1969.
196. Giebisch, G.: Renal potassium excretion. In *The Kidney: Morphology, Chemistry, Physiology,* Vol. 3, Ch. 6. Edited by C. Rouiller and A. F. Muller. Academic Press, New York, 1971.
197. Giebisch, G., Berger, L., and Pitts, R. F.:

The extra-renal response to acute acid-base disturbances of respiratory origin. J. Clin. Invest. *34:* 231–245, 1955.

198. Giebisch, G., Boulpaep, E.L., and Whittembury, G.: Electrolyte transport in kidney tubule cells. Phil. Trans. R. Soc. London *262:* 175–198, 1971.

198a. Giebisch, G., Klose, R. M., and Malnic, G.: Renal tubular potassium transport. In *Proceedings of the Third International Congress of Nephrology,* Vol. 1, pp. 62–75. Edited by J. S. Handler. S. Karger, Basel, 1967.

199. Giebisch, G., MacLeod, M. B., and Pitts, R. F.: Effect of adrenal steroids on renal tubular reabsorption of bicarbonate. Am. J. Physiol. *183:* 377–386, 1955.

200. Giebisch, G., and Malnic, G.: Some aspects of renal tubular hydrogen ion transport. In *Proceedings of the Fourth International Congress of Nephrology, Stockholm, 1969,* Vol. 1, pp. 181–194. S. Karger, Basel, 1970.

201. Giebisch, G., Windhager, E. E., and Pitts, R. F.: Mechanism of urinary acidification. In *Biology of Pyelonephritis,* pp. 277–287. Edited by E. L. Quinn and E. H. Kass. Little, Brown and Co., Boston, 1960.

202. Gill, J. R., Jr., Bell, N. H., and Bartter, F. C.: Impaired conservation of sodium and potassium in renal tubular acidosis and its correction by buffer anions. Clin. Sci. *33:* 577–592, 1967.

203. Gilman, A., and Brazeau, P.: The role of the kidney in the regulation of acid-base metabolism. Am. J. Med. *15:* 765–770, 1953.

203a. Glabman, S., Klose, A., and Giebisch, G.: Micropunture study of ammonia excretion in the rat. Am. J. Physiol. *205:* 127–132, 1963.

204. Goldberg, G., Paul, W., and Gonick, H.: A titrimetric method for determining organic acids and bases in urine. Clin. Chem. *12:* 831–836, 1966.

204a. Goldring, R. M., Cannon, P. J., Heinemann, H. O., and Fishman, A. P.: Respiratory adjustment to chronic metabolic alkalosis in man. J. Clin. Invest. *47:* 188–202, 1968.

205. Goldstein, L., and Neppel, M. J.: Effect of dietary protein concentration on renal ammonia excretion and glutaminase activity in rats (Abstract). Fed. Proc. *25:* 203, 1966.

206. Goldstein, L., Richterich, R., and Dearborn, E. H.: Kidney glutaminases. II: The glutamine-α-ketoacid transamination-deamidation system of the guinea pig. Enzymologia *18:* 261–270, 1957.

207. Goldstein, M. B., Gennari, F. J., and Schwartz, W. B.: The influence of graded degrees of chronic hypercapnia on the acute carbon dioxide titration curve. J. Clin. Invest. *50:* 208–216, 1971.

208. Gonick, H.: Comparison of net mineral acid (NMA) balance in normals and nephritics (Abstract). Clin. Res. *16:* 165A, 1968.

209. Gonick, H. C., Goldberg, G., and Mulcare, D.: Reexamination of the acid-ash content of several diets. Am. J. Clin. Nutr. *21:* 898–903, 1968.

210. Goodman, A. D., Lemann, J., Jr., Lennon, E. J., and Relman, A. S.: Production, excretion and net balance of fixed acid in patients with renal acidosis. J. Clin. Invest. *44:* 495–506, 1965.

211. Goodman, L. S., and Gilman, A., Editors: *The Pharmacological Basis of Therapeutics: A Textbook of Pharmacology, Toxicology, and Therapeutics for Physicians and Medical Students,* Ed. 4. Macmillan Co., London, 1970.

212. Goodyer, A. N., and Seldin, D. W.: The effects of quiet standing on solute diuresis. J. Clin. Invest. *32:* 242–250, 1953.

213. Gottschalk, C. W.: Osmotic concentration and dilution of the urine. Am. J. Med. *36:* 670–685, 1964.

214. Gottschalk, C. W., Lassiter, W. E., and Mylle, M.: Localization of urine acidfication in the mammalian kidney. Am. J. Physiol. *198:* 581–585, 1960.

215. Gowenlock, A. H., Mills, J. N., and Thomas S.: Acute postural changes in aldosterone and electrolyte excretion in man. J. Physiol. (London) *146:* 133–141, 1959.

216. Grollman, A. P., and Gamble, J. L., Jr.: Metabolic alkalosis, a specific effect of adrenocortical hormones. Am. J. Physiol. *196:* 135–140, 1959.

217. Guldberg, C. M., and Waage, P.: Studies on affinity (in Norwegian). Forhandlinger i Videnskabs-Selskabet, Christiania, 1864.

218. Guldberg, C. M., and Waage, P.: Études sur les affinités chimiques. First semester program, University of Christiania, 1867. Ostwald's Klassiker no. 104.

219. Gutman, A., and Yü, T-F.: Uric acid nephrolithiasis. Am. J. Med. *45:* 756–779, 1968.

220. Haag, H. B., and Larson, P. S.: Studies on the fate of nicotine in the body: the effect of pH on the urinary excretion of nicotine by tobacco smokers. J. Pharmacol. Exp. Ther. *76:* 235–244, 1942.

220a. Haag, H. B., Larson, P. S., and Schwartz, J. J.: The effect of urinary pH on the elimination of quinine in man. J. Pharmacol Exp. Ther. *79:* 136–139, 1943.

221. Haggard, H. W., and Henderson, Y.: How oxygen deficiency lowers the blood alkali. J. Biol. Chem. *43:* 15–27, 1920.

221a. Haldane, J. B. S.: Experiments on the regulation of the blood's alkalinity. II. J. Physiol. (London) *55:* 265–275, 1921.

221b. Haldane, J. S., Kellas, A. M., and Kennaway, E. L.: Experiments on acclimatization to reduced atmospheric pressure. J. Physiol. (London) *53:* 181–206, 1919–1920.

222. Haldane, J. S., and Priestley, J. G.: The regulation of the lung-ventilation. J. Physiol. (London) *32:* 225–265, 1905.
222a. Haldane, J. S., and Smith, J. L.: The physiological effects of air vitiated by respiration. J. Pathol. Bacteriol. *1:* 168–186, 1892.
222b. Hallervorden, E.: Über das Verhalten des Ammoniaks im Organismus und seine Beziehung zur Harnstoffbildung. Arch. Exp. Pathol. Pharmakol. *10:* 125–146, 1878.
223. Hallervorden, E.: Aus der med. Klin. zu Königsberg. Über Ausscheidung von Ammoniak im Urin bei pathologischen Zuständen. Arch. Exp. Pathol. Pharmakol. *12–13:* 237–275, 1880–1881.
224. Hallervorden, E.: Über Ammoniakausscheidung. Virchows Arch. *143:* 705–706, 1896.
225. Hammes, G. G.: Very fast reactions in solution. Science *151:* 1507–1511, 1966.
226. Hantzsch, A.: Die Theorie der Ionogenen Bindung als Grundlage der Ionentheorie nach Versuchen über die Natur der nicht ionisierten Säuren. Z. Elektrochem. *29:* 221–246, 1923.
227. Hantzsch, A., and Weissberger, A.: Reaktionskinetische Untersuchungen an starken Säuren. Z. Physik. Chem. *125:* 251–263, 1927.
228. Harrington, J. T., and Lemann, J. R., Jr.: The metabolic production and disposal of acid and alkali. Med. Clin. North Am. *54:* 1543–1554, 1971.
229. Haskins, H. D.: Nitrogenous metabolism as affected by diet and alkaline diuretics. J. Biol. Chem. *2:* 217–229, 1906.
230. Hasselbalch, K. A.: Die Berechnung der Wasserstoffzahl des Blutes aus der freien und gebundenen Kohlensäure desselben, und die Sauerstoffbindung des Blutes als Funktion der Wasserstoffzahl. Biochem. Z. *78:* 112–144, 1916–1917.
231. Hastings, A. B., and Sendroy, J., Jr.: The effect of variation in ionic strength on the apparent first and second dissociation constants of carbonic acid. J. Biol. Chem. *65:* 445–455, 1925.
232. Hastings, A. B., Sendroy, J., Jr., and Van Slyke, D. D.: Studies of gas and electrolyte equilibria in blood. XII. The value of pK′ in the Henderson-Hasselbalch equation for blood serum. J. Biol. Chem. *79:* 183–192, 1928.
233. Häusler, G.: Zur Technik und Spezifität des histochemischen Carboanhydrasenachweises in Modellversuch und in Gewebschnitten von Rattennieren. Histochemie *1:* 29–47, 1958.
234. Hayes, C. P., Jr., Mayson, J. S., Owen, E. E., and Robinson, R. R.: A micropuncture evaluation of renal ammonia excretion in the rat. Am. J. Physiol. *207:* 77–82, 1964.
235. Hempling, H. G.: Kinetic analysis of ammonia equilibration in kidney of dog. J. Appl. Physiol. *31:* 620–621, 1971.
236. Henderson, L. J.: The theory of neutrality regulation in the animal organism. Am. J. Physiol. *21:* 427–448, 1908.
237. Henderson, L. J.: Das Gleichgewicht zwischen Basen und Säuren im tierischen Organismus. Ergeb. Physiol. *8:* 254–325, 1909.
238. Henderson, L. J.: Zur Kenntnis der Ionengleichgewichts im Organismus. III. Messungen der normalen Harnacidität. Biochem. Z. *24:* 40–44, 1910.
239. Henderson, L. J.: A critical study of the process of acid excretion. J. Biol. Chem. *9:* 403–424, 1911.
240. Henderson, L. J.: *The Fitness of the Environment: An Inquiry into the Biological Significance of the Properties of Matter.* Macmillan Co., New York, 1913; reprinted, with an introduction by George Wald, by the Beacon Press, Boston, 1958.
241. Henderson, L. J.: *Blood: A Study in General Physiology.* Yale University Press, New Haven, Conn., 1928.
242. Henderson, L. J., and Palmer, W. W.: On the intensity of urinary acidity in normal and pathological conditions. J. Biol. Chem. *13:* 393–405, 1912–1913.
243. Henderson, L. J., and Palmer, W. W.: On the several factors of acid excretion. J. Biol. Chem. *17:* 305–315, 1914.
244. Henderson, L. J., and Palmer, W. W.: On the several factors of acid excretion in nephritis. J. Biol. Chem. *21:* 37–55, 1915.
245. Henderson, Y., and Haggard, H. W.: Respiratory regulation of the CO_2 capacity of the blood. I. High levels of CO_2 and alkali. J. Biol. Chem. *33:* 333–344, 1918.
245a. Hendrix, B. M., and Sanders, J. P.: The effect of injections of sodium phosphates and sodium hippurate upon the excretion of acid and ammonia by the kidney. J. Biol. Chem. *58:* 503–513, 1923.
246. Hendrix, T. R.: Studies of the pathogenesis of cholera. Trans. Am. Clin. Climatol. Assoc. *82:* 63–70, 1971.
247. Henneman, P. H., Wallach, S., and Dempsey, E. F.: The metabolic defect responsible for uric acid renal stone formation (Abstract). J. Clin. Invest. *37:* 901, 1958.
248. Henneman, P. H., Wallach, S., and Dempsey, E. F.: The metabolic defect responsible for uric acid stone formation. J. Clin. Invest. *41:* 537–542, 1962.
249. Henriques, O. M.: Die Bindungsweise des Kohlendioxyds im Blute. II. Vorläufige Mitteilung. Der experimentell Nachweis schnell reagierenden, gebundenen CO_2 im Hämoglobin. Biochem. Z. *200:* 5–9, 1928.
250. Heymans, C., Boukaert J. J., and Dautreband, L.: Sinus carotidien et réflexes respiratoires. III. Sensibilité des sinus carotidiens aux substances chimiques;

action stimulante respiratoire réflex du sulfure de sodium, du cyanure de potassium, de la nicotine, et de la lobéline. Arch. Int. Pharmacodyn. Ther. *40:* 54–91, 1931.

250a. Hierholzer, K.: Secretion of potassium and acidification in collecting ducts of mammalian kidney. Am. J. Physiol. *201:* 318–324, 1961.

251. Hills, A. G.: A kinetic treatment of renal ammonia balance. In *Renal Metabolism and Epidemiology of Some Renal Diseases,* pp. 119–124. Edited by J. Metcoff. Proceedings of the Fifteenth Annual Conference on the Kidney, October 10 and 11, 1963, Swampscott, Mass. National Kidney Foundation, New York, 1964.

252. Hills, A. G.: Rückdiffusion von CO_2 und NH_3 aus dem Endsegment des Nephrons bei der Harnkonzentrierung. In *3 Symposion der Gesellschaft für Nephrologie, 11–13 September, 1964,* pp. 119–124. Edited by K. J. Ullrich and K. Hierholzer. Hans Huber, Bern and Stuttgart, 1965.

253. Hills, A. G.: Reversible nephropathies. J. Fla. Med. Assoc. *56:* 402–404, 1969.

254. Hills, A. G.: The fitness of carbon dioxide and ammonia to serve acid-base balance: mammalian urine and plasma pH as evolutionary adaptations. Am. Naturalist. *103:* 131–153, 1969.

254a. Hills, A. G.: Pathogenesis of hyperkalemia in hypoaldosteronism. J. Clin. Endocrinol. Metab. *29:* 988–989, 1969.

255. Hills, A. G.: Acid-base balance. Trans. Am. Clin. Climatol. Assoc. *83:* 186–197, 1971.

256. Hills, A. G.: Unpublished observations.

257. Hills, A. G., Kerr, W. D., and Reid, E. L.: Peripheral protein catabolism as the ultimate source of urinary ammonia. Presented to the Fourth International Congress of Nephrology, Stockholm, Abstracts *1:* 221, June, 1969.

258. Hills, A. G., and Reid, E. L.: Renal ammonia balance. A kinetic treatment. Nephron *3:* 221–256, 1966.

259. Hills, A. G., and Reid, E. L.: Renal maintenance of acid-base balance in health. Johns Hopkins Med. J. *120:* 368–379, 1967.

260. Hills, A. G., and Reid, E. L.: pH defended; more on pH (Letters and Comments). Ann. Intern. Med. *65:* 1150–1151, 1966; *66:* 238–239, 1967.

261. Hills, A. G., and Reid, E. L.: P_{CO_2} and P_{NH_3} in mammalian kidney and urinary tract related to urine pH and flow. Am. J. Physiol. *219:* 423–434, 1970.

262. Hills, A. G., and Reid, E. L.: Unpublished studies.

263. Hills, A. G., Reid, E. L., and Kerr, W. D.: Studies of circulatory transport of L-glutamine and of the cellular sources of the urinary ammonia-nitrogen in the fasted mammal. Am. J. Physiol. *223:* 1470–1476, 1972.

264. Hilton, J. G., Capeci, N. E., Kiss, G. T., Kruesi, O. R., Glaviano, V. V., and Wegria, R.: The effect of acute elevation of the plasma chloride concentration on the renal excretion of bicarbonate during acute respiratory acidosis. J. Clin. Invest. *35:* 481–487, 1956.

264a. Höber, R.: Über die Hydroxylionen des Blutes. Pfluegers Arch. Gesamte Physiol. *81:* 522–539, 1900.

265. Höber, R.: *Physikalische Chemie der Zelle und der Gewebe.* W. Engelmann, Leipzig, 1902.

265a. Höber, R.: Über die Hydroxylionen des Blutes. Zweite Mittleilung. Pfluegers Arch. Gesamte Physiol. *98:* 551–576, 1903.

266. Höber, R.: Die Acidität des Harns vom Standpunkt der Ionenlehre. Beitr. Chem. Physiol. Pathol. *3:* 525–542, 1903.

266a. Hodson, C. J.: Radiology of the kidney. In *Renal Disease,* Ed. 2, Ch. 6. Edited by D. A. K. Black. F. A. Davis Co., Philadelphia, 1967.

267. Hofmann, F.: Über den Übergang von freien Säuren durch das alkalische Blut in den Harn. Z. Biol. *7:* 338–353, 1872.

268. Hollander, W., Jr., and Blythe, W. B.: Nephropathy of potassium depletion. In *Diseases of the Kidney,* Ed. 2, Ch. 25. Edited by M. B. Strauss and L. G. Welt. Little, Brown and Co., Boston, 1971.

269. Holmes, A. M., Hesling, C. M., and Wilson, T. M.: Drug-induced secondary hyperaldosteronism in patients with pulmonary tuberculosis. Q. J. Med. *39(NS):* 299–315, 1970.

270. Holmgård, A.: Quantitative analysis of enzymes in normal and diseased kidney tissue. Scand. J. Clin. Lab. Invest. *14* (*suppl. 65*)*:* 1–79, 1962.

271. Hornbein, T. F.: The chemical regulation of ventilation. In *Physiology and Biophysics,* Ed. 19, Ch. 42. Edited by T. C. Ruch and H. D. Patton. W. B. Saunders Co., Philadelphia, 1965.

272. Howe, N. M., and Hawk, P. B.: Studies on water drinking: XIII (Fasting Studies: VIII). Hydrogen ion concentration of feces. J. Biol. Chem. *11:* 129–140, 1912.

273. Howell, D. S., and Davis, J. O.: Relationship of sodium retention to potassium excretion by the kidney during administration of desoxycorticosterone acetate to dogs. Am. J. Physiol. *179:* 359–363, 1954.

274. Hubbard, R. S.: The excretion of ammonia and nitrogen. J. Biol. Chem. *58:* 711–719, 1924.

275. Hubbard, R. S., and Munford, S. A.: The excretion of acid and ammonia. J. Biol. Chem. *54:* 465–479, 1922.

276. Huckabee, W. E.: Abnormal resting blood lactate. I. The significance of hyperlactatemia in hospitalized patients. Am. J. Med. *30:* 833–839, 1961.

277. Huckabee, W. E.: Abnormal resting blood

lactate. II. Lactic acidosis. Am. J. Med. *30:* 840–848, 1961.
278. Huckabee, W. E.: Henderson vs. Hasselbalch. Clin. Res. *9:* 116–119, 1961.
278a. Hudson, J. R., Chobanian, A. V., and Relman, A. S.: Hypoaldosteronism. A clinical study of a patient with an isolated adrenal mineralocorticoid deficiency resulting in hyperkalemia and Stokes-Adams attacks. New Eng. J. Med. *257:* 529–536, 1957.
279. Hunt, J. N.: The influence of dietary sulphur on the urinary output of acid in man. Clin. Sci. *15:* 119–134, 1956.
280. Huth, E. J., Webster, G. D., and Elkinton, J. R.: The renal excretion of hydrogen ion in renal tubular acidosis. III. An attempt to detect latent cases in a family; comments on nosology, genetics and etiology of the primary disease. Am. J. Med. *29:* 586–598, 1960.
281. Ihde, A.: *The Development of Modern Chemistry.* Harper and Row, New York, 1964.
282. Iqbal, K., and Ottaway, J. H.: Glutamine synthetase in muscle and kidney. Biochem. J. *119:* 145–156, 1970.
283. Jacobs, M. H.: The production of intracellular acidity by neutral and alkaline solutions containing carbon dioxide. Am. J. Physiol. *53:* 457–463, 1920.
284. Jacobs, M. H.: The influence of ammonium salts on cell reaction. J. Gen. Physiol. *5:* 181–188, 1922.
285. Jacobs, M. H.: The exchange of material between the erythrocyte and its surroundings. Harvey Lect. *22:* 146–164, 1926–1927.
286. Jacobs, M. H.: Some aspects of cell permeability to weak electrolytes. Cold Spring Harbor Symp. Quant. Biol. *8:* 30–39, 1940.
287. Jacobs, M. H., and Stewart, D. H.: The distribution of penetrating ammonium salts between cells and their surroundings. J. Cell. Comp. Physiol. *7:* 351–365, 1936.
288. Jacquez, J. A., Poppel, J. W., and Jeltsch, R.: Solubility of ammonia in human plasma. J. Appl. Physiol. *14:* 255–258, 1959.
289. Jailer, J. W., Rosenfeld, M., and Shannon, J. A.: The influence of orally administered alkali and acid on the renal excretion of quinacrine, chloroquine, and santoquine. J. Clin. Invest. *26:* 1168–1172, 1947.
290. Jamison, R. L., Buerkert, J., and Lacy, F.: A micropuncture study of collecting tubule function in rats with hereditary diabetes insipidus. J. Clin. Invest. *50:* 2444–2452, 1971.
291. Janicki, R. H., and Goldstein, L.: Glutamine synthetase and renal ammonia metabolism. Am. J. Physiol. *216:* 1107–1110, 1969.
292. Johnell, H. E.: The first dissociation constant for carbonic acid and the solubility of carbon dioxide in human amniotic fluid. Scand. J. Clin. Lab. Invest. *27:* 233–238, 1971.
293. Kahn, M., and Goodridge, F. G.: *Sulfur Metabolism: A Review of the Literature.* Lea and Febiger, Philadelphia, 1926.
294. Kappagoda, C. T., Linden, R. J., and Snow, H. M.: An approach to the problems of acid-base balance. Clin. Sci. *39:* 169–182, 1970.
295. Karnovsky, M. J., and Himmelhoch, S. R.: Histochemical localization of glutaminase I activity in kidney. Am. J. Physiol. *201:* 786–790, 1961.
296. Kassirer, J. P., Berkman, P. M., Lawrenz, D. R., and Schwartz, W. B.: The critical role of chloride in the correction of hypokalemic alkalosis in man . Am. J. Med. *38:* 172–189, 1965.
297. Kassirer, J. B., Lowance, D. C., and Schwartz, W. B.: Aldosterone-induced metabolic alkalosis in man. Abstracts, American Society of Nephrology, Fifth Annual Meeting, 1971, p. 36.
298. Kassirer, J. B., and Schwartz, W. B.: Correction of metabolic alkalosis in man without repair of potassium deficiency. A reevaluation of the role of potassium. Am. J. Med. *40:* 19–26, 1966.
299. Katz, J.: Die minerale Bestandtheile des Muskelfleisches. Pfluegers Arch. Gesamte Physiol. *63:* 1–85, 1896.
300. Kavanau, J. L.: *Water and Solute-Water Interactions.* Holden-Day, San Francisco, 1964.
301. Kaye, G.: Studies in the reaction of urine. Aust. J. Exp. Biol. Med. *6:* 187–214, 1929.
302. Kayser, C.: Le sommeil. J. Physiol. Pathol. Gen. *41:* 1A–60A, 1949.
303. Kempton, R. T.: Differences in the elimination of neutral red and phenol red by the frog kidney. J. Cell. Comp. Physiol. *13:* 73–81, 1939.
304. Kendall, J.: The velocity of the hydrogen ion, and a general dissociation formula for acids. J. Chem. Soc. *101:* 1275–1296, 1912.
304a. Kennedy, T. J., Eden, N., and Berliner, R. W.: Interpretation of urine CO_2 tension (Abstract). Fed. Proc. *16:* 72, 1957.
305. Kennedy, T. J., Jr., Orloff, J., and Berliner, R. W.: Significance of carbon dioxide tension in urine. Am. J. Physiol. *169:* 596–608, 1952.
306. Kennedy, T. J., Jr., Winkley, J. H., and Dunning, M. F.: Gastric alkalosis with hypokalemia. Am. J. Med. *6:* 790–794, 1949.
307. Kerr, W. D., Reid, E. L., and Hills, A. G.: Peripheral glutamine release—a defense against ammonemia (Abstract). Clin. Res. *16:* 345, 1968.
308. Kety, S. S., and Schmidt, C.: The effects of active and passive hyperventilation on cerebral blood flow, cerebral oxygen consumption, cardiac output, and blood pressure of normal young men. J. Clin. Invest. *25:* 107–119, 1946.
309. Kilburn, K. H.: Shock, seizures, and coma

with alkalosis during mechanical ventilation. Ann. Intern. Med. *65:* 977–984, 1966.

309a. Kildeberg, P., Engel, K., and Winters, R. W.: Balance of net acid in growing infants: endogenous and transintestinal aspects. Acta Paediatr. Scand. *58:* 321–329, 1969.

310. Klahr, S., Tripathy, K., and Lotero, H.: Renal regulation of acid-base balance in malnourished man. Am. J. Med. *48:* 325–331, 1970.

311. Klein, W., and Moritz, F., Das Harnammoniak beim gesunden Menschen unter dem Gesichtspunkt einer ausschliesslich neutralisatorischen Funktion desselben, sowie die Bilanzverhältnisse zwischen Säuren and Alkalien im menschlichen Harn bei verschiedener Ernährung. Dtsch. Arch. Klin. Med. *99:* 162–214, 1910.

312. Kleitman, N.: Studies on the physiology of sleep. I. The effects of prolonged sleeplessness on man. Am. J. Physiol. *66:* 67–92, 1923.

313. Kleitman, N.: Studies on the physiology of sleep. III. The effect of muscular activity, rest and sleep on the urinary excretion of phosphorus. Am. J. Physiol. *74:* 225–237, 1925.

314. Kleitman, N.: *Sleep and Wakefulness*, revised enlarged edition. University of Chicago Press, Chicago, 1963.

315. Klümper, J. D., Ullrich, K. J., and Hilger, H. H.· Das Verhalten des Harnstoffs in den Sammelrohren der Säugetierniere. Pfluegers Arch. Gesamte Physiol. *267:* 238–243, 1958.

316. Knieriem, W. von: Beiträge zur Kenntniss der Bildung des Harnstoffs im thierischen Organismus. Z. Biol. *10:* 263–294, 1874.

317. Kong, H. H. P., and Alleyne, G. A. O.: Defect in urinary acidification in adults with sickle-cell anaemia. Lancet *2:* 954–955, 1968.

318. Krebs, H. A.: Metabolism of amino acids. IV. The synthesis of glutamine from glutamic acid and ammonia, and the enzymatic hydrolysis of glutamine in animal tissues. Biochem. J. *29:* 1951–1969, 1935.

319. Krebs, H. A., Eggleston, L. V., and Hems, R.: Synthesis of glutamic acid in animal tissues. Biochem. J. *43:* 406–414, 1948.

320. Kroetz, C.: Über einige stoffliche Erscheinungen bei verlangertem Schlafentzug. Z. Gesamte Exp. Med. *52:* 770–778, 1926.

321. Kuhn, T. S.: *The Structure of Scientific Revolutions.* International Encyclopedia of Unified Science Vol. 2, No. 2. University of Chicago Press, Chicago, 1962.

322. Kuhn, W., and Ramel, A.: Aktiver Salztransport als möglicher (und wahrscheinlicher) Einzeleffekt bei der Harnkonzentrierung in der Niere. Helv. Chim. Acta *42:* 628–661, 1959.

323. Kuhn, W., and Ryffel, K.: Herstellung konzentrierter Lösungen aus verdünnten durch blosse Membranwirkung. Ein Modellversuch zur Funktion der Niere. Hoppe-Seylers Z. Physiol. Chem. *276:* 145–178, 1942.

324. Kunau, R. T., Jr., Frick, A., Rector, F. C., Jr., and Seldin, D. W.: Mircopuncture study of the proximal tubular factors responsible for maintenance of alkalosis during potassium deficiency in the rat. Clin. Sci. *34:* 223–231, 1968.

325. Kurtzman, N. A.: Regulation of renal bicarbonate reabsorption by extracellular volume. J. Clin. Invest. *49:* 586–595, 1970.

326. Lange, F.: Physiologische Untersuchungen über das Verhalten und die Wirkung einiger Ammoniaksalze im thierischen Organismus. Inaugural dissertation, Dorpat, 1874.

327. Lange, N. A.: *Handbook of Chemistry*, Ed. 4, p. 1078. Handbook Publishers, Sandusky, O., 1941.

328. Lassar, O.: Zur Alkalescenz des Blutes. Pfluegers Arch. Gesamte Physiol. *9:* 44–52, 1874.

329. Lathem, W.: Hyperchloremic acidosis in chronic pyelonephritis. New Eng. J. Med. *258:* 1031–1036, 1958.

330. Latta, T.: Malignant cholera. Lancet *2:* 274–277, 1831–1832.

331. Laurence, R., and Marsh, D. J.: Effect of diuretic states on hamster collecting duct potential differences. Am. J. Physiol. *220:* 1610–1616, 1971.

332. Leathem, J. H., Editor.: *Protein Nutrition and Free Amino Acid Patterns.* Rutgers University Press, New Bruswick, N. J., 1968.

333. Leathes, J. B.: Renal efficiency tests in nephritis and the reaction of the urine. Br. Med. J. *2:* 165–167, 1919.

333a. Lemann J., Jr., and Lennon, E. J.: A potential error in the measurement of urinary titratable acid. J. Lab. Clin. Med. *67:* 906–913, 1966.

333b. Lemann, J., Jr., and Lennon, E. J.: Role of diet, gastrointestinal tract, and bone in acid-base homeostasis. Kidney Int. *1:* 275–279, 1972.

334. Lemann, J., Jr., Lennon, E. J., Goodman, A. D., and Relman, A. S.: The role of fixed tissue buffers in acid base regulation. Trans. Assoc. Am. Physicians *77:* 188–195, 1964.

335. Lemann, J., Jr., Litzow, J. R., and Lennon, E. J.: The effects of chronic acid loads in normal man: further evidence for the participation of bone mineral in the defense against chronic metabolic acidosis. J. Clin. Invest. *45:* 1608–1614, 1966.

336. Lemann J., Jr., and Relman, A. S.: The relation of sulfur metabolism to acid-base balance and electrolyte excretion: the effects of DL-methionine in normal man. J. Clin. Invest. *38:* 2215–2223, 1959.

337. Lennon, E. J. and Lemann, J., Jr.: Defense of hydrogen ion concentration in chronic metabolic acidosis. A new evaluation of an

old approach. Ann. Intern. Med. *65:* 265–274, 1966.

338. Lennon, E. J., and Lemann, J., Jr.: Is pH defensible? (Letters and Comments). Ann. Intern. Med. *65:* 1151–1152, 1966.
339. Lennon, E. J., Lemann, J., Jr., and Litzow, J. R.: The effects of diet and stool composition on the net external acid balance of normal subjects. J. Clin. Invest. *45:* 1601–1607, 1966.
340. Lennon, E. J., Lemann, J., and Relman, A. S.: The effects of phosphoproteins on acid balance in normal subjects. J. Clin. Invest. *41:* 637–645, 1962.

340a. Leusen, I.: Regulation of cerebrospinal fluid composition with reference to breathing. Physiol. Rev. *52:* 1–55, 1972.

341. Leusen, I. R.: Chemosensitivity of the respiratory center. Influence of CO_2 in the cerebral ventricles on respiration. Am. J. Physiol. *176:* 39–44, 1954.

341a. Leusen, I. R.: Chemosensitivity of the respiratory center. Influence of changes in the H^+ and total buffer concentration in the cerebral ventricle on respiration. Am. J. Physiol. *176:* 45–51, 1954.

342. Levine, D. Z.: Effect of acute hypercapnia on proximal tubular water and bicarbonate reabsorption. Am. J. Physiol. *221:* 1164–1170, 1971.

342a. Levitt, M. F., Turner, L. B., Sweet, A. Y., and Pandiri, D.: The response of bone, connective tissue, and muscle to acute acidosis. J. Clin. Invest. *35:* 98–106, 1956.

343. Lewis, G. N.: The law of physicochemical change. Proc. Am. Acad. Arts Sci. *37:* 49–69, 1901.
344. Lewis, G. N.: Outlines of a new system of thermodynamic chemistry. Proc. Am. Acad. Arts Sci. *43:* 259–293, 1907.
345. Lewis, G. N.: *Valence and the Structure of Atoms and Molecules.* Chemical Catalog Co., New York, 1923.
346. Lewis, G. N.: Acids and bases. J. Franklin Inst. *226:* 293–313, 1938.
347. Lewis, G. N., and Randall, M.: The thermodynamic treatment of concentrated solutions, and applications to thallium amalgams. J. Am. Chem. Soc. *43:* 233–254, 1921.
348. Liddle, G. W.: Summarization of the effects of hormones on water and electrolyte metabolism. In *Textbook of Endocrinology,* Ed. 4, Ch. 15. Edited by R. H. Williams. W. B. Saunders Co., Philadelphia, 1968.
349. Liebig, J.: Über die Constitution des Harns der Menschen und der fleischfressenden Thiere. Ann. Chem. Pharm. *50:* 161–196, 1844.
350. Liebig, J.: *Animal Chemistry or Organic Chemistry in Its Application to Physiology and Pathology.* Edited from the Author's Manuscript by William Gregory with Additions, Notes and Corrections by Dr. Gregory and John W. Webster, M.D. A Facsimile of the Cambridge Edition of 1842. Johnson Reprint Corp., New York, 1964.
351. Lightwood, R.: Calcium infarction of the kidneys in infants. Proceedings of the Eighth Annual General Meeting, British Paediatric Association. Arch. Dis. Child. *10:* 205–206, 1935.
352. Lightwood, R., Maclagan, N. F., and Williams, J. G.: Persistent acidosis in an infant: cause not yet ascertained. Proc. R. Soc. Med. *29:* 1431–1433, 1936.
353. *The Little and Ives Complete Book of Science,* p. 281. Illustrated. J. J. Little and Ives Co., New York. 1958.

353a. Litzow, J. R., Lemann, J., Jr., and Lennon, E. J.: The effect of treatment of acidosis on calcium balance in patients with chronic azotemic renal disease. J. Clin. Invest. *46:* 280–286, 1967.

354. Lockemann, G.: Scheele. In *Das Buch der Grossen Chemiker,* Vol. 1, pp. 274–290. Edited by G. Bunge. Verlag-Chemie, Weinheim/Bergstrasse, 1929; reprinted 1965.

354a. Loeb, R. F.: The adrenal cortex and electrolyte behavior (Harvey Lecture). Bull. N.Y. Acad. Med. *18:* 263–288, 1942.

355. Loeb, R. F., Atchley, D. W., Richards, D. W., Jr., Benedict, E. M., and Driscoll, M. E.: On the mechanism of nephrotic edema. J. Clin. Invest. *11:* 621–639, 1932.

355a. Loeschke, H. H.: The acid-base status of cerebrospinal fluid and the regulation of breathing. In *Proceedings of the International Union of Physiological Scientists, 1971,* Vol. 7, pp. 231–232. Twenty-fifth International Congress, Munich.

356. Lohrer, J.: Über den Übergang der Ammoniaksalze in den Harn. Inaugural dissertation, Dorpat, 1862.
357. Longson, D., and Mills, J. N.: The failure of the kidney to respond to respiratory acidosis. J. Physiol. (London) *122:* 81–92, 1953.
358. Lotspeich, W. D.: Metabolic aspects of acid-base change. Science *155:* 1066–1075, 1967.
359. Lowenstein, J. M.: Ammonia production in muscle and other tissues: the purine nucleotide cycle. Physiol. Rev. *52:* 382–414, 1972.
360. Lowry, T. M.: The uniqueness of hydrogen. Chem. Ind. *43:* 43–47, 1923.
361. Lowry, T. M.: The electronic theory of valency. Part IV. The origin of acidity. Trans. Faraday Soc. *20:* 13–17, 1924.
362. Luder, W. R., and Zuffanti, S.: *The Electronic Theory of Acids and Bases,* Ed. 2. Dover Publications, New York, 1961.
363. Luke, R. G., and Levitin, H.: Impaired renal conservation of chloride and acid-base changes associated with potassium depletion. Clin. Sci. *32:* 511–526, 1967.
364. Lund, P.: Control of glutamine synthesis in rat liver. Biochem. J. *124:* 653–660, 1971.

365. Lund, P., Brosnan, J. T., and Eggleston, L. V.: The regulation of ammonia metabolism in mammalian tissues. In *Essays in Cell Metabolism*, pp. 167–188. Edited by W. Bartley, H. L. Kornberg, and J. R. Quayle. Wiley-Interscience, London, 1970.
366. Lund, P., and Goldstein, L.: Glutamine synthetase activity in tissues of lower vertebrates. Comp. Biochem. Physiol. *31:* 205–210, 1969.
367. Lusk, G.: *The Elements of the Science of Nutrition*. W. B. Saunders Co., Philadelphia, 1906.
368. Lyon, M. L., and Pitts, R. F.: Species differences in renal glutamine synthesis *in vivo*. Am. J. Physiol. *216:* 117–122, 1969.
369. Mach, R. S., Fabre, J., Duckert, A., Borth, R., and Ducommun, P.: Action clinique et métabolique de l'aldostérone (Electrocortine). Schweiz. Med. Wochenschr. *84:* 407–416, 1954.
370. MacInnes, D. A., and Shedlovsky, T.: The determination of the ionization constant of acetic acid, at 25°, from conductance measurements. J. Am. Chem. Soc. *54:* 1429–1438, 1932.
370a. MacKay, C. M., Wick, A. N., Carne, H. O., and Barnum, C. P.: The influence of alkalosis and acidosis upon fasting ketosis. J. Biol. Chem. *138:* 63–68, 1941.
370b. MacLeod, J. J. R., and Hoover, D. H.: Studies in experimental glycosuria. XII. Lactic acid production in the blood following the injection of alkaline solutions of dextrose or of alkaline solutions alone. Am. J. Physiol. *42:* 460–465, 1917.
371. Magnus-Levy, A.: Aus der med. Klin. zu Strassburg in E. Untersuchungen über die Acidosis im Diabetes melitus und die Säurenintoxication in Coma Diabeticum. Arch. Exp. Pharmakol. Pathol. *45:* 389–434, 1901.
372. Mainzer, F., and Bruhn, M.: Über Löslichkeit, Dissoziation und Spannung der Kohlensäure im Harn. Biochem. Z. *230:* 395–410, 1931.
373. Malnic, G., de Mello Aires, M., and Giebisch, G.: Potassium transport across distal tubules of rat kidney during alterations of acid-base equilibrium. Am. J. Physiol. *221:* 1192–1208, 1971.
374. Malnic, G., de Mello Aires, M., and Giebisch, G.: Micropuncture study of renal tubular hydrogen ion transport in the rat. Am. J. Physiol. *222:* 147–158, 1972.
375. Malnic, G., and Giebisch, G.: Mechanism of renal hydrogen ion secretion. Kidney Int. *1:* 280–296, 1972.
376. Malnic, G., Klose, R. M., and Giebisch, G.: Micropuncture study of renal potassium excretion in the rat. Am. J. Physiol. *206:* 674–686, 1964.
377. Manchester, R. C.: The diurnal rhythm in water and mineral exchange. J. Clin. Invest. *12:* 995–1008, 1933.
378. Mann, T., and Keilin, D.: Sulfanilamide as a specific inhibitor of carbonic anhydrase. Nature (London) *146:* 164–165, 1940.
379. Maren, T. H.: Carbonic anhydrase: chemistry, physiology, and inhibition. Physiol. Rev. *47:* 595–781, 1967.
380. Maren, T. H.: Renal carbonic anhydrase and the pharmacology of sulfonamide inhibitors. In *Handbook of Experimental Pharmacology*. Edited by O. Eichler, A. Farah, H. Herken, and A. D. Welch. Vol. 14: *Diuretica*, pp. 195–256, edited by H. Herken. Springer-Verlag, Berlin, 1969.
381. Marliss, E. G., Aoki, T. T., Pozefsky, T., Most, A. S., and Cahill, G. F., Jr.: Muscle and splanchnic glutamine and glutamate metabolism in post-absorptive and starved man. J. Clin. Invest. *50:* 814–817, 1971.
382. Maron, S. H., and Prutton, C. F.: *Principles of Physical Chemistry*, Ed. 4. Macmillan Co., London, 1965.
383. Marsh, D. J., Ullrich, K. J., and Rumrich, G.: Micropuncture analysis of the behavior of potassium ions in rat renal cortical tubules. Pfluegers Arch. Gesamte Physiol. *277:* 107–119, 1963.
384. Marshall, E. K., Jr.: The effect of loss of carbon dioxide on the hydrogen ion concentration of urine. J. Biol. Chem. *51:* 3–10, 1922.
385. Masoro, E. J., and Siegel, P. D.: *Acid-Base Regulation: Its Physiology and Pathophysiology*. W. B. Saunders Co., Philadelphia, 1971.
386. Massry, S. G., Katz, A. I., Agmon, J., and Toor, M.: Effect of water diuresis on electrolyte and hydrogen ion excretion in hot climate. Nephron *5:* 124–133, 1968.
387. Maurice, P. F., and Henneman, P. H.: Medical aspects of renal stones. Medicine (Baltimore) *40:* 315–346, 1961.
388. McCance, R. A., and Widdowson, E. M.: The response of the kidney to an alkalosis during salt deficiency. Proc. R. Soc. London [Biol.] *120:* 228–239, 1936.
388a. McCollum, E. V.: *A History of Nutrition: The Sequence of Ideas in Nutritional Investigations*. Houghton Mifflin Co., Boston, 1957.
389. McDermott, W. V., Jr.: The role of ammonia intoxication in hepatic coma. Bull. N.Y. Acad. Med. *34:* 357–365, 1958.
390. McGee, L. C., and Hastings, A. B.: The carbon dioxide tension and acid-base balance of jejunal secretions in man. J. Biol. Chem. *142:* 893–904, 1942.
391. McKay, C. N., Wick, A. N., Carne, H. D., and Barnum, C. P.: The influence of alkalosis and acidosis upon fasting ketosis. J. Biol. Chem. *138:* 63–68, 1941.
391a. Meister, A.: *Biochemistry of the Amino Acids*, Ed. 2. Pergamon Press, New York, 1965.
392. Meldrum, N. U., and Roughton, F. J. W.: Carbonic anhydrase. Its preparation and

properties. J. Physiol. (London) *80:* 113–142, 1933.

393. Melick, R. A., and Henneman, P. H.: Clinical and laboratory studies of 207 consecutive patients in a kidney stone clinic. New Eng. J. Med. *259:* 307–314, 1958.
394. Menaker, W.: Buffer equilibria and reabsorption in the production of urinary acidity. Am. J. Physiol. *154:* 174–178, 1948.
395. *The Merck Manual of Diagnosis and Therapy*, Ed. 11. Merck and Co., Rahway, N. J., 1966.

395a. Messeter, K., Pontén, A., and Siesjö, B. K.: The influence of deep barbiturate anesthesia upon the regulation of extra- and intracellular pH in the rat brain during hypercapnia. Acta Physiol. Scand. *85:* 174–182, 1972.

396. Messeter, K., and Siesjö, B. K.: The effect of acute and chronic hypercapnia upon the lactate, pyruvate, α-ketoglutarate, glutamate, and phosphocreatine contents of the rat brain. Acta Physiol. Scand. *83:* 344–351, 1971.
397. Metcalfe-Gibson, A., McCollum, F. M., Morrison, R. B. I., and Wrong, O.: Urinary excretion of hydrogen ion in patients with uric acid calculi. Clin. Sci. *28:* 325–342, 1965.
398. Meyerhof, O.: *Die Chemischen Vorgänge im Muskel und Ihr Zusammenhang mit Arbeitsleistung und Wärmebildung*. Springer-Verlag, Berlin, 1930.
399. Miller, A., Teirstein, A. S., Duberstein, E. L., Chusid, E. L., Bader, M. E., and Bader, R. A.: Use of oxygen inhalation in evaluation of respiratory acidosis in patients with apparent metabolic alkalosis. Am. J. Med. *45:* 513–519, 1968.
400. Mills, J. N., and Stanbury, S. W.: A reciprocal relationship between K^+ and H^+ excretion in the diurnal excretory rhythm in man. Clin. Sci. *13:* 177–186, 1954.
401. Mills, J .N., Thomas, S., and Williamson K. S.: The effects of intravenous aldosterone and hydrocortisone on the urinary electrolytes of the recumbent human subject. J. Physiol. (London) *156:* 415–423, 1961.
402. Milne, M. D.: Renal tubular dysfunction. In *Diseases of the Kidney*, pp. 786–840. Edited by M. B. Strauss and L. G. Welt. Little, Brown and Co., Boston, 1963.
403. Milne, M. D., Scribner, B. H., and Crawford, M. A.: Non-ionic diffusion and the excretion of weak acids and bases. Am. J. Med. *24:* 709–729, 1958.
404. Minkowski, O.: Aus der med. Klin. in Königsberg. Über das Vorkommen von Oxybuttersäure im Harn bei Diabetes mellitus. Ein Beitrag zur Lehre von Coma diabeticum. Arch. Exp. Pathol. Pharmakol. *18:* 35–48, 1884.
405. Miquel, R.: Einiges über die Wirkung der Schwefelsäure auf den thierischen Organismus. Arch. Physiol. Heilk. *10:* 479–482, 1851.
406. Mitchell, H. S., Rynbergen, J. J., Anderson, L., and Dibble, M. V.: *Cooper's Nutrition in Health and Disease*, Ed. 15. J. B. Lippincott Co., Philadelphia, 1968.

406a. Mitchell, J. H., Wildenthal, K., and Johnson, R. L., Jr.: The effects of acid-base disturbance on cardiovascular and pulmonary function. Kidney Int. *1:* 375–389, 1972.

407. Mitchell, R. A., Herbert, D. A., and Corman, C. T.: Acid-base constants and temperature coefficients for cerebrospinal fluid. J. Appl. Physiol. *20:* 27–30, 1965.
408. Molière: *Le Malade Imaginaire*. In *Oeuvres Complètes*, Vol. 6, p. 435. Nelson, Paris, 1900.
409. Montgomery, H., and Pierce, J.: The site of acidification of urine within the renal tubule in the amphibia. Am. J. Physiol. *118:* 144–152, 1937.
410. Mookerjee, B. K., Allen, A. C., and Dossetor, J. B.: Validity of net renal hydrogen ion excretion in the presence of urinary tract infection. J. Lab. Clin. Med. *71:* 582–585, 1968.
411. Morris, R. C., Jr.: Renal tubular acidosis. Mechanisms, classification and implications. New Eng. J. Med. *281:* 1405–1413, 1969.

411a. Morris, R. C., Jr., Sebastian, A., and McSherry, E.: Renal acidosis. Kidney Int. *1:* 322–340, 1972.

412. Morris, R. C., Jr., Yamauchi, H., Palubinskas, A. J., and Howenstine, J.: Medullary sponge kidney. Am. J. Med. *38:* 883–892, 1965.
413. Muldowney, F. P., Carroll, D. V., Donohoe, J. F., and Freaney, R.: Correction of renal bicarbonate wastage by parathyroidectomy. Implications in acid-base homeostasis. Q. J. Med. *160(NS):* 487–498, 1971.
414. Mulrow, P. J.: Aldosterone in hypertension and edema. In *Duncan's Diseases of Metabolism*, Ed. 6, Ch. 20. Edited by P. K. Bondy and L. E. Rosenberg. W. B. Saunders Co., Philadelphia, 1969.
415. Munro, A. F.: Nitrogen excretion and arginine activity during amphibian development. Biochem. J. *33:* 1957–1965, 1939.
416. Munro, H. N.: Evolution of protein metabolism in mammals. In *Mammalian Protein Metabolism*, Vol. 3, pp. 133–182, Edited by H. N. Munro, Academic Press, New York, 1969.
417. Munro, H. N.: A general survey of mechanisms regulating protein metabolism in mammals. In *Mammalian Protein Metabolism*, Vol. 4, pp. 3–130. Edited by H. N Munro. Academic Press, New York, 1970.

417a. Muntwyler, E., and Griffin, G. E.: Effect of potassium on electrolytes of rat plasma and muscle. J. Biol. Chem. *193:* 563–573, 1951.

417b. Muntwyler, E., Griffin, G. E., Samuelson,

G. S., and Griffith, L. G.: The relation of the electrolyte composition of plasma and skeletal muscle. J. Biol. Chem. *185:* 525–536, 1950.
418. Nahas, G. G., Editor: *Current Concepts of Acid-Base Measurement.* Ann. N.Y. Acad. Sci. *133:* 1–274, 1966.
419. Nash, T. P., and Benedict, S. R.: The ammonia content of the blood and its bearing on the mechanism of acid neutralization in the animal organism. J. Biol. Chem. *48:* 463–488, 1921.
420. Nathan, D. G., and Rodkey, F. L.: A colorimetric procedure for the determination of blood ammonia. J. Lab. Clin. Med. *49:* 779–785, 1957.
420a. Needham, J.: *Biochemistry and Morphogenesis.* Cambridge University Press, New York, 1950.
421. Neubauer, C.: Über den Ammoniakgehalt des normalen Harns. J. Prakt. Chem. *64:* 177–187 and 278–282, 1855.
422. Newton, R. F.: Thermodynamics. *Encyclopaedia Brittanica,* Vol. 21, pp. 1015–1026. Encyclopaedia Brittanica, Inc., William Benton, Chicago, 1967.
423. Nollet, J.-A.: Recherches sur les causes du bouillonnement des liquides. Mém. Acad. R. Sci. (Paris), 1748.
424. Nordmann, J., and Nordmann, R.: Organic acids in blood and urine. In *Advances in Clinical Chemistry,* pp. 53–120. Edited by C. P. Stewart and H. Sobotka. Academic Press, New York and London, 1961.
425. Norn, M.: Über Schwankungen der Kalium-, Natrium-, und Chlorid-Ausscheidung durch die Niere im Laufe des Tages. Scand. Arch. Physiol. *55:* 184–210, 1929.
426. Nutbourne, D. M.: The effect of dilution on the titratable acid in urine and acidified phosphate buffer solutions, and the correction for this effect in the determination of the rate of elimination of hydrogen ions from the body by the renal tubules. Clin. Sci. *20:* 263–278, 1961.
427. *Nutritional Data,* Ed. 3. H. J. Heinz Co., Pittsburgh, 1956.
428. Ochwadt, B. K., and Pitts, R. F.: Effects of intravenous infusion of carbonic anhydrase on carbon dioxide tension of alkaline urine. Am. J. Physiol. *185:* 426–429, 1956.
429. Oelert, H., Hills, A. G., Rumrich, G., and Ullrich, K. J.: Corticaler Ammoniakdruck und Ammoniakausscheidung in der Ratteniere. In *Aktuelle Probleme der Nephrologie,* pp. 537–543. IV Symposion der Gesellschaft für Nephrologie. Springer-Verlag, Berlin, 1966.
430. Oelert, H., Uhlich, E., and Hills, A. G.: Messungen des Ammoniakdruckes in den corticalen Tubuli der Rattenniere. Pfluegers Arch. Gesamte Physiol. *300:* 35–48, 1968.
430a. Oelert, H., Uhlich, E., and Hills, A. G.: Transport of ammonia in the mammalian cortical nephron. In *Renal Transport and Diuretics,* pp. 25–27. Edited by K. Thurau and H. Jahrmärker. International Symposium, Feldafing, June 21–23, 1968. Springer-Verlag, Berlin, 1969.
431. Ohman, J. L., Marliss, E. B., Aoki, T. T., Munichoodappa, C. S., Khanna, V. V., and Kozak, G. P.: The cerebrospinal fluid in diabetic ketoacidosis. New Eng. J. Med. *284:* 283–290, 1971.
431a. Oliva, P. B.: Severe alveolar hypoventilation in a patient with metabolic alkalosis. Am. J. Med. *52:* 817–821, 1972.
432. Oliver, J.: The antithesis of structure and function in renal activity. Bull. N.Y. Acad. Med. *37:* 81–118, 1961.
433. Onsager, L.: The motions of ions: principles and concepts. Science *166:* 1359–1364, 1969.
434. Orloff, J., and Berliner, R. W.: The mechanism of the excretion of ammonia in the dog. J. Clin. Invest. *35:* 223–235, 1956.
435. Orloff, J., and Burg, M.: The kidney. Annu. Rev. Physiol. *33:* 83–130, 1971.
436. Orloff, J., and Davidson, D. G.: The mechanism of potassium excretion in the chicken. J. Clin. Invest. *38:* 21–30, 1959.
437. O'Shaughnessy, W. B.: Experiments on the blood in cholera. Lancet *1:* 490, 1831–1932.
438. Østberg, O.: Der Citronensäuregehalt des Harns bei Acidose und Alkalose. Biochem. Z. *226:* 162–163, 1930.
439. Ostwald, W.: Elektrochemische Studien. Dritte Abhandlung. Über den Einfluss der Zusammensetzung und Constitution der Säuren auf ihre elektrische Leitfähigkeit. J. Prakt. Chem. *32(NS):* 300–374, 1885.
440. Ostwald, W.: Über die Affinitätsgrössen organischer Säuren und ihre Beziehungen zur Zusammensetzung und Konstitution derselben. Z. Physik. Chem. *5:* 170–197, 1889.
441. Ostwald, W.: Davy. In *Das Buch der Grossen Chemiker,* Vol. 1, pp. 405–416. Edited by G. Bunge. Verlag-Chemie, Weinheim/Bergstrasse, 1929; reprinted 1965.
442. Owen, E. E., and Robinson, R. R.: Amino acid extraction and ammonia metabolism by the human kidney during the prolonged administration of ammonium chloride. J. Clin. Invest. *42:* 263–276, 1963.
443. Owen, E. E., Tyor, M. P., Flanagan, J. R., and Berry, J. N.: The kidney as a source of blood ammonia in patients with liver disease: the effect of acetazolamide. J. Clin. Invest. *39:* 288–294, 1960.
444. *The Oxford English Dictionary.* Vol. 1, p. 1158; Vol. 13 (suppl.), p. 127–128. Clarendon Press, Oxford, 1933.
445. *The Oxford English Dictionary.* 12 vols. and suppl. Clarendon Press, Oxford, 1933.
446. Palmaer, W.: Arrhenius. In *Das Buch der*

Grossen Chemiker, pp. 443–462. Edited by G. Bunge. Verlag-Chemie, Weinheim/Bergstrasse, 1929; reprinted 1965.

447. Parsons, T. R.: On the reaction of the blood in the body. J. Physiol. (London) *51:* 440–459, 1917.
448. Paschkis, K. E., Rakoff, A. E., Cantarow, A., and Rupp, J. J.: *Clinical Endocrinology*, Ed. 3, Ch. 5. Harper and Row, New York, 1967.
449. Pauling, L.: *The Nature of the Chemical Bond and the Structure of Molecules and Crystals: An Introduction to Modern Structural Chemistry*, Ed. 3. Cornell University Press, Ithaca, N. Y., 1960.
450. Pauling, L., and Hayward, R.: *The Architecture of Molecules.* W. H. Freeman and Co., San Francisco, 1964.
451. Pequin, L.: Dégradation et synthèse de la glutamine chez la carpe (Cyprinus carpio L). Arch. Sci. Physiol. (Paris) *21:* 193–203, 1967.
452. Pequin, L., and Serfaty, F.: La régulation hépatique et intestinale de l'ammoniémie chez la carpe. Arch. Sci. Physiol. (Paris) *22:* 449–459, 1968.

452a. Peters, J. P., Jr.: The response of the respiratory mechanism to rapid changes in the reaction of the blood. Am. J. Physiol. *44:* 84–108, 1917.

453. Peters, J. P.: *Body Water: The Exchange of Fluids in Man.* Charles C Thomas, Springfield, Ill., 1935.
454. Peters, J. P., and Van Slyke, D. D.: *Quantitative Clinical Chemistry*, Ed. 2, Vol. 1: *Interpretations.* Williams & Wilkins Co., Baltimore, 1946.
455. Pfeffer, W. F. P.: *Osmotische Untersuchungen.* W. Englemann, Leipzig, 1887.
456. Pflüger, E.: Über die Ursache der Atombewegungen, sowie der Dyspnoë und Apnoë. Pfluegers Arch. Gesamte Physiol. *1:* 61–106, 1868.

456a. pH of intravenous solutions; pH vs. acid load (Letters to the Editor). New Eng. J. Med. *280:* 332; 900–902; and 1480–1481, 1969.

457. Piéron, H.: *Le Problème Physiologique du Sommeil.* Masson et Cie, Paris, 1913.
458. Pilkington, L. A., Young, T.-K., and Pitts, R. F.: Properties of renal luminal and antiluminal transport of plasma glutamine. Nephron *7:* 51–60, 1970.
459. Pines, K. L., and Mudge, G. H.: Renal tubular acidosis with osteomalacia. Am. J. Med. *11:* 302–311, 1951.
460. Pitts, R. F.: The comparison of urea with urea + ammonia clearances in acidotic dogs. J. Clin. Invest. *15:* 571–575, 1936.
461. Pitts, R. F.: Renal excretion of acid. Fed. Proc. *7:* 418–426, 1948.
462. Pitts, R. F.: Mechanisms for stabilizing the alkaline reserves of the body. In *The Harvey Lectures*, 1952–1953, pp. 172–209. Academic Press, New York.
463. Pitts, R. F.: Present status of our knowledge concerning ammonium excretion: biochemical mechanisms and physiologic regulation. In *Proceedings, Fifteenth Annual Conference on the Kidney, Swampscott, 1963*, pp. 32–48. Edited by J. Metcoff. National Kidney Foundation, New York, 1964.
464. Pitts, R. F.: *Physiology of the Kidney and Body Fluids: An Introductory Text*, Ed. 1. Year Book Medical Publishers, Chicago, 1963.
465. Pitts, R. F.: Renal production and excretion of ammonia. Am. J. Med. *36:* 720–742, 1964.
466. Pitts, R. F.: The renal metabolism of ammonia. Physiologist *9:* 97–109, 1966.
467. Pitts, R. F.: *Physiology of the Kidney and Body Fluids: An Introductory Text*, Ed. 2. Year Book Medical Publishers, Chicago, 1968.
468. Pitts, R. F.: Renal metabolism of amino acids and ammonia. In *Renal Transport and Diuretics*, pp. 11–24. Edited by K. Thurau and H. Jahrmärker. International Symposium, Feldafing, June 21–23, 1968. Springer-Verlag, Berlin, 1969.
469. Pitts, R. F.: Non-ionic diffusion and ammonia secretion. In *Proceedings of the Fourth International Congress of Nephrology, Stockholm, 1969*, Vol. 1: *Embryology, Ultrastructure, Physiology*, pp. 195–205. Edited by N. Alwall, F. Berglund, and B. Josephson S. Karger, Basel, 1970.
470. Pitts, R. F.: The role of ammonia production and excretion in regulation of acid-base balance. New Eng. J. Med. *284:* 32–38, 1971.

470a. Pitts, R. F.: Control of renal production of ammonia. Kidney Int. *1:* 297–305, 1972.

471. Pitts, R. F., and Alexander, R. S.: The nature of the renal tubular mechanism for acidifying the urine. Am. J. Physiol. *144:* 239–254, 1945.
472. Pitts, R. F., Ayer, J. L., and Schiess, W. A.: The renal regulation of acid-base balance in man. III. The reabsorption and excretion of bicarbonate. J. Clin. Invest. *28:* 35–44, 1949.
473. Pitts, R. F., and Lotspeich, W. D.: Bicarbonate and the renal regulation of acid-base balance. Am. J. Physiol. *147:* 138–154, 1946.
474. Pitts, R. F., and Stone, W. J.: Renal metabolism and excretion of ammonia. In *Proceedings of the Third International Congress of Nephrology*, Vol. 1: *Physiology*, pp. 123–135. Edited by J. S. Handler. S. Karger, Basel, 1967.
475. Pledge, H. T.: *Science Since 1500: A Short History of Mathematics, Physics, Chemistry, and Biology.* H. M. Stationery Office, London, 1939; reprinted 1959, Harper Torchbook Science Library, Harper and Row, New York.

475a. Plum, F.: Hyperpnea, hyperventilation, and brain dysfunction. Ann. Intern. Med. *76:* 328, 1972.
476. Pollak, V. E., and Mattenheimer, H.: Glutaminase activity in the kidney in gout. J. Lab. Clin. Med. *66:* 564–570, 1965.
477. Pollak, V. E., Mattenheimer, H., de Bruin. H., and Weinman, K.: Experimental metabolic acidosis. The enzymatic basis of ammonia production by the dog kidney. J. Clin. Invest. *44:* 169–181, 1965.
478. Poppell, J. W., Cuajunco, F., Jr., Horsley, J. S. III, Randall, H. T., and Roberts, K. E.: Renal arteriovenous ammonium difference and total ammonium production in normal, acidotic, and alkalotic dogs (Abstract). Clin. Res. Proc. *4:* 137, 1956.
479. Portwood, R. M., Seldin, D. W., Rector, F. C., Jr., and Cade, R.: The relation of urinary CO_2 tension to bicarbonate excretion. J. Clin. Invest. *38:* 770–776, 1959.
480. Preuss, H. G., Bise, B. B., and Schreiner, G. E.: The determination of glutamine in plasma and urine. Clin. Chem. *12:* 329–337, 1966.
481. Preuss, H. G., Davis, B. B., Maher, J. F., Bise, B. W., and Schreiner, G. E.: Ammonia metabolism in renal failure. Ann. Intern. Med. *65:* 54–61, 1966.
482. Prutton, C. F., and Maron, S. H.: *Fundamental Principles of Physical Chemistry.* Macmillan Co., New York, 1944.
483. Purkerson, M. L., Lubowitz, H., White, R. W., and Bricker, N. S.: On the influence of extracellular fluid volume expansion on bicarbonate reabsorption in the rat. J. Clin. Invest. *48:* 1754–1760, 1969.
484. Raoult, F.-M.: Loi générale des tensions de vapeur des dissolvants. Compt. Rend. *104:* 1430–1433, 1887.
485. Rapoport, A., Crassweller, P. O., Husdan, H., From, G. L. A., Zweig, M., and Johnson M. D.: The renal excretion of hydrogen ion in uric acid stone formers. Metabolism *16:* 176–188, 1967.
486. Rector, F. C., Jr.: Micropuncture studies on the mechanism of urinary acidification. In *Renal Metabolism and Epidemiology of Some Renal Diseases*, pp. 9–24. Edited by J. Metcoff. Proceedings of the Fifteenth Annual Conference on the Kidney. National Kidney Foundation, New York, 1964.
487. Rector, F. C., Jr.: Renal secretion of hydrogen. In *The Kidney: Morphology, Chemistry, Physiology*, Vol. 3, pp. 209–252. Edited by C. Rouiller and A. F. Muller. Academic Press, New York, 1971.
487a. Rector, F. C., Jr.: Introduction, symposium on acid-base homeostasis. Kidney Int. *1:* 273–274, 1972.
488. Rector, F. C., Jr., Buttram, H., and Seldin, D. W.: An analysis of the mechanism of the inhibitory influence of K^+ on renal H^+ secretion. J. Clin. Invest. *41:* 611–617, 1962.
489. Rector, F. C., Jr., Carter, N. W., and Seldin, D. W.: The mechanism of bicarbonate reabsorption in the proximal and distal tubules of the kidney. J. Clin. Invest. *44:* 278–290, 1965.
490. Rector, F. C., Jr., Seldin, D. W., Roberts, A. D., Jr., and Smith, J. S.: The role of plasma CO_2 tension and carbonic anhydrase activity in the renal reabsorption of bicarbonate. J. Clin. Invest. *39:* 1706–1721, 1960.
491. Refsum, H. E.: Hypokalemic alkalosis with paradoxical aciduria during artificial ventilation of patients with pulmonary insufficiency and high plasma bicarbonate concentration. Scand. J. Clin. Lab. Invest. *13:* 481–488, 1961.
492. Refsum, H. E.: Acid-base status in patients with chronic hypercapnia and hypoxaemia. Clin. Sci. *27:* 407–415, 1964.
493. Refsum, H. E.: Acid-base disturbances in chronic pulmonary disease. Ann. N.Y. Acad. Sci. *133:* 142–159, 1966.
494. Reid, E. L.: Exaltation and abatement of urinary ammonia excretion (Abstract). J. Clin. Invest. *44:* 1089–1090, 1965.
495. Reid, E. L., and Hills, A. G.: Diffusion of CO_2 out of the distal nephron during antidiuresis in man. Clin. Sci. *28:* 15–28, 1965.
496. Reid, E. L., and Hills, A. G.: Renal ammonia balance (Abstract). Fed. Proc. *25:* 203, 1966.
497. Reid, E. L., and Hills, A. G.: Generalized first-order equations to describe renal ammonia balance in diuretic mammals (Abstract). Clin. Res. *17:* 54A, 1969.
498. Reid, E. L., and Hills, A. G.: The effect of delayed dehydration of carbonic acid on renal bicarbonate clearance and its significance for acid-base balance. Clin. Sci. *37:* 381–393, 1969.
499. Relman, A. S.: Renal acidosis and renal excretion of acid in health and disease. In *Advances in Internal Medicine*, Vol. 12, pp. 1–375. Edited by William Dock and I. Snapper. Year Book Medical Publishers, Chicago, 1964.
500. Relman, A. S.: The acidosis of renal disease. Am. J. Med. *44:* 706–713, 1968.
501. Relman, A. S.: The control of acid secretion. In *Proceedings of the Fourth International Congress of Nephrology, Stockholm, 1969*, Vol. 1: *Embryology, Ultrastructure, Physiology*, pp. 175–180. S. Karger, Basel, 1970.
501a. Relman, A. S.: Metabolic consequences of acid-base disorders. Kidney Int. *1:* 347–359, 1972.
501b. Relman, A. S., Efsten, B., and Schwartz, W. B.: The regulation of renal bicarbonate reabsorption by plasma carbon dioxide tension. J. Clin. Invest. *32:* 972–978, 1953.

502. Relman, A. S., Lennon, E. J., and Lemann, J., Jr.: Endogenous production of fixed acid and the measurement of the net balance of acid in normal subjects. J. Clin. Invest. *40:* 1621–1630, 1961.
503. Relman, A. S., and Schwartz, W. B.: The effect of DOCA on electrolyte balance in normal man and its relation to sodium chloride intake. Yale J. Biol. Med. *24:* 540–558, 1952.
504. Relman, A. S., and Schwartz, W.: Effects of electrolyte disorders on renal structure and function. In *Renal Disease*, Ed. 2, Ch. 29. Edited by D. A. K. Black. F. A. Davis Co., Philadelphia, 1967.
505. Reynolds, T. B.: Observations on the pathogenesis of renal tubular acidosis. Am. J. Med. *25:* 503–515, 1958.
506. Rhorer, L. v.: Die Bestimmung der Harnacidität auf elektrometrischem Wege. Pfluegers Arch. Gesamte Physiol. *86:* 586–602, 1901.
507. Richterich, R. W., and Goldstein, L.: Distribution of glutamine metabolizing enzymes and production of urinary ammonia in the mammalian kidney. Am. J. Physiol. *195:* 316–320, 1958.
508. Riggs, D. S.: *The Mathematical Approach to Physiological Problems: A Critical Primer*. Williams & Wilkins Co., Baltimore, 1963.
509. Ringer, W. E.: Zur Acidität des Harns. Z. Physiol. Chem. *60:* 341–363, 1909.
510. Robert, P.: *Dictionnaire Alphabétique et Analogue de la Langue Française*, Vol. 6, p. 464. Société du Nouveau Littré, Paris, 1966.
511. Roberts, K. E., Magida, M. G., and Pitts, R. F.: Relationship between potassium and bicarbonate in blood and urine. Am. J. Physiol. *172:* 47–54, 1953.
512. Roberts, K. E., Randall, H. T., Sanders, H. L., and Hood, M.: Effects of potassium on renal tubular reabsorption of bicarbonate. J. Clin. Invest. *34:* 666–672, 1955.
513. Robbins, O., Jr.: *Ionic Reactions and Equilibria*. Macmillan Co., New York, 1967.
514. Robin, E. D.: Abnormalities of acid-base regulation in chronic pulmonary disease, with special reference to hypercapnia and extracellular alkalosis. New Eng. J. Med. *268:* 917–922, 1963.
515. Robin, E. D., and Bromberg, P. A.: Claude Bernard's *milieu intérieur* extended: intracellular acid-base relationships. Am. J. Med. *27:* 689–692, 1959.
516. Robinson, H. W., Price, J. W., and Cullen, G. E.: Studies of the acid-base condition of the blood. III. The value of pK_1 in the Henderson-Hasselbalch equation for human and dog sera determined with the Simms electrode. J. Biol. Chem. *106:* 7–27, 1934.
517. Robinson, J. R.: *Fundamentals of Acid-Base Regulation*, Ed. 3. Blackwell Scientific Publications, Oxford; reprinted 1969.
518. Rodriguez-Soriano, J., and Edelmann, C. M., Jr.: Renal tubular acidosis. Annu. Rev. Med. *29:* 363–382, 1969.
519. Roos, A.: Intracellular pH and buffering power of rat muscle. Am. J. Physiol. *221:* 182–188, 1971.
520. Rosado, A., Flores, G., Mora, J., and Soberón, G.: Distribution of an ammonia load in the normal rat. Am. J. Physiol. *203:* 37–42, 1962.
521. Röse, C.: Die Beziehungen zwischen Eiweissmindestbedarf und Basengehalt der menschlichen Nahrung. Z. Gesamte Exp. Med. *96:* 793–798, 1935.
522. Rosenbaum, J. D., Ferguson, B. C., Davis, R. K., and Rossmeisl, E. C.: The influence of cortisone upon the diurnal rhythm of renal excretory function. J. Clin. Invest. *31:* 507–520, 1952.
523. Roth, D. G., and Gamble, J. L., Jr.: Deoxycorticosterone-induced alkalosis in dogs. J. Physiol. (London) *208:* 90–93, 1965.
524. Rotheram, E. B., Jr., Safar, P., and Robin, E. D.: CNS disorder during mechanical ventilation in chronic pulmonary disease. J.A.M.A. *189:* 993–996, 1964.
525. Roughton, F. J. W.: The kinetics and rapid thermochemistry of carbonic acid. J. Am. Chem. Soc. *63:* 2930–2934, 1941.
526. Roughton, F. J. W.: Some recent work on the chemistry of carbon dioxide transport by the blood. Harvey Lect. *39:* 96–142, 1943–1944.
527. Roughton, F. J. W.: Respiratory function of the blood. In *Respiratory Physiology in Aviation*, Ch. 5, pp. 51–102. Edited by W. M. Boothby, U.S. Air Force School of Aviation Medicine, 1954.
528. Roughton, F. J. W.: Transport of oxygen and carbon dioxide. In *Handbook of Physiology*, Sect. 3: *Respiration*, Vol. 1, Ch. 31. American Physiological Society, Washington, D. C., 1964.
528a. Roxe, D. M., DiSalvo, J., and Balagura-Baruch, S.: Renal glucose production in the intact dog. Am. J. Physiol. *218:* 1676–1681, 1970.
529. Royer, P., and Broyer, M.: L'acidose rénale au cours des tubulopathies congénitales. In *Actualités Néphrologiques de l'Hopital Necker*, pp. 73–92. Éditions Médicales Flammarion, Paris, 1967.
530. Rubini, M. E., Blythe, W. B., Herndon, E. G., and Meroney, W. H.: Influence of potassium deficiency in response to an acidifying salt (Abstract). Clin. Res. Proc. *5:* 193, 1957.
531. Rubinstein, R., Howard, A. V., and Wrong, O. M.: *In vivo* dialysis of faeces as a method of stool analysis. IV. The organic anion component. Clin. Sci. *37:* 549–564, 1969.
532. Ruderman, W. B., and Lund, P.: Amino acid metabolism in skeletal muscle. Regu-

lation of glutamine and alanine release in the perfused rat hindquarter. Isr. J. Med. Sci. *8:* 295–302, 1972.
533. Ruszkowski, M., Arasimovicz, C., Knapowski, J., Steffen, J., and Weiss, K.: Renal reabsorption of amino acids. Am. J. Physiol. *203:* 891–896, 1962.
534. Ryberg, C.: Some investigations on the carbon dioxide tension of the urine in man. Acta Physiol. Scand. *15:* 123–139, 1948.
535. Ryberg, C. T.: The physiological variations in the reaction of the human urine. Acta Physiol. Scand. *6:* 271–278, 1943.
536. Salkowski, E.: Über die Möglichkeit der Alkalientziehung beim lebenden Thier. Virchows Arch. *58:* 1–34, 1873.
537. Salkowski, E.: Bemerkungen über die Wirkung der unorganischen Säuren und der Fleischnahrung. Virchows Arch. *76:* 368–373, 1879.
538. Salkowski, E., und Munk, I.: Über die Beziehungen der Reaction des Harns zu seinem Gehalt an Ammoniaksalzen. Virchows Arch. *71:* 500–508, 1877.
538a. Schenker, S., McCandless, D. W., Brophy, E., and Lewis, M. S.: Studies on the intracerebral toxicity of ammonia. J. Clin. Invest. *46:* 838–848, 1957.
539. Schiess, W. A., Ayer, J. L., Lotspeich, W. D., and Pitts, R. F.: The renal regulation of acid-base balance in man. II. Factors affecting the excretion of titratable acid by the normal human subject. J. Clin Invest. *27:* 47–64, 1948.
540. Schloeder, F. X., and Stinebaugh, B. J.: Defect of urinary acidification during fasting. Metabolism *15:* 17–25, 1966.
541. Schlösing, T.: Über die Bestimmung des Ammoniaks und namentlich über die Bestimmung desselben im Tabak. J. Prakt. Chem. *52:* 372–382, 1851.
542. Schmidt, C.: *Charakteristik der Epidemischen Cholera gegenüber verwandten Transudationsanomalieen: Eine Physiologisch-Chemische Untersuchung.* G. A. Reyher, Leipzig and Mitau, 1850.
543. Schmidt-Nielsen, B.: Urea excretion in mammals. Physiol. Rev. *38:* 139–168, 1958.
543a. Schmidt-Nielsen, B., and Kerr, D. N. S., Editors: *Urea and the Kidney.* Excerpta Medica Foundation, Amsterdam, 1970.
544. Schoenheimer, R.: The dynamic state of body constituents. *Harvard University Monographs in Medicine and Public Health.* Harvard University Press, Cambridge, Mass., 1942.
545. Schwartz, W. B., Bank, N., and Cutler, R. W. P.: The influence of urinary ionic strength on phosphate pK_2' and the determination of titratable acid. J. Clin. Invest. *38:* 347–356, 1959.
546. Schwartz, W. B., Brackett, N. C., Jr., and Cohen, J. J.: Defense of the hydrogen ion concentration during acute and chronic hypercapnia; the response to progressive elevation of carbon dioxide tension. Trans. Assoc. Am. Physicians *77:* 182–187, 1964.
547. Schwartz, W. B., Brackett, N. C., and Cohen, J. J.: The response of extracellular hydrogen ion concentration to graded degrees of chronic hypercapnia. The physiologic limits of the defense of pH. J. Clin. Invest. *44:* 291–301, 1965.
548. Schwartz, W. B., Hall, P. W., Hays, R. M., and Relman, A. S.: On the mechanism of acidosis in chronic renal disease. J. Clin. Invest. *38:* 39–52, 1959.
548a. Schwartz, W. B., Jensen, R. L., and Relman, A. S.: The disposition of acid administered to sodium-depleted subjects: the renal response and the role of the whole body buffers. J. Clin. Invest. *33:* 587–597, 1954.
549. Schwartz, W. B., Jensen, R. L., and Relman A. S.: Acidification of the urine and increased ammonium excretion without changes in acid-base equilibrium: sodium reabsorption as a stimulus to the acidifying process. J. Clin. Invest. *34:* 673–680, 1955.
549a. Schwartz, W. B., Ørning, K. J., and Porter, R.: The internal distribution of hydrogen ions with varying degrees of metabolic acidosis. J. Clin. Invest. *36:* 373–382, 1957.
550. Schwartz, W. B., van Ypersele de Strihou, C., and Kassirer, J. P.: Role of anions in metabolic alkalosis and potassium deficiency. New Eng. J. Med. *279:* 630–639, 1968.
551. Seegmiller, J. E., Schwartz, R., and Davidson, C. S.: The plasma "ammonia" and glutamine content in patients with hepatic coma. J. Clin. Invest. *33:* 984–988, 1954.
551a. Seldin, D. W., Carter, N. W., and Rector, F. C., Jr.: Consequences of renal failure and their management. In *Diseases of the Kidney*, Ed. 2, Ch. 6. Edited by M. B. Strauss and L. G. Welt. Little, Brown and Co., Boston, 1971.
552. Seldin, D. W., Portwood, R. M., Rector, M., Rector, F. C., Jr., and Cade, R.: Characteristics of renal bicarbonate reabsorption in man. J. Clin. Invest. *38:* 1663–1671, 1959.
552a. Seldin, D. W., and Rector, F. C.,Jr.: The generation and maintenance of metabolic alkalosis. Kidney. Int. *1:* 306–321, 1972.
552b. Seldin, D. W., Welt, L. G., and Cort, J. H.: The role of sodium salts and adrenal steroids in the production of hypokalemic alkalosis. Yale J. Biol. Med. *29:* 229–247 1956.
553. Seldin, D. W., and Wilson, J. D.: Renal tubular acidosis. In *The Metabolic Basis of Inherited Disease*, Ed. 2, Ch. 54, pp. 1230–1246. Edited by J. B. Stanbury, J. B. Wyngarden, and D. S. Frederickson. Blakiston Division, McGraw-Hill Book Co., New York, 1966.

554. Sellards, A. W.: *The Principles of Acidosis and Clinical Methods for Its Study.* Harvard University Press, Cambridge, Mass., 1919.
555. Severinghaus, J. W.: Blood gas concentrations. In *Handbook of Physiology,* Sect. 3: *Respiration,* Vol. 2, Ch. 61. Edited by W. O. Fenn and H. Rahn. American Physiological Society, Washington, D. C., 1965.
556. Severinghaus, J. W., Stupfel, M., and Bradley, A. F.: Variations of serum carbonic acid pK′ with pH and temperature. J. Appl. Physiol. *9:* 197–200, 1956.
557. Shaffer, P.: On the quantitative determination of ammonia in urine. Am. J. Physiol. *8:* 330–354, 1903.
558. Shalhoub, R., Weber, W., Glabman, S., Canessa-Fischer, M., Klein, J., de Haas, J., and Pitts, R. F.: Extraction of amino acids from and their addition to renal blood plasma. Am. J. Physiol. *204:* 181–191, 1963.
559. Shannon, J. A.: Glomerular filtration and urea excretion in relation to urine flow in dog. Am. J. Physiol. *117:* 206–225, 1936.
560. Shannon, J. A.: Renal tubular excretion. Physiol. Rev. *19:* 63–93, 1939.
561. Shannon, J. A., and Fisher, S.: The renal tubular reabsorption of glucose in the normal dog. Am. J. Physiol. *122:* 765–774, 1938.
562. Sherman, H. C.: *Chemistry of Food and Nutrition,* Ed. 8. Macmillan Co., New York, 1952.
563. Sherman, H. C., and Gettler, A. O.: The balance of acid-forming elements in foods, and its relation to ammonia metabolism. J. Biol. Chem. *11:* 323–338, 1912.
564. Sherman, H. C., and Sinclair, J. E.: The balance of acid-forming and base-forming elements in foods. J. Biol. Chem. *3:* 307–309, 1907.
565. Shires, G. T., and Holman, J.: Dilution acidosis. Ann. Intern. Med. *28:* 557–559, 1948.
566. Shohl, A. T.: Mineral metabolism in relation to acid-base equilibrium. Physiol. Rev. *3:* 509–543, 1923.
567. Shohl, A. T., and Sato, A.: Acid-base metabolism. I. Determination of base balance. J. Biol. Chem. *58:* 235–255, 1924.
567a. Siesjö, B. K.: The regulation of cerebrospinal fluid pH. Kidney Int. *1:* 360–374, 1972.
567b. Siesjö, B. K., and Messeter, K.: Factors determining intracellular pH. In *Ion Homeostasis of the Brain,* pp. 244–269. Edited by B. K. Siesjö and S. C. Sørensen. Alfred Benson Symposium III. Munksgaard, Copenhagen, 1971.
568. Siggaard-Andersen, O.: The first dissociation exponent of carbonic acid as a function of pH. Scand. J. Lab. Clin. Invest. *14:* 587–596, 1962.
569. Siggaard-Andersen, O.: *The Acid-Base Status of the Blood,* Ed. 2. Williams & Wilkins Co., Baltimore, 1964.
569a. Siggaard-Andersen, O.: Titratable acid or base of body fluids. In *Current Concepts of Acid-Base Measurement,* pp. 41–58. Edited by G. G. Nahas. Ann. N.Y. Acad. Sci. *133:* 1–274, 1966.
570. Simpson, D. P.: Control of hydrogen ion homeostasis and renal acidosis. Medicine (Baltimore) *50:* 503–541, 1971.
571. Simpson, G. E.: Diurnal variations in the rate of urine excretion for two hour intervals: some associated factors. J. Biol. Chem. *59:* 107–122, 1924.
572. Simpson, G. E.: The effects of sleep on urinary chlorides and pH. J. Biol. Chem. *67:* 505–516, 1926.
573. Simpson, G. E.: Changes in composition of urine brought about by sleep and other factors. J. Biol. Chem. *84:* 393–411, 1929.
574. Singer, R. B., Clark, J. D., Barker, E. S., Crosley, A. P., Jr., and Elkinton, J. R.: The acute effects in man of rapid intravenous infusion of hypertonic sodium bicarbonate solution. I. Changes in acid-base balance and distribution of the excess buffer base. Medicine (Baltimore) *34:* 51–95, 1955.
575. Singer, R. B., and Hastings, A. B.: An improved clinical method for the estimation of disturbances of the acid-base balance of human blood. Medicine (Baltimore) *27:* 223–242, 1948.
576. Skramlik, E. v.: Über Harnacidität. Z. Physiol. Chem. *71–72:* 290–310, 1911.
577. Smith, H. W.: *The Kidney: Structure and Function in Health and Disease.* Oxford University Press, New York, 1951.
578. Smith, H. W.: I. Theory of solutions. Circulation *21:* 808–817, 1960.
579. Smith, L. H., Jr., and Williams, H. E.: Kidney stones. In *Diseases of the Kidney,* Ed. 2, Ch. 26. Edited by M. B. Strauss and L. G. Welt. Little, Brown and Co., Boston, 1971.
580. Smith, M.: The minimum endogenous nitrogen metabolism. J. Biol. Chem. *68:* 15–31, 1926.
581. Sørensen, S. P. L.: Études enzymatiques. II. Sur la mesure et l'importance de la concentration des ions hydrogène dans les réactions enzymatiques. Compt. Rend. Lab. Carlsberg *8:* 1–168, 1909.
582. Sörensen, S. P. L.: Enzymstudien. II. Mitteilung. Über die Messung und die Bedeutung der Wasserstoffionenkonzentration bei enzymatischen Prozessen. Biochem. Z. *21:* 131–304, 1909.
583. Speakman, J. C.: *An Introduction to the Electronic Theory of Valency,* Ed. 3. Edward Arnold, London, 1955; reprinted 1956, 1959, 1962.
584. Spiro, K., und Pemsel, W.: Über Basen- und Säurecapacität des Blutes und der

Eiweisskörper. Z. Physiol. Chem. *26:* 233–271, 1898–1899.

585. Stadelmann, E.: Über die Ursachen der pathologischen Ammoniakausscheidung beim Diabetes mellitus und des coma diabeticum. Arch. Exp. Pathol. Pharmakol. *17:* 419–444, 1883.
586. Stadelmann, E., Editor: *Über den Einfluss der Menschlichen Stoffwechsel. Experimentel-Klinische Untersuchungen.* F. Enke, Stuttgart, 1890.
587. Stadie, W. C., and O'Brien, N.: The catalysis of the hydration of carbon dioxide and dehydration of carbonic acid by an enzyme isolated from red blood cells. J. Biol. Chem. *103:* 521–529, 1933.
588. Stahl, J.: Studies of blood ammonia in liver disease. Its diagnostic, prognostic, and therapeutic singificance. Ann. Intern. Med. *56:* 1–24, 1963.
589. Stanbury, S. W.: Calcium and phosphorus metabolism in renal failure. In *Diseases of the Kidney*, Ed. 2, Ch. 8. Edited by M. B. Strauss and L. G. Welt. Little, Brown and Co., Boston, 1971.
590. Stanbury, S. W., and Lumb, G. A.: Metabolic studies of renal osteodystrophy. Medicine (Baltimore) *41:* 1–34, 1962.
591. Stanbury, S. W., and Thomson, A. E.: Diurnal variations in electrolyte excretion. Clin. Sci. *10:* 267–293, 1951.
592. Stanbury, S. W., and Thomson, A. E.: The renal response to respiratory acidosis. Clin. Sci. *11:* 357–374, 1952.
593. Stillman, E., Van Slyke, D. D., Cullen, G. E., and Fitz, R.: Studies of acidosis. VI. The blood, urine and alveolar air in diabetic acidosis. J. Biol. Chem. *30:* 405–456, 1917.
594. Stokes, R. H.: Solutions. *Encyclopaedia Brittanica*, Vol. 20, pp. 884–892. Encyclopaedia Brittanica, Inc., William Benton, Chicago, 1967.
595. Stone, W. J., Balagura, S., and Pitts, R. F.: Diffusion equilibrium for ammonia in the kidney of the acidotic dog. J. Clin. Invest. *46:* 1603–1608, 1967.
596. Stone, W. J., and Pitts, R. F.: Pathways of ammonia metabolism in the intact functioning kidney of the dog. J. Clin. Invest. *46:* 1141–1150, 1967.
597. Strassburg, G.: Die Topographie der Gasspannungen im thierischen Organismus. Pfluegers Arch. Gesamte Physiol. *6:* 65–96, 1872.
598. Strieck, F.: Metabolic studies in a man who lived for years on a minimum protein diet. Ann. Intern. Med. *11:* 643–650, 1937.
599. Strunz, F.: Paracelsus. In *Das Buch der Grossen Chemiker*, Vol. 1, pp. 85–98. Edited by G. Bunge. Verlag-Chemie, Weinheim/Bergstrasse, 1929; reprinted 1965.
600. Sullivan, L. P., and McVaugh, M.: Effect of rapid and transitory changes in blood and urine pH or NH_4 excretion. Am. J. Physiol. *204:* 1077–1085, 1963.
601. Swan, R. C., and Pitts, R. F.: Neutralization of infused acid by nephrectomized dogs. J. Clin. Invest. *34:* 205–212, 1955.
602. Tannen, R. L.: The relationship between urine pH and acid excretion—the influence of urine flow rate. J. Lab. Clin. Med. *74:* 757–769, 1969.
603. Tannen, R. L.: The response of normal subjects to the short ammonium chloride test: the modifying influence of renal ammonia production. Clin. Sci. *41:* 583–595, 1971.
604. Thomas, S.: Some effects of change of posture on water and electrolyte excretion by the human kidney. J. Physiol (London) *139:* 337–352, 1957.
605. Thomas, S.: Effects of change of posture on the diurnal renal excretory rhythm. J. Physiol. (London) *148:* 489–506, 1959.
606. Thomas, S.: Solute excretion in man during changing urine flow occurring spontaneously and induced by vasopressin injection. J. Clin. Invest. *43:* 1–10, 1964.
607. Thompson, D. D., and Barrett, M. J.: Renal reabsorption of bicarbonate. Am. J. Physiol. *176:* 201–206, 1954.
608. Thompson, D. W.: *On Growth and Form*, Ed. 2. Cambridge University Press, Cambridge, Mass., 1942; reprinted 1959.

608a. Thorn, G. W.: *The Diagnosis and Treatment of Adrenal Insufficiency*, Ed. 2. Charles C Thomas, Springfield, Ill., 1951.

609. Thurau, K., Deetjen, P., und Kramer, K.: Hämodynamik des Nierenmarks. II. Mitteilung. Wechselbeziehung zwischen vaskulärem und tubulärem Gegenstromsystem bei arteriellen Drucksteigerungen, Wasserdiurese und osmotischer Diurese. Pfluegers Arch. Gesamte Physiol. *270:* 270–285, 1960.
610. Tigerman, H., and MacVicar, R.: Glutamine, glutamic acid, ammonia administration, and tissue glutamine. J. Biol. Chem. *189:* 793–799, 1951.

610a. Tobin, R. B.: Plasma, extracellular, and muscle responses to acute metabolic acidosis. Am. J. Physiol. *186:* 131–138, 1956.

611. Toor, M., Massry, S., Katz, A. I., and Agmon, J.: The effect of fluid intake on the acidification of urine during rest and exercise in hot climate. Clin. Sci. *27:* 259–270, 1964.
612. Trager, W., and Hutchinson, M. D.: The influence of the ammonium ion on the plasma atabrine level and on the urinary excretion of atabrine. J. Clin. Invest. *25:* 694–700, 1946.
613. Tranquada, R. E.: Lactic acidosis. Calif. Med. *101:* 450–461, 1964.
614. Travell, J.: Influence of pH on absorption of nicotine from urinary bladder and subcutaneous tissues. Proc. Soc. Exp. Biol. Med. *45:* 552–556, 1940.
615. Trenchard, D., Noble, M. I., and Guz, A.: Serum carbonic acid pK′ abnormalities in

patients with acid-base disturbances. Clin. Sci. *32:* 189–200, 1967.
616. *Trübners Deutsches Wörterbuch,* Vol. 5, p. 225. Walter De Gruyter and Co., Berlin, 1954.
616a. Tuller, M. A., and Mehdi, F.: Compensatory hypoventilation and hypercapnia in primary metabolic alkalosis. Report of three cases. Am. J. Med. *50:* 281–290, 1971.
617. Tyor, M. P., Owen, E. E., Berry, J. N., and Flanagan, J. G.: The relative role of extremity, liver, and kidney as ammonia receivers and donors in patients with liver disease. Gastroenterology *39:* 420–424, 1960.
618. Uhlich, E., Baldamus, C. A., und Ullrich, K. J.: Verhalten von CO_2-Druck und Bicarbonat im Gegenstromsystem des Nierenmarks. Pfluegers Arch. Gesamte Physiol. *303:* 31–48, 1968.
619. Ullrich, K. J.: Die Nierenphysiologie auf dem Weg zur molekularen Betrachtungsweise. Jahrbuch 1969 der Max-Planck-Gesellschaft zur Förderung der Wissenschaften e.v., pp. 153–174.
619a. Ullrich, K. J., and Eigler, F. W.: Sekretion von Wasserstoffionen in den Sammelrohren der Säugetierniere. Pfluegers Arch. Gesamte Physiol. *267:* 491–496, 1958.
620. Ullrich, K. J., Hilger, H. H., and Klümper, J. D.: Sekretion von Ammoniumionen in den Sammelrohren der Säugetierniere. Pfluegers Arch. Gesamte Physiol. *267:* 244–250, 1958.
621. Ullrich, K. J., Kramer, K., and Boylan, J. W.: Present knowledge of the countercurrent system in the mammalian kidney. In *Heart, Kidney, and Electrolytes,* pp. 1–37, Edited by C. K. Friedberg. Grune and Stratton, New York, 1962.
622. Ullrich, K. J., Radtke, H. W., and Rumrich G.: The role of bicarbonate and other buffers on isotonic fluid absorption in the proximal convolutions. Pfluegers Arch. Gesamte Physiol. *330:* 149–161, 1971.
622a. *Utilization of Nonprotein Nitrogen.* American Institute of Nutrition Symposium. Fed. Proc. *31:* 1151–1193, 1972.
623. Ussing, H. H.: Transport of electrolytes and water across epithelia. Harvey Lect. *59:* 1–30, 1963.
624. Ussing, H. H., Widdas, W. F., Smyth, D. H., Wilson, T. H., Davson, H., Oldendorf, W. H., Burgen, A. S. V., and Lathe, G. H.: Symposium on membrane transport. Proc. R. Soc. Med. *60:* 317–336, 1967.
625. VanderWerf, C. A.: *Acids, Bases, and the Chemistry of the Covalent Bond.* Reinhold Publishing Corp., New York, 1961.
625a. Van Slyke, D. D.: The carbon dioxide carriers of the blood. Physiol. Rev. *1:* 141–176, 1921.
626. Van Slyke, D. D.: Some points of acid-base history in physiology and medicine. In *Current Concepts of Acid-Base Measurement,* 5–14, Edited by G. G. Nahas. Ann. N.Y. Acad. Sci. *133:* 1–274, 1966.
627. Van Slyke, D. D., and Cullen, G. E.: Studies of acidosis. I. The bicarbonate concentration of the blood plasma; its significance, and its determination as a measure of acidosis. J. Biol. Chem. *30:* 289–346, 1917.
627a. Van Slyke, D. D., Hastings, A. B., Murray C. D., and Sendroy, J., Jr.: Studies of gas and electrolyte equilibria in blood. VII. The distribution of hydrogen, bicarbonate, and chloride ions in oxygenated and reduced blood. J. Biol. Chem. *65:* 701–728, 1925.
627b. Van Slyke, D. D., and Meyer, G. M.: The fate of protein digestion products in the body. V. The effects of feeding and fasting on the amino-acid content of the tissues. J. Biol. Chem. *16:* 231–233, 1913–1914.
628. Van Slyke, D. D., and Palmer, W. W.: Studies of acidosis. XVI. The titration of organic acids in urine. J. Biol. Chem. *41:* 567–585, 1920.
629. Van Slyke, D. D., Phillips, R. A., Hamilton P. B., Archibald, R. M., Futcher, P. H., and Hiller, A. Y.: Glutamine as a source material of urinary ammonia. J. Biol. Chem. *150:* 481–482, 1943.
630. Van Slyke, D. D., Sendroy, J., Jr., Hastings, A. B., and Neill, J. M.: Studies of gas and electrolyte equilibria in blood. X. The solubility of carbon dioxide at 38° in water, salt solution, serum, and blood cells. J. Biol. Chem. *78:* 765–799, 1928.
631. Van Slyke, D. D., Wu, H., and McLean, F. C.: Studies of gas and electrolyte equilibria in the blood. V. Factors controlling the electrolyte and water distribution in the blood. J. Biol. Chem. *56:* 765–849, 1923.
631a. van't Hoff, J. H.: Die Grenzebene, ein Beitrag zur Kenntniss der Esterbildung. Ber. Dtsch. Chem. Ges. *10:* 669–678, 1877.
632. van't Hoff, J. H.: The function of osmotic pressure in the analogy between solutions and gases. Phil. Mag., Series 5, *26:* 81–105, 1888.
633. Vermeulen, R.: Isopropyl alcohol and diabetes. Pa. Med. *69:* 53–54, 1966.
634. Vieira, F. L., and Malnic, G.: Hydrogen ion secretion by rat renal cortical tubules as studied by an antimony microelectrode. Am. J. Physiol. *214:* 710–718, 1968.
634a. Vinay, P., and Lemieux, G.: Renal hemodynamics and ammoniagenesis. Abstracts, Fifth International Congress of Nephrology, Mexico City, 1972, No. 438.
635. Vitti, T. G., Vukmirovich, R., and Gaebler, O. H.: Utilization of ammonia nitrogen, administered by intragastric, intraperitoneal, and subcutaneous routes: effects of growth hormone. Arch. Biochem. Biophys. *106:* 475–482, 1964.

636. Waddell, W. J., and Bates, R. G.: Intracellular pH. Physiol. Rev. *49:* 285–329, 1969.
637. Wald, G.: Life and light. Sci. Am. *201:* 92–108, 1959.
638. Wald, G.: Life in the second and third periods, or why phosphorus and sulfur for high-energy bonds? In *Horizons in Biochemistry*, pp. 127–142. Edited by M. Kasha and B. Pullman. Academic Press, New York, 1962.
639. Wald, G.: The origins of life. Proc. Natl. Acad. Sci. U.S.A. *52:* 595–611, 1964.
640. Walker, A. J.: Ammonia formation in the amphibian kidney. Am. J. Physiol. *131:* 187–194, 1940.
641. Wallace, W. M., and Hastings, A. B.: The distribution of the bicarbonate ion in mammalian muscle. J. Biol. Chem. *144:* 637–649, 1942.
642. Walser, M.: Sodium excretion. In *The Kidney: Morphology, Biochemistry, Physiology*, Vol. 3, Ch. 3, pp. 127–207. Edited by C. Rouiller and A. F. Muller. Academic Press, New York, 1971.
643. Walser, M., and Mudge, G. H.: Renal excretory mechanisms. In *Mineral Metabolism*, pp. 287–336. Edited by C. L. Comer and F. Bronner. Academic Press, New York, 1960.
644. Walshe, J. M.: Effect of penicillamine on failure of renal acidification in Wilson's disease. Lancet *1:* 775–778, 1968.
645. Walter, F.: Untersuchungen über die Wirkung der Säuren auf den thierischen Organismus. Arch. Exp. Pathol. Pharmakol. *7:* 148–178, 1877.
646. Warburg, E. J.: XXII. Studies on carbonic acid compounds and hydrogen ion activities in blood and salt solutions. Biochem. J. *16:* 153–340, 1922.
647. Watson, J. F., Clapp, J. R., and Berliner, R. W.: Micropuncture study of potassium concentration in proximal tubule of dog, rat, and Necturus. J. Clin. Invest. *43:* 595–605, 1964.
648. Webster, L. T., Jr., and Gabuzda, G. J.: Ammonium uptake by the extremities and brain in hepatic coma. J. Clin. Invest. *37:* 414–424, 1958.
649. Weil-Malherbe, H.: The metabolism of ammonia in brain. In *Consciousness and the Chemical Environment of the Brain*, pp. 50–55. Ross Laboratories, Columbus, O., 1958.
650. Weiss, M. B., and Longley, J. B.: Renal glutaminase I distribution and ammonia excretion in the rat. Am. J. Physiol. *198:* 223–226, 1960.
651. Wesson, L. G., Jr.: Electrolyte excretion in relation to diurnal cycles of renal function. Plasma electrolyte concentrations and aldosterone secretion before and during salt and water balance changes in normotensive subjects. Medicine (Baltimore) *43:* 547–592, 1964.
652. Wesson, L. G., Jr.: *Physiology of the Human Kidney*. Grune and Stratton, New York, 1969.
653. White, H. L., Rosen, I. T., Fischer, S. S., and Wood, G. H.: The influence of posture on renal activity. Am. J. Physiol. *78:* 185–200, 1926.
654. Whittam, R., and Wheeler, K. P.: Transport across cell membranes. Annu. Rev. Physiol. *32:* 21–60, 1970.
655. Wilde, P.: Disquisitiones quaedam de alcalibus per urinam excretis. Dissertation. Dorpat, 1855.
656. Wilhelmy, L.: Über das Gesetz, nach welchem die Einwirkung der Säuren auf dem Rohrzucker stattfindet. Ann. Phys. Chem. *81:* 413–428 and 499–526, 1850. Ostwald's Klassiker, no. 29.
656a. Willis, L. R.: Schneider, E. G., Lynch, R. E. and Knox, F. G.: Effect of chronic alteration of Na balance on reabsorption by proximal tubule of the dog. Am. J. Physiol. *223:* 34–39, 1972.
657. Wilson, D. L.: Direct effects of adrenal cortical steroids on the electrolyte content of rabbit leucocytes. Am. J. Physiol. *190:* 104–108, 1957.
657a. Winderlich, R.: Justus Liebig. In *Das Buch der Grossen Chemiker*, Vol. 2, pp. 1–30. Edited by G. Bunge. Verlag-Chemie, Weinheim/Bergstrasse, 1929; reprinted 1965.
658. Windhager, E. E.: Säureexkretion und Ammoniakausscheidung. In *Drittes Symposium der Ges. für Nephrologie*, Berlin, 1964, pp. 37–61. Edited by K. J. Ullrich and K. Hierholzer. Hans Huber, Bern, 1965.
659. Windhager, E. E.: *Micropuncture Techniques and Nephron Function*. Appleton-Century-Crofts, New York, 1968.
660. Winkler, A. W., and Smith, P. K.: Renal excretion of potassium salts. Am. J. Physiol. *138:* 94–103, 1942.
661. Winterberg, H.: Zur Theorie der Säurevergiftung. Z. Physiol. Chem. *25:* 202–235, 1898.
662. Winters, R. W., Lowder, J. A., and Ordway N. K.: Observations on carbon dioxide tension during recovery from metabolic acidosis. J. Clin. Invest. *37:* 640–645, 1958.
663. Winters, R. W., Scaglione, P. R., Nahas, G. G., and Verosky, M.: The mechanism of acidosis produced by hyperosmotic infusions. J. Clin. Invest. *43:* 647–658, 1964.
664. Winterstein, H.: Die Regulierung der Atmung durch das Blut. Pfluegers Arch. Gesamte Physiol. *138:* 167–184, 1911.
665. Winterstein, H.: Chemical control of pulmonary ventilation. I. The physiology of the chemoreceptors. II. Hypoxia and respiratory acclimatization. III. The "reaction theory" of respiratory control.

New Eng. J. Med. *225:* 216–233, 272–278, 331–337, 1956.
666. Wirz, H., Hargitay, B., and Kuhn, W.: Lokalisation des Konzentrierungsprozesses in der Niere durch direkte Kryoskopie. Helv. Physiol. Acta *9:* 196–207, 1951.
667. Wissbrun, K. F., French, D. M., and Patterson, A., Jr.: The true ionization constant of carbonic acid in aqueous solution from 5 to 45°. J. Phys. Chem. *58:* 693–695, 1954.
668. Woeber, K. A., Reid, E. L., Kiem, I., and Hills, A. G.: Diffusion of gases out of the distal nephron-segment in man. I. NH_3. J. Clin. Invest. *42:* 1689–1704, 1963.
669. Woeber, K. A., Ricca, L., and Hills, A. G.: Pathogenesis of uric acid urolithiasis. Clin. Res. *10:* 45, 1962.
670. Wohlgemuth, J.: Über die Herkunft der schwefelhaltigen Stoffwechselprodukte im tierischen Organismus. Mitteilung I. Z. Physiol. Chem. *40:* 81–100, 1903.
671. Wolf, A. V.: Renal regulation of water and some electrolytes in man with special reference to their relative retention and excretion. Am. J. Physiol. *148:* 54–68, 1947.
672. Woodbury, J. W.: Regulation of pH. In *Physiology and Biophysics*, Ed. 19, Ch. 46. Edited by T. C. Ruch and H. D. Patton. W. B. Saunders Co., Philadelphia, 1965.
673. Wrong, O., and Davies, H. E. F.: The excretion of acid in renal disease. Q. J. Med. *28(NS):* 259–313, 1959.
674. Wu, C.: Glutamine synthetase. I. A comparative study of its distribution in animals and its inhibition by DL-allo-δ-hydroxylysine. Comp. Biochem. Physiol. *8:* 335–351, 1963.
675. Wu, C., Bollman, J. L., and Butt, H. R.: Changes in free amino acids in the plasma during hepatic coma. J. Clin. Invest. *34:* 845–849, 1955.
676. Yoshimura, H., Fujimoto, M., Okumura, O., Sugimoto, J., and Kuwada, T.: Three-step regulation of acid-base balance in body fluid after acid load. Jap. J. Physiol. *11:* 109–125, 1961b.

ILLUSTRATION CREDITS

For permission to reproduce illustrations published elsewhere, acknowledgment is made to the following.

Figure I-2.2. Reproduced from Davis and Day (122). © 1961, Educational Services, Inc. Reprinted by permission of Doubleday and Company, Inc., Garden City, N. Y.

Figure I-2.3. Reproduced from Pauling and Hayward (450). © 1964, W. H. Freeman and Company, San Francisco.

Table I-3.1, Table I-3.2, and Figure I-4.1. Modified from Prutton and Maron (382). © 1944, Macmillan Company, New York.

Table I-5.1. Values taken from Edsall and Wyman (137). © 1966, Academic Press, Inc., New York.

Figures I-5.3 and I-5.4. Modified from Clark (104). © 1948, The Williams & Wilkins Company, Baltimore.

Figure I-6.1. Modified from Hills and Reid (259). © 1967, Johns Hopkins University Press, Baltimore.

Figure I-6.2. Reproduced from Gamble (182). © 1964, Harvard University Press, Cambridge, Mass.

Figure I-6.3. Reproduced from Hills and Reid (258). © 1966, S. Karger, Basel.

Figure II-1.1. Reproduced from Elliot, Sharp, and Lewis (150). © 1959, The Williams & Wilkins Company, Baltimore.

Figure II-6.1. Reproduced from Hills and Reid (258). © 1966, S. Karger, Basel.

Figure II-6.2. Slightly modified from Hills (254). © 1969, University of Chicago Press, Chicago.

Figure II-6.3. Reproduced from Hills (255). © 1971, American Clinical and Climatological Association, Nashville, Tenn.

Figure II-7.2. Reproduced from Shannon (560). © 1939, American Physiological Society, Washington, D. C.

Figure II-7.3. Reproduced from Rector, Seldin, Roberts, and Smith (490). © 1960, American Society for Clinical Investigation, Salt Lake City, Utah.

Figure II-8.1. Reproduced from Hills (255). © 1971, American Clinical and Climatological Association, Nashville, Tenn.

Figure II-8.2. Data obtained from Reid and Hills (495): © 1965, Blackwell Scientific Publications, Oxford; and from Woeber, Reid, Keim, and Hills (668): © 1963, American Society for Clinical Investigation, Salt Lake City, Utah.

Figures II-8.3 and II-8.4. Reproduced from Reid and Hills (498). © 1969, Blackwell Scientific Publications, Oxford.

Figure II-9.6. Reproduced from Wrong and Davies (673). © 1959, Clarendon Press, Oxford.

Figures II-9.13, II-9.14, and II-9.15. Reproduced from Hills, Reid, and Kerr (263). © 1972, American Physiological Society, Washington, D. C.

Figures II-10.1, II-10.2, and II-10.3. Reproduced from Hills and Reid (261). © 1970, American Physiological Society, Washington, D.C.

Figure II-10.7. Reproduced from Hills and Reid (259). © 1967, Johns Hopkins University Press, Baltimore.

Figure III-1.3. Reproduced from Singer and Hastings (575). © 1948, The Williams & Wilkins Company, Baltimore.

Figure III-1.4. Reproduced from Siggaard-Andersen (569a). © 1966, New York Academy of Sciences, New York.

Table III-3.4. Data obtained from Schwartz, Hall, Hays, and Relman (548). © 1959, American Society for Clinical Investigation, Salt Lake City, Utah.

INDEX